Source Testing For Air Pollution Control

by

Hal B. H. Cooper, Jr., M.S.

and

August T. Rossano, Jr., Sc.D.

University of Washington
Dept. of Civil Engineering
Air Resources Program
Seattle, Washington 98105

McGraw-Hill Book Company
New York   St. Louis   San Francisco
London   Montreal   Singapore

Source Testing

For

Air Pollution Control

Copyright © 1971 by McGraw-Hill, Inc.
All rights reserved. Printed in the
United States of America. No part
of this publication may be reproduced,
stored in a retrieval system, or
transmitted, in any form or by any
means, electronic, mechanical,
photocopying, recording, or otherwise,
without the prior written permission
of the publisher.

Library of Congress Catalog Card
Number: 70-160153 0-07-012760-3

Republished in 1974 by McGraw-Hill
Book Company.

Cover design by Charlotte Newman

TABLE OF CONTENTS

Preface iv
Acknowledgements iv
List of Figures 219
List of Tables 221

Chapter 1. Introduction 1
   I. Importance 1
   II. Procedure 2
   III. References 3

Chapter 2. Terminology 7
   I. Definitions 7
   II. Units 7
   III. Symbols 8
   IV. Reference Conditions 9
   V. Performance Testing 12
   VI. References 12

Chapter 3. Principles 15
   I. Gas Laws 15
   II. Fluid Motion 17
   III. Particle Dynamics 18
   IV. Generalities 20
   V. References 20

Chapter 4. Gas Flow Measurements 23
   I. Sampling Points 23
   II. Flow Conditions 24
   III. Velocity Determination 33
   IV. Total Gas Flow Rate 37
   V. Low Velocity Measurement 38
   VI. Total Volume Measurement 40
   VII. Special Problems 41
   VIII. References 42

Chapter 5. Sampling Trains 47
   I. Introduction 47
   II. Components 48
   III. Sample Flow Measurement 49
   IV. Flow Calibration Devices 51
   V. Flow Corrections 52
   VI. References 52

Chapter 6. Principles of Particulate Sampling 53
   I. Introduction 53
   II. Isokinetic Sampling 53
   III. Statistical Considerations 67
   IV. References 72

Chapter 7. Methodology of Particulate Sampling 77
   I. Sampling Procedure 77
   II. Sample Calculations 77
   III. Sampling Equipment 80
   IV. Sampling Trains 83
   V. Collection Devices 84
   VI. References 99

Chapter 8. Particulate Sampling Trains 105
   I. Specific Organization 105
   II. Specific Contaminants 105
   III. Combustion Sources 110
   IV. Industrial Sources 113
   V. Performance Testing 118
   VI. Particle Size Analysis 120
   VII. References 128

Chapter 9. Gaseous Sampling 137
   I. Introduction 137
   II. Sample Collection Methods 139
   III. Wet Chemical Methods 146
   IV. Instrumental Analyses 152
   V. Sample Problem 154
   VI. References 156

Chapter 10. Continuous Monitoring 161
   I. Introduction 161
   II. Particulate Monitoring 161
   III. Gaseous Monitoring 168
   IV. Design of Systems 179
   V. References 181

Chapter 11. Special Applications 187
   I. Introduction 187
   II. Additional Measuring Techniques .. 187
   III. Mobile Sources 191
   IV. Additional Sources 194
   V. Bibliographies 200
   VI. References 200

Chapter 12. Appendices 203
   Appendix A. Pertinent Source Test Data ... 203
   Appendix B. Derivations 208
   Appendix C. Sources of Equipment 210
   Appendix D. Addresses 218

Index 223
   Author Index 223
   Subject Index 226

PREFACE

Pollution of our atmospheric resources by the activities of man has become a problem of great national and worldwide concern. Determination of the character and quantity of emissions from individual sources is an essential step in any program to control and minimize these emissions.

The purpose of this work is to present a discussion of principles and methods used for testing of gaseous and particulate materials being emitted from industrial, combustion and other sources. It is designed to be utilized by both industrial and governmental personnel actively involved in source testing, and by students, engineers, and others desiring an introductory or background knowledge of the subject. Necessary tables and charts are to be provided for ready reference and to simplify calculations.

The material is organized to give the reader a logical presentation of the steps taken in source testing. It is designed to emphasize sampling methods with a general description of analytical procedures. However, it is not designed as a so-called "cookbook" of analytical procedures as these are available in other works. The introductory material is involved with purposes, terminology, units, preliminary evaluations, and theoretical principles. Determination of rates and conditions of gas flow in ducts is necessary for accurate determination of emission levels. One section is devoted to a brief description of the devices used and requirements of sampling train construction, with special emphasis on collection devices.

An extensive discussion of the equipment, methodology, sampling and analytical techniques in use for gaseous and particulate materials is presented. Continuous monitoring techniques for gaseous and particulate materials from stationary sources is given special attention because of its considerable importance in enforcement of emission standards. A final section describes methods for determination of odor threshold levels, mobile source monitoring, and construction of mobile laboratories for field studies. Useful background information is included in the Appendix.

In addition to a discussion of current source sampling technology, the book attempts to point out shortcomings and areas where additional effort is needed. It is felt by the authors that techniques for source testing are not perfected and there is a need for considerable work in certain areas.

The book is particularly timely in that it coincides with increasingly stringent State and local air pollution regulations, and with the forthcoming implementation of National emission standards. It will prove to be of value to any persons interested in the area of enforcement, compliance, or study of these Standards.

ACKNOWLEDGEMENTS

The authors are indebted to the following persons and organizations for their assistance in the preparation of this manuscript. Messrs. Murl Miller and Richard Abrams of the Scott Paper Company made a thorough evaluation of the wet impingement method of particulate sampling. Messrs. James S. Leonard and Gary N. Thoen of the Weyerhaeuser Company provided useful information on continuous monitoring for gases and particulates. Messrs. Dennis Moody, Fred Martin, and Vincent Tretter of the Georgia Pacific Corp., Mr. James S. Walther of the Crown Zellerbach Corp., and Mr. Neil Gansler of the Longview Fibre Company provided useful techniques for flow measurements and continuous monitoring. Messrs. Leon Duncan and Joseph Megy added helpful data on mobile laboratory construction and particulate sampling methodology.

The following people from governmental regulatory agencies added helpful information regarding sampling techniques. Mr. Howard Devorkin of the Los Angeles County Air Pollution Control District supplied useful information on techniques employed by that agency. Mr. John T. Donovan of the Bay Area Air Pollution Control District, and Mr. Mario Rivera of the Dade County Pollution Control Authority, provided detailed information regarding particulate sampling techniques in use by their respective agencies. Mr. Walter S. Smith of the U.S. Public Health Service made several very helpful suggestions during the initial preparation stage of the manuscript. Mr. Albert S. Moore of the U.S. Bureau of Mines allowed use of materials relating to maintenance of isokinetic sampling conditions on a continuous, automated basis.

The following firms were kind enough to allow use of their materials in the course of preparing this book: Western Precipitation Corp. (Mr. Harold Haaland), Buffalo Forge Co., Plibrico Company, Research-Cottrell, Inc., Koch Engineering Co., and Airflow Developments, Ltd. (A. Connor Wilson and Brian F. Cornwall).

A special acknowledgement is extended to Mr. H. B. H. Cooper's former employers at the National Council of the Paper Industry for Air and Stream Improvement, Inc., and its member companies for allowing him to gain the experience necessary to be able to write this book. Without extensive experience in real-world source test problems, it would never have been possible to complete this work.

Last, the sincere appreciations of the authors to Mr. H. B. H. Cooper's parents, Dr. and Mrs. Hal B. H. Cooper, and to Mr. George Cuthill for providing the friendly, peaceful, and conducive surroundings under which the major portion of the manuscript was prepared.

Dr. Robert C. Haring of Chem Tech Services is also to be commended for the fine job in preparing the manuscript for publication.

CHAPTER 1

INTRODUCTION

I. IMPORTANCE

Pollution of our atmospheric resource by the activities of man has become a problem of growing national and worldwide concern with the realization that this resource is not unlimited. A polluted atmosphere is one where humans, animals, plants, or materials are exposed to air contaminants at such levels and durations that adverse effects occur. The three essential elements of air pollution are sources, transport, and receptors.

The presence of any contaminant material in the atmosphere is the result of being emitted from one or more sources. Materials emitted to the atmosphere may be gaseous, liquid, or solid; organic or inorganic; and of many different physical, chemical, and biological properties. The nature and quantities of materials emitted vary with the type of source and process being performed; the sources may be either stationary or mobile. A summary of major man-made emissions from combustion, industrial, and other sources is shown in Table 1 [1].

A. Reasons:

It is necessary to know the types and amounts of materials being emitted from a source (or group of sources) if a potential or existing air pollution problem is to be evaluated and ultimately controlled. There are six major reasons for obtaining source emission level information. First, it may be necessary to determine whether a given process unit is in compliance with existing or proposed emission regulations of a governmental agency. Second, it may be desired to determine the economic impact of material or product losses from a source. Third, the requirements and inlet loading for a proposed collection device may be needed as a part of the engineering design. Fourth, the efficiency of an existing collection device can be evaluated in terms of its inlet and outlet loadings. A fifth reason involves continuous or frequent observation of one or more constituents in the exit stream of a particular unit as a means of process control. An example is the use of oxygen and combustibles monitors for combustion control on fossil fuel-fired boilers [2]. Sixth, enforcement agencies may require reliable source and emission data as a basis upon which to develop aerometric activities, control regulations, and air resource management programs.

B. Approaches:

There are at least three ways for making independent estimates of emission rates. Making a material balance calculation across all solid, liquid, and gaseous streams entering and leaving an entire process or particular process unit is the first method. This can give a rough approximation of materials and economic losses and usually is used as a check against sampling results for possible gross errors.

Second, it may be possible to make use of published emission factors which express the amount of contaminant generated per unit of fuel burned, raw material processed, or product removed [3]. The emission factors are developed from either material balance calculations, chemical theory, or actual sampling results. They provide a rapid and useful method for estimating pollutant emissions, and as a check against sampling results, but should be used with caution. A thorough listing of emission factors has recently been published by Ozolins and Smith [4], and a sample developed from sample results and used for estimating emissions from refuse incineration is shown in Table 2 [1].

The third and most direct method for determining contaminant emission levels from a particular source is to make an actual test. It is desired to measure the amount of material being emitted per unit volume or mass of flue gas. Measurement of the emissions from the total gas flow is not usually practical, so a small portion of the gas stream is withdrawn from the flue for subsequent collection and analysis for the materials of interest. The three essential elements of source testing at a point are: 1) the total amount or volume of flue gas emitted; 2) the total amount or volume of gas sample withdrawn from the flue gas; and 3) the amount of specific contaminant material(s) collected from the gas sample withdrawn. It is then possible to make an analysis of the materials present. An additional complication is presented in the rate of sample withdrawal for particulate sampling, but this will be discussed in a later chapter. All of the above measurements are based on a given time interval, to determine concentrations and emission rates for gaseous and particulate materials. It is usually necessary to obtain as nearly as possible the exact flue gas and sample gas conditions. This includes determination of duct velocity, flow rates, equivalent volumes, temperatures, pressures, moisture contents, and gas compositions and densities at both duct and meter conditions. The approach used in illustrated in Figure 1.

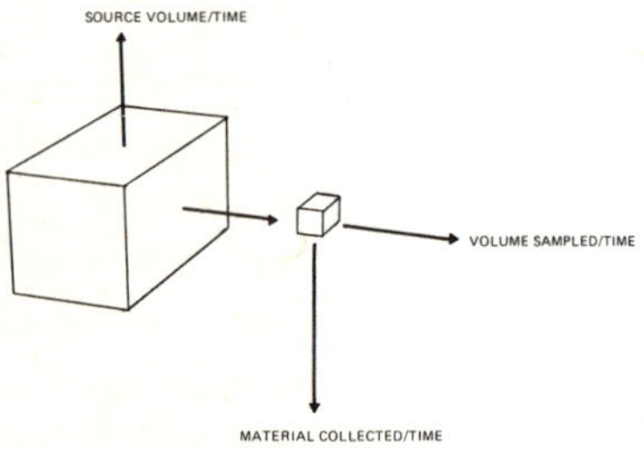

Fig. 1.

C. Requirements:

The objective in making a test of any existing or potential source of pollutants is to get an accurate, precise, and reliable determination of the types, concentrations, and emission rates of materials being released [5]. The gas volume sampled must represent *in toto* or a known portion of the total gas volume to obtain concentration. It is also necessary to know the total flue gas flow rate to compute emission rates for materials. A third requirement is that the sample taken be representative of the total flue gas. The method should be reproducible if the pollutant concentration remains constant. It is desired to allow collection and subsequent detection by a sufficiently sensitive analytical technique where all material removed is measured, or else the losses known. Provisions for dealing with interfering substances are also important.

It should be emphasized that the sampling methods should be as simple as possible and still allow accurate determination. They also should be of sufficient flexibility to allow sampling a maximum number of sources with a minimum of changes, and with a minimal amount of hazardous or dangerous equipment. The analytical methods used should be as accurate, simplified, and reproducible as possible. All equipment and methods used must also be able to function reliably for extended time periods under adverse working conditions.

An additional requirement is that samples taken may later be used as evidence in courts of law. It is then doubly important that all flow measurements, gas samples, analyses, and calculations be made accurately, exactly according to established procedures, and that all results be checked thoroughly and well documented.

II. PROCEDURE

The program for determining the gaseous or particulate levels from a source involves seven basic steps [5]. These can be relatively simple or complicated, depending on the type, purpose, and extent of the particular study. These steps include the following: 1) planning, 2) preliminary evaluation, 3) gas flow measurement, 4) sample collection, 5) sample analysis, 6) calculations, 7) report preparation.

A. Planning:

Preliminary planning is an important element of a successful source testing program. It requires a knowledge of operating characteristics of the process unit to be tested, background as to the character and quantity of emissions expected, potential problems to be encountered, and the reasons for obtaining the information. The objectives, sampling locations, accessibility equipment, power requirements, manpower, timing and scheduling, and transportation to the site must be worked out before starting the test. The persons performing the source test should study the process or equipment to be evaluated before beginning the study.

The first step in planning a source test is to define the objective (or objectives), such as determining compliance with emission regulation, performance of an existing collection device, or the potential need for one. This includes estimation of the: (a) types and amounts of pollutants present by use of emission factors [4] or a material balance, (b) flow rates, and (c) problems to be encountered.

It is necessary to choose one or more suitable sampling locations so as to obtain accurate and representative samples and gas flow measurements. Sample ports, platforms, scaffolding, electric power, and water must be provided if necessary. Safety and minimizing exposure to adverse weather conditions should be given utmost consideration to reduce accident hazards and increase work efficiency.

Sampling equipment to be used for a particular study must be prepared for use, repaired if necessary, flow measuring units calibrated, and all equipment collected and packaged for transportation to the field. A check list of all items required for a study should be utilized. Special equipment for particular tests must be designed, fabricated, and prepared for shipment. Equipment should be packed securely in lightweight containers such as large cardboard boxes, and packed so as to facilitate ease of transportation and minimize the possibility of breakage.

The scheduling requirements for a particular test should be worked out in advance. A process unit is selected, and arrangements made to have it operated at a given set of conditions. In cyclical and other nonuniform operations, the times for sampling must be determined prior to sampling, and may be designed to take an average of these varying conditions. It is possible to specify tests for a number of operating conditions to establish a range of emission levels.

Time and manpower requirements are a necessary part of the plan for a source testing program. At least two persons are required for source testing at a single point because of the number of jobs to be done simultaneously; it is best to have three for particulate sampling. Four to six man-hours off the stack should be allowed for each man-hour involved in sample collection, to allow for study preparation, flow measurement, analyses, calculations, and report preparation. A full day is usually required for a two- or three-man crew to collect three of four particulate samples. It is best to make separate gaseous and particulate determinations so as to avoid confusion resulting from excessive activity. One man should be placed in charge of the entire source testing operation with final authority and responsibility to minimize the amount of administrative confusion.

Scheduling of the source test program should be made through the manager of the appropriate person in authority at a particular industrial or other installation. Persons performing a source testing operation should make special efforts to meet with supervisory and operating personnel in areas where they will be working, to explain the objectives. This provides for an interchange of ideas to utilize the experience and knowledge of both groups [5], a better and more courteous working relationship, and a more effective and efficient test program.

Additional items in performing source tests are transportation, costs, and attire. It is necessary to transport personnel and equipment to and from a given test site. A panel truck or specially-fitted trailer can serve this purpose very well [6]. All equipment should be designed and packaged for easy and safe shipment in small, lightweight containers. Studies should be made as rapidly and with as simple equipment and procedures as possible, to minimize costs. Consideration should be given to continuous monitoring systems for sources where a large number of tests may be required, as a means of minimizing labor costs.

In connection with attire, consideration should be given to practicality, comfort, weather, and safety. It is advisable to wear work clothes, a hard hat, and safety shoes with heavy soles when sampling because test locations are usually dirty, hazardous, and often above warm ducts. Jackets with a number of pockets are handy for storing stopwatches, sliderules, disconnects, and other items. Special rubber boots and rainsuits should be worn in rainy or monsoon weather. Asbestos or rubber gloves should be worn when working near hot ducts or following electrostatic precipitators.

B. Preliminary Evaluation:

Upon arrival at the test site, the sampling equipment is unpacked and prepared for use. The arrangements for sampling facilities, such as parts, platforms, and power outlets, should be checked on and installed in place if necessary. It is usually necessary to allow the first day of any extensive source sampling study for organizing sample equipment, testing ports, facilities, and organizing the test program in general. It is also necessary to make arrangements for laboratory analyses of samples, if these are to be done on the site. Spending sufficient time for organization pays for itself in rapid and accurate sampling results.

A preliminary determination of gas flow rates and conditions at points to be sampled, prior to the actual test, provides knowledge of the conditions to be expected and problems to be faced. It is often useful to take an exploratory sample at a point to be sampled so as to establish the order of magnitude of emission levels

[7]. This is particularly useful for sources and processes for which there is no available knowledge or previous sampling experience. Correlation of preliminary gas flow and emission data with process unit operating parameters in terms of loading and production rates is recommended, to establish familiarity with the particular process. Ranges of operation and potential problems to be encountered should be discussed and coordinated with operating personnel.

A third part of the preliminary evaluation is to calibrate, prepare, and check out all sampling equipment to assure reliable and accurate operation and freedom from leaks and equipment malfunctions. It may also be necessary to make changes or modifications in sampling methods and equipment as a result of the preliminary survey. This may be the result of physical and chemical properties of the materials being tested, conditions in the duct such as excessive positive or negative pressure, or the presence of corrosive or toxic gases. Limitations in working space, excessive temperatures at work sites, and accessibility are additional considerations [5].

C. Gas Flow Measurement:

Accurate measurement of the rate of gas flow and flow conditions in a duct at a point of sampling is essential in computing both concentrations and emission rates of the respective gaseous and particulate materials. Measurements of velocity, temperatures, static pressure, gas composition and density, cross-sectional area, and rate of flow are necessary to establish accurate gas flow parameters. Methodology of flue gas flow measurement is discussed in a later section. Flow measurements are usually made just prior to sample collection.

D. Sample Collection:

The sample collection equipment is assembled and put in place near the source of interest. A known volume of source gas is removed from the flue at the point of sampling and drawn through the sampling train. It is necessary to know the metered amount of gas withdrawn, the equivalent volume at source conditions, and the amount of material collected. Sample collection procedures for both gaseous and particulate materials are covered in subsequent sections.

E. Sample Analysis:

The collected sample is usually enclosed and transported to a laboratory for analysis. It is normally a good idea to analyze samples at the site as soon after collection as possible to minimize the chances for losses or interactions. Samples should be analyzed according to specified procedures as accurately as possible; it may be necessary to evaluate methods for possible losses and interferences, which may require an extensive laboratory study. Accurate and simplified analytical methods should be used, which involve as little time as posible, to reduce time requirements. The use of continuous monitoring systems with combined sample collection and analysis should be considered as a means of saving time and effort. Brief discussions of gaseous and particulate analytical methods are presented; it is not a purpose of this publication to present a synopsis of analytical methods as this is presented in other texts [5] [8] [9] [10] [11] [12] [13] [14] [15].

F. Calculations:

Calculations of the appropriate concentrations and emission rates for particulate and gaseous materials in flue gas streams are made following sample collection and analysis. These computations are performed from knowledge of gas flow rate and conditions, equivalent gas volume sampled, and the amount of material collected. Sample problems are included in each section where appropriate.

G. Report:

A report is prepared at the conclusion of a source test study. Its purpose is to document what was done and the reasons, the findings, and possible corrective actions. Source test reports are usually divided into six basic parts as follows: [5].

(1) The purpose of the study is to describe why a test is being undertaken. Possible reasons have been previously listed.

(2) A description of the process or the process unit or control device should be given. A discussion of the effect of pertinent operating parameters, and a flow diagram, should be included.

(3) The pertinent facts regarding the test should be included for future reference. These include date, time, location, establishment and section, persons participating, and persons contacted.

(4) The sampling and analytical methods

and the calculations used should be described with regard to equipment, flow rates, collection mechanism(s), chemical reactions, selectivity, sensitivity, and potential interferences. A sketch of the sampling train used is usually necessary, and a diagram of analytical apparatus is very helpful.

(5) The results obtained should be presented and discussed in terms of significant trends, predicted or theoretical results, and sampling anomalies.

(6) The conclusions and recommendations, made as a result of data evaluation and limitations, should be included.

It is often a good idea to include a brief summary at the beginning of a report, particularly if the report describes an extensive source test study.

III. REFERENCES

1. Rossano, A. T., Jr., and Cooper, H. B. H., Jr., "Sampling and Analysis for Air Pollution Control," *Chemical Engineering*, Vol. 75, No. 22, pp. 142-146, October 14, 1968.
2. Barnard, C. H., "Increased Efficiency and Decreased Smoke with Boiler Instruments and Controls," *Air Repair*, Vol. 4, No. 1, pp. 20-25, May 1954.
3. Rossano, A. T., Jr., and Schell, N. E., "Procedures for Making an Inventory of Air Pollution Emissions," *Journal of the Air Pollution Control Association* (APCA), Vol. 8, No. 2, pp. 147-152, August 1958.
4. Ozolins, G., and Smith, R., "Rapid Survey Technique for Estimating Community Emissions", U. S. Public Health Service Publication No. 999-AP-29, Cincinnati, Ohio, 1966.
5. Devorkin, H., Chass, R. L., Fudurich, A., and Kanter, C. V.; Holmes, R. G., ed. "Source Testing Manual," Los Angeles County Air Pollution Control District, Los Angeles, California, 1965.
6. Walther, J. E., and Amberg, H. R., "Experience with a Mobile Laboratory in Source Sampling Kraft Mill Emissions," *Tappi*, Vol. 51, No. 11 pp. 126A-129A, November 1968.
7. Haaland, H. H., ed., "Methods for Determination of Velocity, Volume, Dust and Mist Content of Gases," Bulletin WP-50, Western Precipitation Corp., Los Angeles, California, 1968.
8. Jacobs, M. B., *The Chemical Analysis of Air Pollutants*, Interscience Publishers, Inc., New York, New York, 1960.
9. Clayton, G. D., ed., "Stack Testing," Ch. 9 in *Air Pollution Manual, Part 1 - Evaluation*, American Industrial Hygiene Association, Detroit, Michigan, 1961.
10. Bloomfield, B. D. "Source Testing," Ch. 28 in Stern, A. C., ed., *Air Pollution*, Vol. 2, pp. 487-536, Academic Press, Inc. New York, New York, 1968.
11. "Standard Methods for the Examination of Water and Wastewater," 12th ed., American Public Health Association, New York, New York, 1965.
12. "Selected Methods for the Measurement of Air Pollutants," U. S. Public Health Service Publication, No. 999-AP-11, Cincinnati, Ohio, May 1965.
13. "Recommended Methods in Air Pollution Studies," California State Department of Public Health, Air and Industrial Hygiene Laboratory, Berkeley, California, 1962.
14. Ruch, W. E., ed., *Chemical Detection of Gaseous Pollutants*, Ann Arbor Science Publishers, Inc., Ann Arbor, Michigan, 1967.
15. *1965 Book of Industrial Standards*, Part 23, "Industrial Water; Atmospheric Analysis," American Society for Testing and Materials, Philadelphia, Pennsylvania, 1965.

Table 1. Summary of Air Pollution Sources [1]

Operation	Category	Examples	Pollutants for all categories
Material Handling	Crushing, Grinding, Screening	Road mix plants	Mineral particles
	Demolition	Urban renewal	Organic particles
	Milling	Grain elevators	
Combustion Processes	Fuel Burning	Home heating and power plants	Oxides of nitrogen
			Oxides of sulfur
	Motor Vehicles	Autos, trucks, buses	Carbon monoxide
			Smoke and flyash particles
	Refuse Burning	Municipal, apartment and home incinerators	Organic vapors
			Odorous gases
		Open burning	Metallic oxide particles
Industrial Operations	Metallurgical Plants	Smelters, steel mills, aluminum plants	Metal fumes (Zn, As, Pb, Cu)
			Fluoride materials
			SO_2 and H_2S
	Chemical Plants	Petroleum refinery,	Oxide of sulfur & H_2S
		Pulp & paper mills,	Oxides of nitrogen
		Fertilizer plants,	Organic vapors
		H_2SO_4, HNO_3, and organic processing plants	Particles
			Odorous gases
	Cement Plants	Wet and dry process	Oxides of nitrogen
			Mineral particles
	Waste Recovery	Metal scrap yards,	Smoke & soot particles
		Auto body burning,	Organic vapors
		Rendering plants	Odorous gases
Organic Materials	Spray Painting	Automobile assembly,	Hydrocarbon vapors
		Furniture and appliance finishing	Organic vapors
	Inking	Photogravure & printing	"
	Solvent Cleaning	Dry cleaning, degreasing	
Agricultural Operations	Crop Spraying & Dusting	Pest and weed control	Organic phosphates
			Chlorinated hydrocarbons
			Organic lead and arsenic materials
	Field Burning	Stubble and slash burning	Smoke, flyash, and soot
			Carbon monoxide
			Oxides of nitrogen
	Frost Damage Control	Smudge pot and orchard heaters	Smoke and soot
			Sulfur and nitrogen oxides
			Carbon monoxide
Radioactive Materials	Ore Preparation	Crushing, grinding, and screening	Uranium dust
			Beryllium dust
	Fuel Fabrication	Gaseous diffusion	Fluoride materials
	Nuclear Fission	Nuclear reactors	Argon 41
	Spent Fuel Processing	Chemical separation	Iodine 131
	Nuclear Testing	Atmospheric explosions	Radioactive fallout
			Sr 90, Cs 137, C 14, Kr 85

Table 2. Emission Factors For Refuse Incineration [1]
lb material per ton refuse burned

Pollutant	Municipal Multiple Chamber	Industrial and Commercial Single Chamber	Multiple Chamber
Aldehydes	1.1	5-64	0.3
Carbon monoxide	0.7	20-200	0.5
Hydrocarbons	1.4	20-50	0.3
Oxides of Nitrogen	2.1	1.6	2.0
Oxides of Sulfur	1.9	n.a.	1.8
Ammonia	0.3	n.a.	n.a.
Organic Acids	0.6	n.a.	n.a.
Particulates	6-12	20-25	4.0

Note: n.a. = not available

CHAPTER 2

TERMINOLOGY

It is important to place the terminology on a uniform basis in source testing to minimize the possibility of confusion. Definitions, symbols, and units frequently used are listed, along with sections describing reference conditions and performance testing.

I. DEFINITIONS

Definition is made by reference to source test manuals of the Dade County [1], Los Angeles County [2], and Bay Area [3] Air Pollution Control districts, to ASME Power Test Codes 21 [4] and 27 [5], and other references.

Aerosol is a dispersion of solid or liquid particles suspended in a gaseous phase.

Air contaminants are solid, liquid, or gaseous materials or combinations thereof, which by their presence may produce undesirable effects on humans, animals, plants, surfaces, or materials.

Air pollution is the presence in the outdoor atmosphere of one or more air contaminants in combinations, in such quantities, and of such duration as to be potentially injurious to human, plant, or animal life, or to property, or which unreasonably interferes with the comfortable enjoyment of life or property, or the conduct of business.

Collection device is the unit (or units) used to collect the particulate or gaseous material of interest.

Dust particles may be organic or inorganic in nature, and are formed from operations other than combustion processes.

Emission is the act of passing air contaminants into the atmosphere by either a gas stream or other means.

Flue is any duct, breeching, stack, or other passageway for air, gases, or airborne materials to be conducted to the atmosphere. These terms are used, within limitations, somewhat interchangeably.

Flyash is the noncombustible, inorganic residue remaining after the combustion of organic materials, and which has become entrained in a gas stream. It includes such substances as ash, cinders, and sand.

Fumes are fine solid particles formed by condensation or sublimation of vapors from volatilized or molten solid materials, formed at high temperatures, and which have become suspended in a gas stream.

Gases are formless, dimensionless fluids which occupy any confined space where the molecules have a tendency to separate from each other. They are changed to the liquid or solid state by increased pressures at controlled temperatures, or by decreased temperatures at controlled pressures. Materials in the gaseous state at standard conditions are presumed to be gases.

Liquids are substances which "have free flow and movement of molecules among themselves, but without the tendency to separate from each other" [6]. Substances present in the liquid state at standard conditions are presumed to be liquids.

Mists are suspensions of finely divided liquid droplets in a gas or gaseous stream.

Particulate matter is any material or substance emitted into, or present in, the atmosphere in a finely divided form of solid or liquid state or combination thereof, normally at standard conditions. It does not include chemically or physically uncombined water vapor.

Prime mover is the device used to provide the means for withdrawing flue gas from the duct into the sample train.

Run is a subdivision of a test which corresponds to one-date observation at a single point in a sample location.

Sampling train is a series of devices into which a flue gas stream of known volume is withdrawn to facilitate sample collection and analysis.

Smoke is the assortment of solid particles, usually organic in nature, produced by incomplete combustion of organic material, e.g., tar, soot and carbon particles.

Solids are substances which are rigid and definite in form without free movement between molecules, and without the tendency for molecules to become separated from each other. Substances present in the solid state at standard conditions are presumed to be solids.

Soot particles are dark, organic particles which are formed from the incomplete combustion of organic fuels.

Source is any point or place from which materials are being emitted which are (or may become) air contaminants.

Stack gas is the total aggregate of gaseous, liquid, and solid materials being emitted from a source.

Standard conditions are used as a reference point for compressible fluids, at which results are calculated and reported. Unless otherwise specified, it refers to a temperature of 60°F (15.6°C), 1.0 atmosphere total pressure at dry gas conditions. The standard atmospheric pressure at sea level is 29.92 inches of mercury (760.0 mm Hg). Barometric conditions other than these are to be corrected to standard pressure at sea level.

Test means a complete single sample collection at a sample location with a complete set of observations and recorded data [5].

Vapor is the gaseous phase of a substance normally in the liquid or solid state.

Water vapor is liquid water which has become evaporated at source or ambient conditions. It is not ordinarily considered to be a contaminant *per se*.

II. UNITS

The following units find common application in calculations involving air pollution measurements. More complete tabulations are found in the *Handbook of Air Pollution* [7] and Himmelblau [8].

Area:
$1.0 \text{ ft}^2 = 144.0 \text{ in.}^2 = 930.3 \text{ cm}^2 = 0.093 \text{ m}^2$

Concentration (Volume/Volume):
$1.0 \text{ ppm} = 10^3 \text{ ppb} = 10^{-3} \text{ ppt}$
$1.0 \text{ ppm} = 1.0 \text{ ft}^3 \text{ gas}/10^6 \text{ ft}^3 \text{ diluent}$

Concentration (Weight/Volume):
$1.0 \text{ gr/ft}^3 = 2.28 \text{ gm/m}^3$

Emission Rate:
$1.0 \text{ lb/hour} = 24.0 \text{ lb/day} = 7.56 \text{ gm/min}$
$1.0 \text{ lb/day} = 4.86 \text{ gr/min}$

Length:
$1.0 \text{ ft} = 12.0 \text{ in.} = 30.5 \text{ cm} = 0.305 \text{ meter}$

Mass:
$1.0 \text{ lb} = 454.0 \text{ gm} = 7{,}000 \text{ grains} = 7{,}000 \text{ gr}$
$15.43 \text{ grains} = 1.0 \text{ gm} = 1{,}000 \text{ mg}$
$1.0 \text{ gm} = 1{,}000 \text{ mg} = 10^6 \text{ }\mu\text{g}$

Power:
$1.0 \text{ hp} = 0.746 \text{ kw} = 550 \text{ ft-lb/sec}$
$= 0.707 \text{ BTU/sec}$

Pressure:
$1.0 \text{ atm} = 29.92 \text{ in. Hg} = 33.8 \text{ ft H}_2\text{O}$
$1.0 \text{ atm} = 14.7 \text{ lb/in.}^2 = 76.0 \text{ cm Hg}$
$= 760.0 \text{ mm Hg}$

Temperature:
$°R = °F + 460; °K = °C + 273;$
$°F = 1.8(°C) + 32$

Time:
$1.0 \text{ hr} = 60.0 \text{ min} = 3{,}600 \text{ sec}$
$1.0 \text{ day} = 24.0 \text{ hr} = 1{,}440 \text{ min} = 86{,}400 \text{ sec}$

Volume:
$1{,}000 \text{ ml} = 1.0 \text{ liter} = 10^{-3} \text{m}^3$
$28.32 \text{ liter} = 1.0 \text{ ft}^3 = 7.48 \text{ gal}$
$35.3 \text{ ft}^3 = 1.0 \text{ m}^3$

III. SYMBOLS

A. Subscripts

b = Barometric
c = Condensed water
d = Dry conditions
db = Dry bulb
g = Gaseous
i = Individual constituent
m = Meter conditions
n = Sample nozzle or probe
O = Orifice conditions
o = Standard conditions
      Initial conditions in time problems
p = Particulate
s = Stack conditions
st = Saturation conditions
T = Total mixture
w = Water vapor
wb = Wet bulb

B. Terms

Symbol	Description	Unit
A	Cross-sectional area of a duct, tube, or nozzle	ft²
C	Concentration of material in a gaseous stream at a certain point	

D	1.) Diameter 2.) Characteristic dimension of a duct	
d	Density of a gas at a certain set of conditions	lb/ft³
E	Removal efficiency on a weight basis of a collection device for a given material	Percent
ER	Emission rate of material in a gas stream at a certain point	lb/hour
F_{PT}	Correction factor for an S-type pitot tube	none
g	Acceleration due to gravity	ft/sec²
H	Humidity	
L	Volumetric liquid flow rate for a certain set of conditions	gal/min
L_c	Liquid volume of water condensed	milliliters
MW	Molecular weight of a gas or gaseous mixture	$\frac{\text{lb}}{\text{lb-mole}}$
m	Mass of a gas present in a given volume at a certain set of conditions	lbs
N	Number of sample or traverse points utilized during a test	none
n	Number of moles of a gas present	lb-mole
ΔP	Pressure differential between impact and static pressures when using a pitot tube for gas flow measurements	in. H₂O
P	Absolute pressure at a point	in. Hg
p	Partial pressure exerted by a constituent gas in a gaseous mixture	in. Hg
Q	Volumetric rate of gas flow at a certain set of conditions	ft³/min
R	Universal gas constant expressed in convenient units	$\frac{(\text{ft}^3)(\text{in. Hg})}{(\text{lb-mole})(°R)}$
T	Temperature at a point	°R
t	Time elapsed in a test	minutes
U	Velocity at a single point	ft/min
	in a gas stream at the conditions of flow	
V	Volume occupied by a gas at a particular set of conditions	ft³
v	Partial volume occupied by a single constituent in a gaseous mixture	ft³
W	Weight of material collected in collection device during a performance test	lbs
w	Weight of material collected in sampling train	gr or gm

C. Abbreviations

Symbol	Description	Unit
CF	Volume of gas sampled at actual conditions	ft³
CFM	Volumetric rate of flow at actual conditions	ft³/min
FPM	Velocity of a gas at actual conditions	ft/min
SDCF	Volume of gas at standard conditions of 60°F, 29.92 in. Hg, dry	ft³
SDCFM	Volumetric rate of gas flow at standard conditions of 60°F, 29.92 in. Hg, dry	ft³/min

D. Useful Numbers

Number	Description	Units
28.84	Molecular weight of dry air	lb/lb-mole
62.43	Weight of liquid water at standard conditions	lb/ft³
32.16	Acceleration due to gravity	ft/sec²
379.0	Volume occupied by one lb-mole of gas at 60°F, 29.92 in. Hg, dry	ft³/lb-mole

IV. REFERENCE CONDITIONS

A. Standard Conditions

The purpose of "standard conditions" in air pollution work is to provide a common base for defining allowable concentrations in compressible fluids. It is necessary to *specify* which "standard condition" is meant when reporting any stack concentration value to minimize the possibility of confusion or misinterpretation.

Some of the more common "standard conditions" in use are listed in Table 3. Several codes do not specify moisture conditions, providing for an element of confusion in interpreting regulations. Use of dry standard gas conditions is strongly encouraged to eliminate nonuniformities resulting from varying moisture contents in gas streams.

Unless otherwise specified, all future notations in this book to standard conditions refer to dry gas at 60°F and 1.0 atmosphere (29.92 inches of mercury at sea level). Conditions obtained at other than sea level should be corrected from the appropriate barometric pressures to sea level to avoid being unduly penalized by emission standards.

The variation of the relative density and barometric pressure for air with altitude is shown in Table 4. Corrections for barometric pressures to sea level are usually available from local meteorological data services.

B. Specified Conditions

Limitations are often set on the allowable emissions from combustion process exit gases by adjusting concentrations to a uniform basis to prevent circumventing emission standards by mere dilution with air. It is a common practice to require that stack concentrations be corrected to either 12 percent by volume carbon dioxide or 50 percent excess air, where both are expressed on a dry gas basis. These are nominal values that have been observed in flue gases from coal-fired boilers. An alternative correction is to 6 percent oxygen on a dry gas basis [1], which has also been found to correspond to approximately 50 percent excess air in incinerator, and coal and oil-fired boiler exit gases. It is a common provision in codes which make use of specified conditions to eliminate the carbon dioxide contribution from auxiliary fuel prior to making calculations, particularly for incinerators. The derivation for each relationship used is presented in the appendix.

C. Twelve Percent Carbon Dioxide

A stack gas concentration is determined at standard conditions of 60°F, 29.92 inches Hg, dry. The average carbon dioxide content in the flue gas during the test is determined on a dry basis. The concentration of the material is then adjusted by considering that sufficient air is either added or removed from the flue gas so that its carbon dioxide content is 12 percent by volume on a dry basis. The method for concentration correction to 12 percent carbon dioxide is shown in the following equation, and a graph showing correction factors at varying CO_2 levels is shown in Figure 2.

$$F_{CO_2} = \frac{12.00}{(\% CO_2)} \quad (1)$$

$$C_{12} = (F_{CO_2})(C_o) = \frac{12.00}{(\% CO_2)}(C_o) \quad (2)$$

where:

C_o = Measured concentration of constituent at standard conditions

C_{12} = Measured concentration of constituent at standard conditions when corrected to 12% CO_2 by volume on a dry basis

F_{CO_2} = Correction factor for constituent concentration when adjusting to 12% CO_2 by volume on a dry basis

$\% CO_2$ = Percent carbon dioxide by volume on a dry basis

This correction is favored over the others because of its simplicity, since it requires only a single Orsat determination for carbon dioxide from the flue gas.

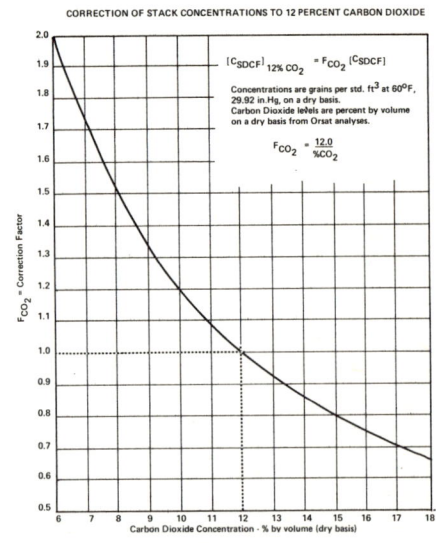

Fig. 2.

D. Six Percent Oxygen

The procedure is very similar to carbon dioxide except that a different reference substance from the combustion process is used as the basis for adjusting concentrations. The concentration is determined at standard conditions as before, and the oxygen content of the flue gas determined on a dry basis. The concentration is then adjusted to six percent oxygen by considering the addition or removal of sufficient air. The equation for correcting concentrations to six percent oxygen is shown below, and a chart of corrections for varying oxygen levels is shown in Figure 3.

$$F_{O_2} = \frac{0.15}{0.21 - (\% O_2/100)} \quad (3)$$

$$C_6 = (F_{O_2})(C_o) \quad (4)$$

C_o = Measured concentration of constituent at standard conditions

C_6 = Measured concentration of constituent at standard conditions when corrected to 6% oxygen by volume on a dry basis

F_{O_2} = Correction factor for constituent concentration when adjusting to 6% oxygen by volume on a dry basis

$\% O_2$ = Percent oxygen on a dry basis

The use of oxygen as a reference material requires two Orsat determinations, but may be easier if there is a continuous oxygen monitor on the exit gases from a combustion unit.

E. Fifty Percent Excess Air

The third and most complicated procedure is to correct flue gas concentrations to 50 percent excess air over the theoretical amount required for combustion. Excess air may be determined from either a fuel or flue gas analysis, where the amount of excess air is defined as follows:

% Excess Air
$$= \frac{\text{Air-Supplied} - \text{Theoretical Air}}{\text{Theoretical Air}} \times 100 \quad (5)$$

However, this concept requires an analysis of the fuel burned, a very questionable procedure when burning such highly nonuniform fuels as waste wood (hogged fuel) or refuse. An *approximation* of the excess air involves only an Orsat analysis and considers the volumetric ratio of nitrogen to oxygen in the flue gas. However, the method is limited because of its inherent complexity, the presumption of complete combustion, and the possible presence of significant amounts of oxygen or nitrogen in certain fuels. Therefore, it should be used with considerable caution. An Orsat analysis gives the respective CO_2, O_2, and CO contents, and the nitrogen content is then calculated as follows, where all concentrations are expressed in percent by volume on a dry basis.

$$\%(N_2) = 100 - [(\%CO_2) + (\%O_2) + (\%CO)] \quad (6)$$

The ratios of nitrogen to oxygen for conditions of ambient air, theoretical combustion air, and certain intermediate conditions are shown in Table 5.

The necessary corrections are determined by consideration of addition or removal of sufficient air to provide for nitrogen/oxygen ratios of 11.3, corresponding to 50 percent excess air. Once the nitrogen/oxygen ratio is determined from Orsat analyses, the percent excess air is determined from the left vertical axis shown in Figure 4. The correction factor to be applied is determined from the right vertical axis and may be computed by using the following expression:

$$F_{EA} = \frac{11.3}{(N_2/O_2)} \quad (7)$$

$$(C)_{\substack{50\% \\ EA}} = \frac{11.3}{(N_2/O_2)}(C_o) = (F_{EA})(C_o) \quad (8)$$

In most cases it is easiest to use 12 percent

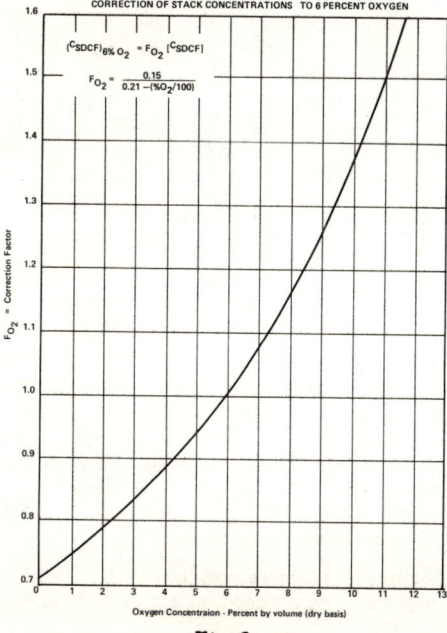

Fig. 3.

carbon dioxide as the specified condition for fossil-fuel fired boilers and incinerators because of the simplicity in determination.

The oxygen correction may be used as an alternative procedure, but may require two Orsat measurements. The use of 50 percent excess air is more complicated than the other corrections and is subject to limitations in its application.

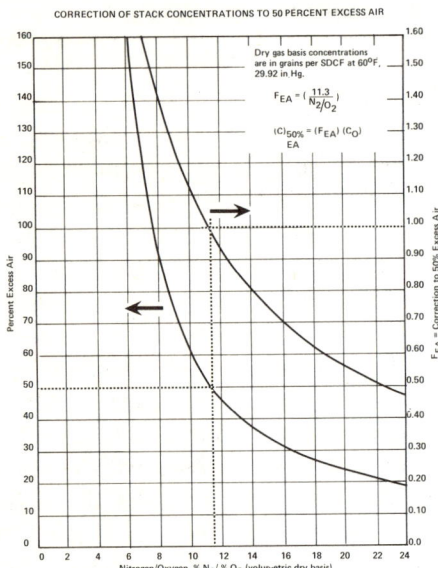

Fig. 4.

V. PERFORMANCE TESTING

Performance testing involves determination of the collection efficiency of an air pollution control device. By nature, it is involved with sampling both inlet and outlet to determine emission levels, preferably simultaneously. The collection efficiency is defined as the ratio of the amount of material entering the device which is collected, to the amount of material entering the device, where both are based on the same time interval. The expression is given by the following equation:

$$E = \frac{W}{ER_{In}} \cdot 100 = \frac{(ER_{In} - ER_{out})}{ER_{In}} \cdot 100 \quad (9)$$

where:
- E = Collection efficiency of collector in percent by weight
- W = Amount of material collected in lbs per hour
- ER_{In} = Amount of material entering collector in lbs per hour
- ER_{out} = Amount of material leaving collector in flue gas in lbs per hour

It is a good idea to perform a material balance by checking source testing results with the material collected to aid accuracy. It is sometimes difficult to gain sample access to the collector inlet, and a material balance is then necessary. In this case:

$$ER_{In} = W + ER_{out} \quad (10)$$

The above equations can apply to total particulates, particle breakdown by size range, and for different gases. More thorough treatments on the subject are available [5] [10].

VI. REFERENCES

1. Rivera, M. F., "Incinerator Source Testing Manual," Dade County Pollution Control Authority, Miami, Florida, September 1968.
2. Devorkin, H., Chass, R. L., Fudurich, A., and Kanter, C. V., Holmes R. G. ed.; Source Testing Manual" Los Angeles County Air Pollution Control District, Los Angeles, California, 1965.
3. Wolfe, E. A., ed., "Source Test Methods," Bay Area Air Pollution Control District, San Francisco, California, 1961.
4. "Dust Separating Apparatus," Performance Test Code 21-1941, American Society of Mechanical Engineers, New York, New York, 1941.
5. "Determining Dust Concentration in a Gas Stream," Performance Test Code 27-1957, American Society of Mechanical Engineers, New York, New York, 1957.
6. *Webster's New Collegiate Dictionary*, G. & C. Merriam Company, Springfield, Massachusetts, 1956.
7. Sheehy, J., Achinger, W., and Simon, R., "Handbook of Air Pollution," U. S. Public Health Service, Division of Air Pollution, Cincinnati, Ohio, 1966.
8. Himmelblau, D. M., *Basic Principles and Calculations of Chemical Engineering*, 2nd ed., Prentice-Hall, Inc., Englewood Cliffs, New Jersey, 1967.
9. Jorgensen, R., ed., *Fan Engineering*, Buffalo Forge Co., Buffalo, New York, 1961.
10. Caplan, K. J., ed., "Performance Testing," Ch 12 in *Air Pollution Manual, Part 2–Control Equipment*, American Industrial Hygiene Association, Detroit, Michigan, 1968.

Table 3. Summary of Standard Conditions.

Agency	Temp. °F	Pressure in. Hg	Moisture Content % by vol.
1. Puget Sound, Bay Area	60	29.92	0.0 (dry)
2. STP (Gas Laws), ASME	32	29.92	0.0 (dry)
3. South Carolina	70	29.92	(1)
4. San Bernardino	60	29.92	(1)
5. Los Angeles	60	29.92	(2)
6. Florida	68	29.92	(1)

Notes:
1. Moisture content is not specified.
2. Treat water as a noncondensable fluid and include all water.

Table 4. Changes in Density and Barometric Pressure of Air with Altitude.

Altitude feet	Relative Density	Barometric Pressure inches Hg
Sea level	1.000	29.92
500	0.982	29.38
1,000	0.964	28.86
1,500	0.947	28.33
2,000	0.930	27.82
2,500	0.913	27.32
3,000	0.896	26.82
3,500	0.880	26.32
4,000	0.864	25.84
4,500	0.848	25.36
5,000	0.832	24.89
6,000	0.801	23.98
7,000	0.772	23.09
8,000	0.743	22.22
9,000	0.715	21.39
10,000	0.688	20.58
15,000	0.564	16.89

Courtesy of Buffalo Forge Company, *Fan Engineering*, p. 32 [9]

Table 5. Nitrogen/Oxygen Ratios for Varying Excess Air Levels.

Excess Air %	Nitrogen %T.A./100	Oxygen %T.A./100	N_2/O_2 Ratio
0.0 (T.A.)	79.0	0.0	$79/0 = \infty$
10.0	(1.10)(79.0)	(0.1)(21.0)	$87/2.1 = 41.4$
25.0	(1.25)(79.0)	(0.25)(21.0)	$99/5.2 = 18.9$
50.0	(1.50)(79.0)	(0.50)(21.0)	$118/10.5 = 11.3$
100.0	(2.00)(79.0)	(1.00)(21.0)	$158/21 = 7.5$
200.0	(3.00)(79.0)	(2.00)(21.0)	$247/42 = 6.1$
500.0	(6.00)(79.0)	(5.00)(21.0)	$474/105 = 4.5$
Ambient Air (2)	(79.0)	(21.0)	$79/21 = 3.76$

Notes:
1. T.A. = Theoretical combustion air required.
2. Percent excess air is infinity.

CHAPTER 3

PRINCIPLES

Several theoretical principles are useful in source testing calculations to define the behavior of gas streams, water vapor, and dynamic properties of gases and particles. Consideration will be given to: 1) the expansion and contraction characteristics of gases, 2) dynamic aspects of gas stream flow, 3) psychrometric properties of water vapor-air mixtures, and 4) the dynamics of particle motion in fluids.

I. GAS LAWS

The ideal gas laws define the volume occupied by an imaginary gas in terms of its temperature, pressure, and mass of the gas molecules. However, the individual molecules are at great enough distances from each other that the interactive forces between them are negligible; the volume occupied by the molecules themselves is negligible so that compressibility effects are insignificant in the total system. These laws are not followed by any actual gases for all conditions, but are definitely useful for air and most other lighter gases sufficiently above their boiling points at low pressures. However, water vapor, sulfur dioxide, and certain other gases deviate significantly from ideal behavior near their boiling points and at high pressures [1]. Several other equations of state have been proposed to correct for these deviations from ideality, but these involve added complexity. The assumptions used with the ideal gas laws have been found to be sufficiently accurate and simple to find wide usage in correcting from one set of conditions to another and in computing mass emission rates.

The first three laws describe behavior of a total ideal gas volume, while the others describe behavior of constituents in gaseous mixtures.

A. Boyle's Law

Boyle's law states that the volume occupied by a gas at constant mass and temperature is inversely proportional to pressure, or that the product of pressure and volume is a constant for such conditions.

$$PV = \text{Constant} \quad (11)$$

thus:

$$P_1 V_1 = P_2 V_2 \quad (12)$$

where: P = Pressure (in. Hg)
V = Volume (ft³)

B. Charles' Law

Charles' law states that the volume of a gas at constant mass and pressure is directly proportional to temperature.

$$\frac{V}{T} = \text{Constant} \quad (13)$$

thus:

$$\frac{V_1}{T_1} = \frac{V_2}{T_2} \quad (14)$$

where: T = Temperature = °R

C. Perfect Gas Law

The perfect, or ideal, gas law combines the findings of the two above laws, defining volume changes in terms of simultaneous changes in temperature and pressure. The product of pressure and volume divided by temperature is a constant for a given amount of gas. These equations allow correction from one set of conditions to another, as follws:

$$PV = KT, \text{ where } K \text{ is a constant} \quad (15)$$

thus:

$$\frac{P_1 V_1}{T_1} = \frac{P_2 V_2}{T_2} \quad (16)$$

The value for the above constant "K" is the product of the number of moles (n) of gas present and the so-called "Universal Gas Constant," or proportionality constant, R, for the gas laws.

$$K = nR = \frac{m}{MW} R \quad (17)$$

where: n = Moles of gas
m = Mass of air
MW = Average molecular weight of gas
R = Universal Gas Constant

Values are listed in Perry [2] for this constant in several systems of units. Care must be taken to use consistent units in gas law computations, where frequently used values are:

$$R = 21.85 \text{ in units of } \frac{(ft^3)(in\ Hg)}{(°R)(lb\text{-mole})};$$

$$R = 82.06 \frac{(cm^3)(atm)}{(gm\text{-mole})(°K)}.$$

Avogadro's Number expresses the number of molecules present in one mole of any substance, or 6.023×10^{23} molecules per gram-mole in the metric system. There are $454 \times 6.023 \times 10^{23}$ molecules per pound-mole in the English system. The mass of material present is the weight per mole (or molecular weight) times the number of moles (a fixed number of molecules). The term mole merely provides a more convenient form of expression of the equivalent number of molecules in one molecular weight of a substance.

$$m = (n)(MW); \quad n = \frac{m}{MW} \quad (18)$$

The Perfect Gas Law also states the volume occupied by any gas is independent of its molecular weight. The perfect gas law may then be used to define the volume of gas present in terms of mass, temperature and pressure.

$$PV = nRT = \frac{m}{MW} RT \quad (19)$$

It is then possible to determine the amount of gaseous material present, in terms of volume, to allow calculation of concentrations and emission rates.

D. Volume per Mole

Useful relationships are the volumes per mole at conditions of standard temperature and pressure (STP) or 32°F (0°C) and 1.0 atmosphere pressure (see Table 6). These allow direct conversion of amounts of gaseous materials collected to equivalent gaseous phase volumes, in computing gas concentrations and emission rates in flue gases.

E. Gas Density

Gas density (ρ) is defined as the mass per unit volume, and is commonly used to calculate gas concentrations on weight basis, and in computing emission levels in terms of weight of material emitted per unit weight of flue gas.

Repeating equation 19:

$$PV = \frac{m}{MW} RT \quad (20)$$

Thus:

$$\rho = \frac{m}{V} = \left(\frac{m}{MW}\right)\left(\frac{P}{T}\right) \quad (21)$$

The density of dry air at 70°F and 29.92 in. Hg is 0.075 pounds per cubic foot, and 0.076 at 60°F and 29.9 in. Hg. The densities of most flue gases, particularly from combustion processes, are similar to ambient air and require only minor corrections. The presence of large quantities of water vapor in the flue gas tends to reduce the density while carbon dioxide has the opposite effect. Gas density figures should always be referred to a particular set of temperature and pressure conditions to minimize the possibility for confusion.

F. Dalton's Law

Dalton's law states that the summation of partial pressures exerted by the individual gaseous components is equal to the total pressure in a gaseous mixture [1]. The partial pressure of each constituent gas is determined by considering that it alone occupies the given volume at the same temperature and pressure conditions.

$$P_T = \sum_i p_i = p_1 + p_2 + p_3 + \cdots + p_n \quad (22)$$

where: i = Individual components (1, 2, 3, etc.).
T = Total mixture

If both the total gas mixture and the individual constituents follow the ideal gas laws, it can then be shown that the ratio of partial pressure of a constituent to the total pressure is the same as the number of molecules or moles of the constituent to the total number of molecules or moles in the gaseous mixture. At constant temperature and volume, the following relationships apply:

Constituent: $\quad p_i V = n_i RT \quad (23)$

Total: $\quad P_T V = n_T RT \quad (24)$

Ratio: $\quad \dfrac{p_i v}{P_T v} = \dfrac{n_i RT}{n_i RT}$

which reduces to:

$$\frac{p_i}{P_T} = \frac{n_i}{n_T} \quad (25)$$

The partial pressure of each gas is then proportional to its molar concentration in the mixture, and Dalton's law may be used for calculating moisture content of gas streams, particularly when the static pressure at a point is substantially different than 29.92 in. Hg. It is necessary to have knowledge of the vapor pressure of water at different temperatures to use this method and a chart is provided in the appendix for this purpose.

G. Amagat's Law

Amagat's law states that the summation of the partial volumes contributed by each constituent of a gaseous mixure is equal to the total volume, at the same temperature and pressure.

$$V_T = \sum_i V_i = V_1 + V_2 + V_3 + \cdots + V_n \quad (26)$$

$$n_T = \sum_i n_i = n_1 + n_2 + n_3 + \cdots + n_n \quad (27)$$

It can then be shown that the ratio of the partial volume occupied by a constituent gas to the total volume is the same as the ratio of moles of the constituent gas to the total number of moles in the system. The above statements apply for constant temperature and pressure, and the volume percent and mole percent are then equivalent.

Constituent: $\quad PV_i = n_i RT \quad (28)$

Total: $\quad PV_T = n_T RT \quad (29)$

Ratio: $\quad \dfrac{PV_i}{PV_T} = \dfrac{n_i RT}{n_T RT}$

which reduces to:

$$\frac{V_i}{V_T} = \frac{n_i}{n_T} \quad (30)$$

Amagat's law is then used for calculation of moisture content with the psychrometric chart, and in Orsat analyses of flue gases.

II. FLUID MOTION

A. Conservation of Energy

The total energy at any section in a duct at steady flow is equal to the total energy at any other section in the direction of flow plus the energy loss due to friction between the two sections. Bernoulli's equation states that the total head (pressure head + potential head + velocity head) at any section in a duct equals the total head at any other downstream section plus the head loss resulting from friction [3] where all terms are in consistent units of height of the fluid flowing. There is no accumulation or loss of mass or energy in the system.

$$\frac{P_1}{\rho_1 g} + Z_1 + \frac{U_1^2}{2g} = \frac{P_2}{\rho_2 g} + Z_2 + \frac{U_2^2}{2g} + h_f \quad (31)$$

where:

- Z = Elevation above reference plane
- U = Velocity (ft/sec)
- h = Velocity head (ft of fluid)
- g = 32.2 (ft/sec^2) (gravity constant)
- h_f = Energy loss resulting from friction

The kinetic energy of flow at a point is expressed by the term $\dfrac{(U^2)}{(2g)}$. It applies for uniform velocity in a duct, and holds approximately for turbulent flow conditions [2]. The kinetic energy head, or pressure, resulting from velocity at any point is given by the following equation:

$$h = \frac{U^2}{2g} \quad \text{or} \quad U = \sqrt{2gh} \quad (32)$$

The velocity head is usually measured in convenient units such as equivalent inches of water pressure differential, since direct measurement in terms of feet of gas is too cumbersome. These relationships are then used to determine velocity in ducts and stacks using standard or Stausscheibe-type pitot tubes. The same basic relationship also applies to other volumetric measuring devices, such as orifice flow meters and venturi meters.

B. Conservation of Mass

The principle of conservation of mass states that the total amount of material present in a duct remains constant if no material is added or removed.

$$m = \rho_1 A_1 U_1 = \rho_2 A_2 U_2 \quad (33)$$

$$Q = A_1 U_1 = A_2 U_2 \quad \text{if} \quad \rho_1 = \rho_2 \quad (34)$$

where A = The cross-sectional area of duct in ft.^{2}

Q = Volumetric rate of flow in ft^3 per min.

Q = Volumetric rate of flow in ft^3/min.

This principle is used to calculate isokinetic sampling rates for particulate sampling and total gas flows to obtain mass emission rates of materials.

C. Psychrometry

Psychrometric charts are constructed from vapor pressure data, Dalton's and Amagat's law, and enthalpy properties for air-water vapor mixtures; they express the amount of water vapor per unit amount of air at various temperatures. They are usually constructed for a standard atmospheric pressure of 29.92 inches of Hg. Thorough treatments describing their construction and use are available in other references [2] [4] [5]. They provide a rapid and convenient method for estimating moisture contents in ambient air and flue gas streams.

Several terms must first be defined when making use of psychrometric charts. The *dry bulb* temperature is the measured temperature of air or the flue gas stream. The *wet bulb* temperature is the temperature at which a mixture of air or flue gas with water vapor would become saturated. It is the temperature at which heat is transferred from the air-water mixture by convection and conduction at the same rate that heat is being withdrawn from the wick by evaporation of water. It is measured by wrapping a cloth wick around the tip of a thermometer, and saturating it with water. The thermometer is then introduced into the gas stream and the temperature observed. For optimum results, the velocity of the gas stream should be 1,000 to 2,000 feet per minute in passing the wick [4]. The *dew point* temperature is the temperature at which the air-water mixture becomes saturated and condensation begins. It is equal to the wet bulb temperature at saturation and is defined from the partial vapor pressure data for water given in the appendix. The *specific humidity* of an air-water mixture is the weight ratio of water vapor to dry air, expressed as grains of water per pound of dry air. Knowledge of wet and dry bulb temperature allows determination of the moisture contents of gaseous streams using a psychrometric chart.

The equation for vapor pressures of air-water mixtures at conditions other than saturation was developed by Carrier in 1911 [6]. It applies from the dew point temperature to approximately 400°F with sufficient accuracy for use.

$$P_w = P_{wb} - \left[\frac{(P_s - P_{wb})(T_s - T_{wb})}{2{,}830 - 1.44\, T_{wb}}\right] \quad (35)$$

T_s = Dry bulb temperature of flue gas in °F

T_{wb} = Wet bulb temperature of flue gas in °F

P_s = Static pressure of flue gas in inches of Hg

P_w = Actual vapor pressure exerted by water in inches of Hg

P_{wb} = Saturation vapor pressure of water at wet bulb temperature in inches of Hg.

The moisture content may then be calculated by Dalton's law as follows:

$$MC = \frac{P_{H_2O}}{P_T} \times 100 \quad (36)$$

MC = Moisture content of flue gas in percent by volume.

Corrections in the specific humidity by weight are necessary when carbon dioxide is present in the flue gas. However, at levels of 6 to 15 percent CO_2 normally encountered in flue gases this correction is not significant except at high temperatures [2]. An additional correction is to modify the moisture content in terms of barometric pressure, as is evident from Dalton's law. Decreases in total static pressure, due to elevation change or vacuum, raise the equivalent moisture content at a given temperature. A chart of the necessary corrections to be added to or subtracted from specific humidity values are shown in Hougen and Watson [5].

III. PARTICLE DYNAMICS

Solid and liquid particles being emitted from sources have many different properties in terms of size, shape, density, chemical composition, cohesiveness, and structure [7]. These variations present added complexities when sampling for particles in flue gases, as will be discussed in a later section. Particle size distribution in source gases is particularly important when evaluating the needs for collection devices. The Stanford Research Institute has recently published a chart of size characteristics for particles from different sources, as shown in Figure 5. Detailed descriptions of particle size distribution, their significance, and methods for determination are discussed in other publications [8] [9].

The dynamics of particle motion is of interest in source sampling because particle motion does not always coincide with gas stream motion, as discussed by Hawksley [10]. Particles smaller

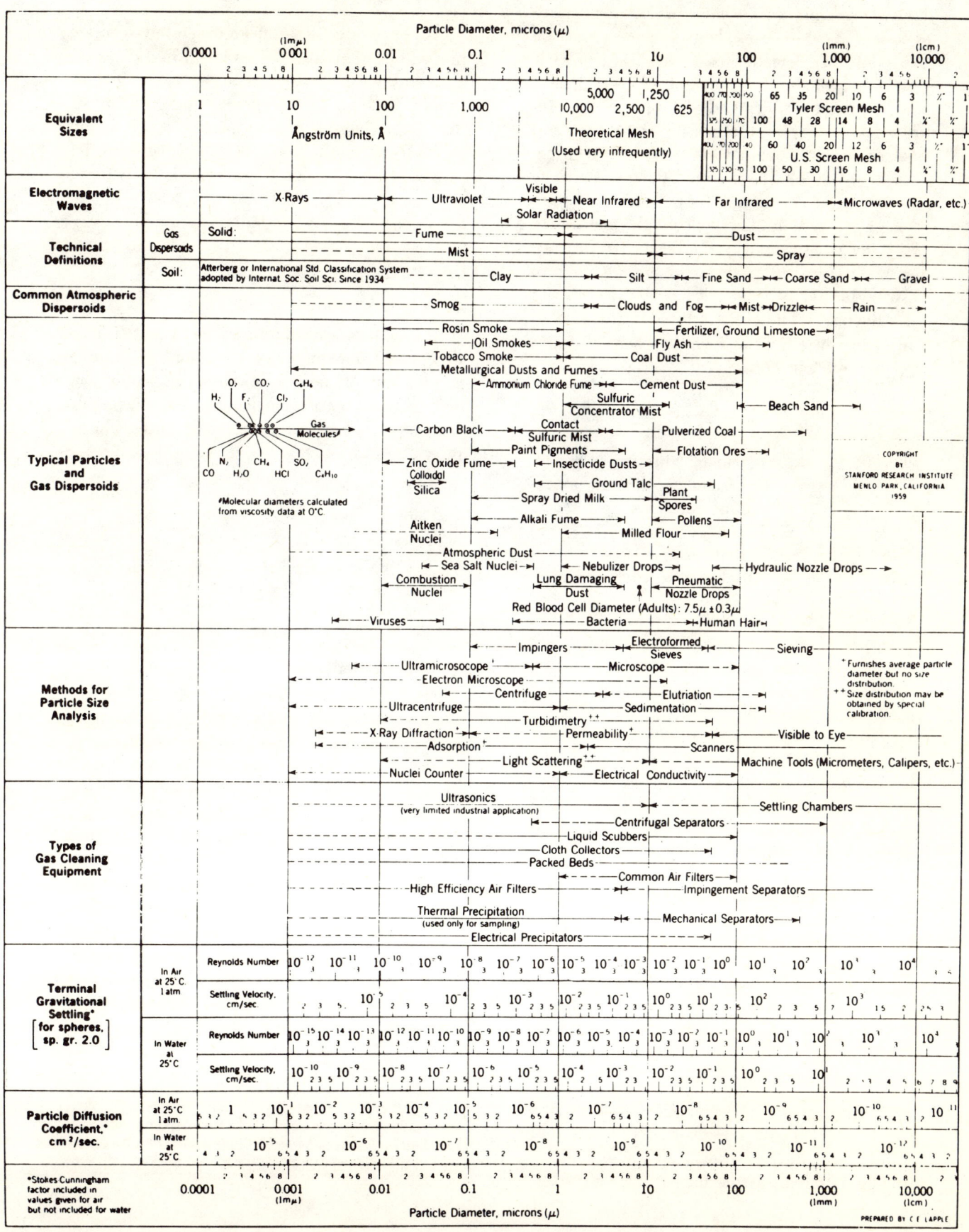

Fig. 5.

Courtesy of Stanford Research Institute.

than three to five microns in size will follow gas flow streamlines irrespective of changes in direction [11]. However, larger particles can be expected to deviate from flow streamlines, particularly in flowing around elbows, bends, and constrictions. This deviation can be predicted for particles in terms of size, shape, and specific gravity, but not always with precision.

Large particles have more difficulty than the gas stream in negotiating turns because of their greater mass; thus, they selectively migrate to the outer edge following the bends. Turbulence tends to return the particles to a more uniform distribution at varying distances downstream [9]. Reynold's number expresses the degree of turbulence in a duct as follows, where all terms are in consistent units.

$$N_{Re} = \frac{\rho V D}{\mu} \quad (37)$$

where: N_{Re} = Reynolds' Number
D = Characteristic length dimension of duct
V = Velocity
ρ = Density of gas
μ = Absolute viscosity of gas

Reynolds' numbers usually are well into the turbulent region in most source test applications, normally ranging from 2.0×10^5 to 10×10^6. This tends to promote mixing and make for a relatively uniform particle distribution [9] and temperature profile across a duct.

An additional consideration is involved with flow in horizontal ducts. Larger particles tend to settle out in the duct with a settling velocity described by the Stokes equation, where all terms are in consistent units.

$$V = \frac{1}{18\mu}(\rho_p - \rho_g)gD^2 \quad (38)$$

where:
ρ_p = Density of the particle
ρ_g = Density of the flue gas
μ = Absolute viscosity of the gas
g = Acceleration due to gravity (gravity constant)
D = Stokes diameter of the particle

This settling of larger particles in long horizontal ducts then tends to produce a non-uniform concentration profile. Therefore, it is usually better to sample for particles in vertical ducts where this selective settling problem does not exist.

It is not possible to make exact and accurate predictions of particle behavior in flue gases for all conditions. Factors such as swirling flow profiles in ducts and tangential gas entry into a duct represent additional problems. However, it can be said that sampling in locations with relatively stable and uniform flow patterns where the particle distribution is relatively uniform across the duct is highly desirable.

IV. GENERALITIES

A. Murphy's Law

A useful rule in source testing is that "if anything can go wrong it will." Therefore, it is a good idea to take all available precautions during sampling and analysis to assure accurate and representative results, and to check all procedures and data.

B. Peter Principle

Rising to one's "level of incompetence" on high stacks at altitudes reaching 300 feet or more can be an extremely dangerous phenomenon [12]. Therefore, in any field tests due caution and prior thought should be given before making any actions or motions.

C. Bluford's Law

For large organizations of any kind, doubling the size of the hierarchy tends to increase its complexity by a factor of four. It is important in planning and performing any source test to always be sure to contact the person who can and will be of the most assistance in making a prospective study. Normally, the support of a person high enough in responsibility to have sufficient authority is very helpful in minimizing problems involved in source tests. Unfortunately, the best of technical expertise can be easily undone by political actions at a somewhat lower performance level. Dealing with the people involved must always be considered in any source test study.

V. REFERENCES

1. Himmelblau, P. M., *Basic Principles and Calculations of Chemical Engineering*, 2nd ed., Prentice-Hall, Inc., Englewood Cliffs, New Jersey, 1967.
2. Perry, J. H., *Chemical Engineers' Handbook*, 4th ed., McGraw-Hill Book Co., New York, New York, 1963.

3. Streeter, V. L., *Fluid Mechanics,* 3rd ed., McGraw-Hill Book Co., New York, New York, 1962.
4. Hays, A. D., "Primer on Psychrometrics," Reprinted from Series in *Air Engineering,* 1966.
5. Hougen, O. A., Watson, K. M., and Ragatz, R., *Chemical Process Principles, Part 1, Material and Energy Balances,* John Wiley and Sons, Inc., New York, New York, 1959.
6. Carrier, W. H., "Rational Psychrometric Formulae," *Transactions of the American Society of Mechanical Engineers,* Vol. 33, pp. 1309-1350, 1911.
7. Lapple, C. E., "Particle Size Analysis and Analyzers," *Chemical Engineering,* Vol. 75, No. 11, pp. 149-156, May 20, 1968.
8. Cadle, R. D., *Particle Size Determination,* Interscience Publishers, Inc., New York, New York, 1955.
9. Orr, C., and DallaValle, J. M., *Fine Particle Measurement,* MacMillan and Company, New York, New York, 1959.
10. Hawksley, P. G. W., Badzioch, S., and Blackett, J. H., *"Measurement of Solids in Flue Gases,* British Coal Utilization Research Association, Leatherhead, Surrey, England, 1961.
11. Bloomfield, B. D., "Source Testing," Ch. 28 in Stein, A. C., ed., *Air Pollution,* Vol. 2, pp. 487-536, Academic Press, Inc., New York, New York, 1968.
12. Peter, L. J., and Hull, R., *The Peter Principle,* Bantam Books, Inc., New York, New York, 1969.

Table 6. Volume per Mole for Gases.

Volume/Mole,	Units	Temp	Pressure
22.4	liter/gm-mole	0°C	1.0 atm
359.0	cf/lb-mole	32°F	1.0 atm
379.0	cf/lb-mole	60°F	1.0 atm

CHAPTER 4

GAS FLOW MEASUREMENTS

Measurement of gas flow rates and other parameters in ducts is important in determining both concentrations and emission rates for gaseous and particulate materials. The variables to be considered in measuring flue gas flows include temperature, pressure, moisture content, dry gas composition and density, velocity, and total volumetric rate of flow. It is always desirable to have the process unit being tested operate at constant flow conditions throughout the test period so as to minimize the number of variables.

I. SAMPLING POINTS

It is first necessary to locate sampling ports in ducts to facilitate making flow measurements and taking samples from flue gases. This involves consideration of the location of the sampling port in terms of accessibility and representative sampling conditions, the arrangement of sampling ports at a given location, and the size and design of the actual port.

A. Sampling Location

When testing for particulates, sampling ports should be located in regions with stable flow patterns, if possible. This will provide the most representative sampling and flow measuring conditions. They should ideally be located eight to ten diameters downstream, and three to five diameters upstream from any bends, elbows, junctions, or other constrictions. This is to minimize problems associated with nonuniform and unstable gas flow profiles, and nonuniform particle concentration patterns by providing distance for adequate mixing to occur, as source gas streams are normally in the turbulent region of flow [1].

However, this is not always possible, so the optimum locations should be chosen, or flow-straightening vanes added to the ducts [2]. Unfortunately, flow-straightening and honeycomb vanes are not always effective in making both concentration and flow profiles uniform, and they may become corroded or plugged, thereby creating a maintenance problem.

It is necessary to consider the accessibility, the need for ladders with safety loops, platforms, and railings, and the availability of water and electric power in locating sampling ports. Shelters to protect equipment and workers from adverse weather conditions are very helpful in making for better working conditions and more accurate results. A sufficient amount of room must be available for inserting pitot tubes and long probes between a duct and the nearest wall or obstruction. Safety considerations in both design of sampling ports and during testing should be kept in mind. It is also a good idea to have a block and tackle arrangement mounted on the sampling platform for lifting equipment to any elevated location.

B. Arrangement

Sampling ports are arranged so as to obtain access to a maximum area of the duct. Sampling ports in round ducts should be located so that there are two located perpendicular to each other in the same plane. It is also possible to place a third port at 45° to the other two, as shown in Figure 6. Rectangular ducts should have the ports located parallel to each other, in the same plane, as shown in Figure 7.

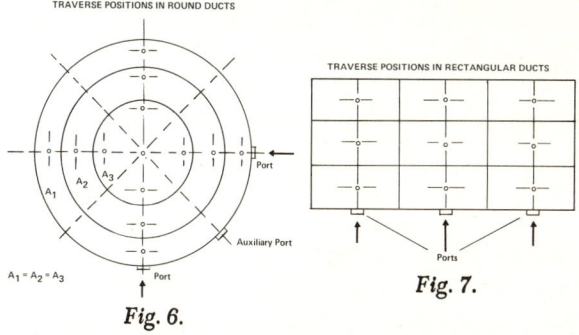

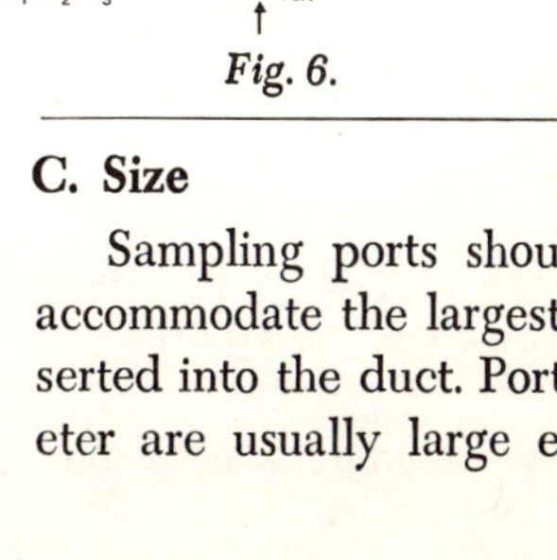

Fig. 6.

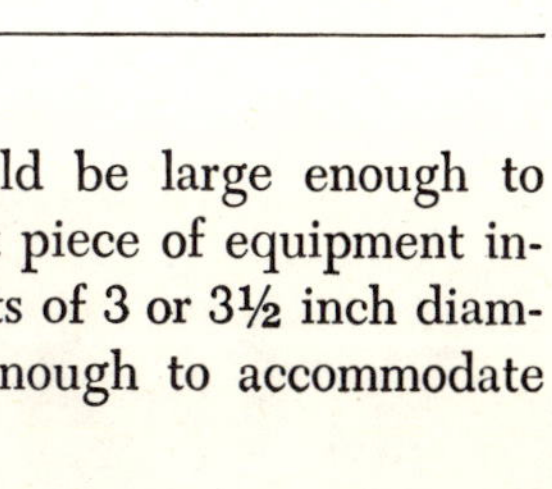

Fig. 7.

C. Size

Sampling ports should be large enough to accommodate the largest piece of equipment inserted into the duct. Ports of 3 or 3½ inch diameter are usually large enough to accommodate sampling equipment. Stainless steel flapper fittings, with rubber gasket seals, provide easily accessible, corrosion-resistant sampling ports. Typical port fittings are shown in Figure 8. Other sizes may be used to fit individual situations. A diagram of sample ports on a round stack is shown in Figure 9. Extensions are placed in line with the sample port to help support pitot tubes or long probes.

Sampling points under positive pressures greater than one to three inches of water can present a definite danger as well as a nuisance to source testing personnel. Research-Cottrell, Inc., has developed a modified sample port assembly which utilizes a closed tube with a valve at one end [3]. It isolates the sample probe or pitot tube prior to introduction into the duct, as shown in Figure 10.

Inclined or vertical ducts may be sampled using a probe similar to that shown in Figure 11. The gas stream is withdrawn and the condensed moisture is allowed to collect in the outer tube section. It may then be removed at the end of the test.

II. FLOW CONDITIONS

It is necessary to define the conditions of gas flow in terms of temperature, pressure, moisture content, and gas composition. This allows gas volumes sampled to be converted from meter to source or standard conditions when computing pollutant concentrations in flue gases.

A. Temperature

Temperature is a measure of the kinetic energy of the molecules in a gas stream, and is measured by several different devices [4]. *Thermometers* measure the height of rise resulting from thermal expansion of a liquid (usually mercury) in a sealed tube, as a function of temperature. They are normally used in the temperature range from −30 to +500°F. *Bimetallic thermometers* are useful in stack measurements because of their ruggedness and dial-type readout. They operate on the principle of differences in thermal expansion of two attached dissimilar metal strips, and are useful for temperatures up to 500°F. *Thermocouples* operate on the principle of the voltage differences produced at a junction between two different metal conductors as a function of temperature. Operating ranges vary with the combination of metals used, but

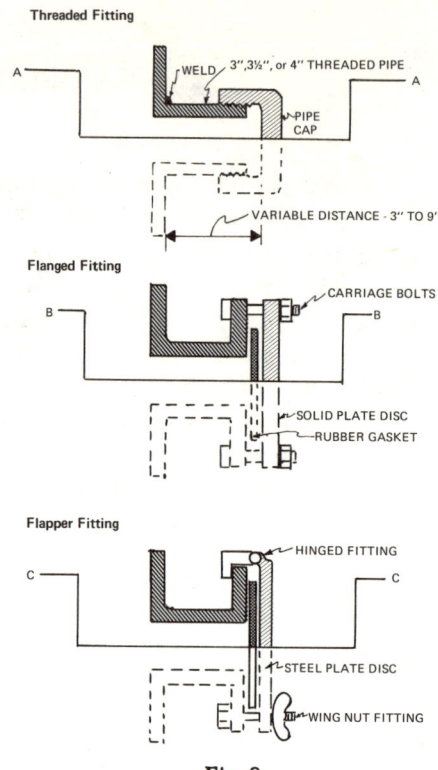

Fig. 8.

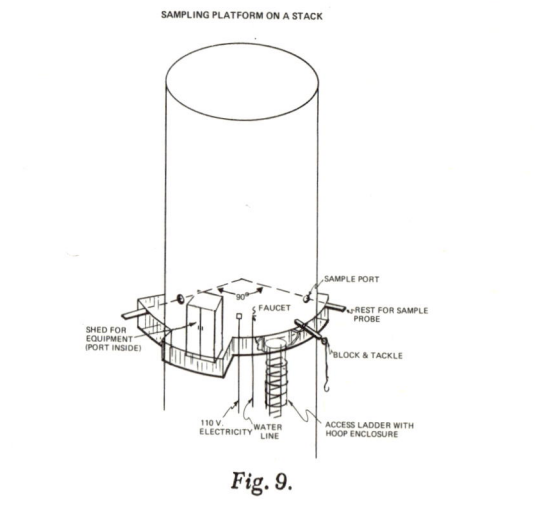

Fig. 9.

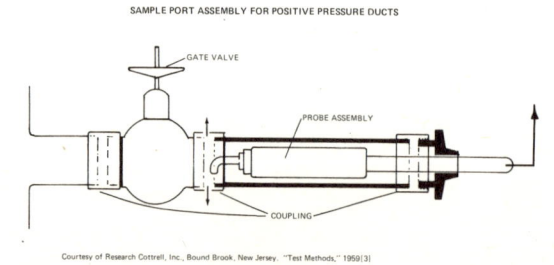

Fig. 10.

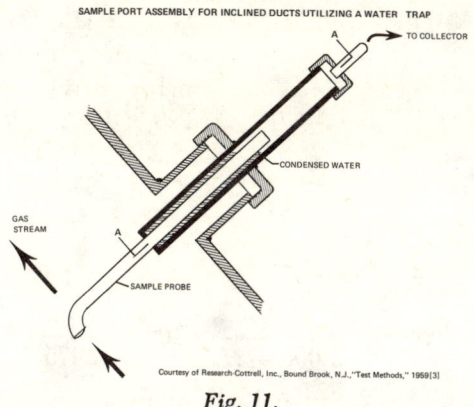

Fig. 11.

are normally from −200°F to 2,500°F [5]. *Thermistors* operate on the principle of changes in electrical resistance with temperature and are useful in the temperature range from −200°F to +1,000°F. *Optical pyrometers* are useful in measuring high temperatures exceeding 3,000°F, and measure the amount of radiant energy emitted from a body or surface. A more thorough discussion of temperature measurement is presented in Perry [5].

Dry bulb temperature is determined by inserting a thermometer, thermocouple, or pyrometer into a stack gas and allowing it to reach equilibrium. It may be measured intermittently or continuously during an extended test. The temperature probe should be inserted sufficiently far into the stack and the port covered to prevent cooling and resultant low readings, usually by stuffing a cloth or asbestos mat over the port opening.

B. Pressure

Pressure is the force per unit area exerted by the motion of gas molecules. *Barometric* pressure is the atmospheric pressure at the particular sampling location; standard atmospheric pressure is defined as 29.92 inches of mercury at sea level. Atmospheric pressure is usually measured with a barometer [5]. The variation of density and barometric pressure with altitude was shown previously in Table 4.

Static pressure in a duct is measured by comparison with barometric pressure, using a manometer or pressure gauge by connecting it to a straight tipped tube inserted into the duct. However, pressure gauges should be checked frequently for accuracy. Very precise pressure measurements can be made by using a micromanometer, whose use is described in Perry [5]

and *Fan Engineering* [6]. An S-type pitot tube can also be used to measure static pressure in a duct by turning it so that the tube opening faces perpendicular to the flow. One tube is disconnected, and the differential is read with the draft gauge, as shown in Figure 12. It is necessary to plug the port opening to prevent errors in measurement.

C. Moisture Content

The moisture content of gas streams may be determined either by measurement of the wet and dry bulb temperature or by condensation of the moisture present. Both techniques involve use of either a psychrometric chart or water vapor pressure tables. The psychrometric chart has been previously explained; it provides a method for graphical determination of moisture contents of flue gases from direct measurement of wet and dry bulb temperatures. However, the chart applies at only one absolute pressure. The saturation vapor pressure for water at different dry bulb temperatures is given in the table in the appendix. Moisture contents may then be computed from equations (35) and (36). Problems can occur in using wet and dry bulb measurements if sufficient quantities of sulfur trioxide are present because sulfuric acid mist is formed, which may raise the dew point substantially [7].

1. Wet and Dry Bulb Method

The wet and dry bulb temperatures are first determined by direct measurement in the duct, and the specific humidity (grains of water per pound of dry air) is observed from the psychrometric chart. The weight ratio is converted to a molar ratio by using the molecular weights of water and the flue gas, respectively. Mole percent and volume percent are equivalent from

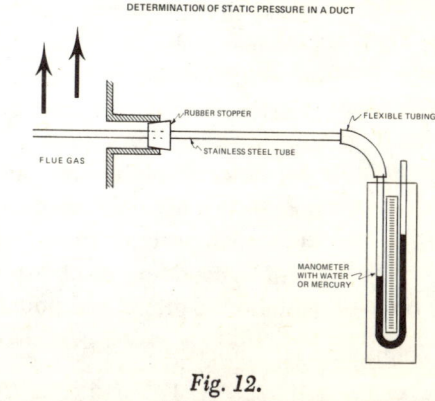

Fig. 12.

Amagat's law and it is then possible to compute moisture content as the ratio of volume of water vapor present to volume of total moist air (dry air plus water vapor). Computation of moisture content from wet and dry bulb temperature measurements using the psychrometric charts proceeds as follows:

$$H_v = \frac{28.84 \text{ lb/lb mole dry air}}{18.00 \text{ lb/lb mole } H_2O} \cdot \frac{1}{7000 \text{ gr/lb}} \cdot (H_m) \quad (39)$$

$$H_v = 2.29 \times 10^{-4} \cdot (H_m)$$

$$MC = \frac{H_v}{1.000 + H_v} \cdot 100 \quad (40)$$

where:

H_m = specific humidity in grains H_2O per lb dry air (from psychrometric chart at T_s, T_{wb})

H_v = volumetric humidity in lb-mole H_2O per lb-mole dry air (or ft³ H_2O per ft³ dry air)

MC = moisture content in % by volume

Two sample problems are given below, to illustrate the details of calculation by the psychrometric chart method or by alternative vapor pressure method.

Sample Problem No. 1—Moisture Content Determination—Psychrometric Chart Method

The dry and wet bulb temperatures in a stack were measured as 175°F and 160°F, respectively. Pressure in the duct was measured as 29.70 inches of mercury (neglect the difference from 29.92 in. Hg). Calculate the moisture content of the flue gas by making use of the psychrometric chart.

Solution:

H_m = Mass humidity in grains of water/pound dry air (from chart)

H_v = Volumetric humidity in ft³ of water/ft³ dry air

MC_s = Moisture content of flue gas in percent water by volume

28.84 = Molecular weight of air in pounds/pound-mole of air

18.00 = Molecular weight of water in pounds/pound-mole of water

7,000 = Number of grains per pound.

Volumetric Humidity:

Measured: $T_{db} = 175°F; T_{wb} = 160°F$

Observed: H_m = 2,070 gr H_2O/lb dry air (from psychrometric chart)

$H_v = 2.29 \times 10^{-4} \cdot (H_m)$
$H_v = (2.29 \times 10^{-4})(2,070 \text{ gr } H_2O/\text{lb da})$
$H_v = 0.475 \text{ ft}^3$

Moisture Content:

$$MC = \frac{H_v}{1.000 + H_v} \cdot 100 = \frac{0.475}{1.475} \cdot 100$$

$MC = 32.2\%$ by volume

An alternative calculation procedure involves use of the Carrier equation expressing the vapor pressure exerted by water vapor at other than saturation conditions. It is more versatile than using the psychrometric chart method because the chart applies at only one static pressure and is particularly useful when the dew point of gas streams is above 180°F. The moisture content is calculated from the vapor pressure of water as follows:

$$P_w = P_{wb} - \left[\frac{(P_s - P_{wb})(T_s - T_{wb})}{(2,830 - 1.30 T_{wb})}\right] \quad (41)$$

$$MC = \frac{P_w}{P_s} \cdot 100 \quad (42)$$

where:

P_w = Actual vapor pressure exerted by water vapor in duct in inches of mercury

P_{wb} = Vapor pressure of water at wet bulb temperature in inches of mercury

P_s = Absolute static pressure in duct in inches of mercury

T_s = Dry bulb temperature of flue gas in °F

T_{wb} = Wet bulb temperature of flue gas in °F.

The vapor pressure exerted by water at different temperatures is listed in the appendix. Calculation of moisture content for a flue gas stream proceeds as shown in Sample Problem 2.

Sample Problem No. 2—Moisture Content Determination—Vapor Pressure Method

The dry and wet bulb temperatures measured in a duct were 175°F and 160°F, respectively. The static pressure in the duct was found to be 2.0 inches of water vacuum at a barometric pressure of 29.92 inches of mercury. Compute the moisture content of the flue gas from the vapor pressure relation-

ship for water and the vapor pressure tables.
Solution:
- ΔP = Pressure difference between static and barometric pressures in inches of mercury
- 13.6 = Inches of water per inch of mercury.

Observed: $T_s = 175°F$; $T_{wb} = 160°F$
$P_b = 29.92$ in. Hg; $P_{wb} = 9.65$ in. Hg (from Tables)

Static Pressure in Duct:

$$P_s = P_b - \Delta P = 29.92 - \frac{2.0}{13.6} = 29.92 - 0.15$$

$$P_s = 29.77 \text{ in. Hg}$$

Vapor Pressure of Water:

$$P_w = P_{wb} - \frac{(P_s - P_{wb})}{(2830 - 1.44\,T_{wb})}(T_s - T_{wb})$$

$$P_w = 9.65 - \frac{(29.77 - 9.65)}{2830 - 1.44(160)}(175 - 160)$$

$$P_w = 9.65 - \frac{20.12}{2600}(15) = 9.65 - 0.12$$

$$P_w = 9.53 \text{ in. Hg}$$

Moisture Content:

$$MC = \frac{P_w}{P_s} \times 100 = \frac{9.53}{29.77} \times 100$$

$$MC = 32.2\% \text{ by volume}$$

a. Temperature below 212°F:

Direct measurement of wet and dry bulb temperatures in the ducts is possible, along with a determination of absolute static pressure. The moisture content can then be computed directly by use of either the psychrometric chart or vapor pressure methods.

b. Temperatures above 212°F:

Complications develop when attempting to measure wet bulb temperatures greater than 212°F because of the tendency of the wick to dry at elevated temperature. The first technique is to simultaneously measure wet and dry bulb temperatures in a duct every thirty seconds. The dry bulb temperature will normally rise rapidly to an equilibrium value. The wet bulb temperature will rise to reach

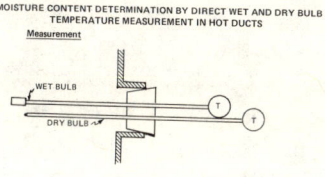

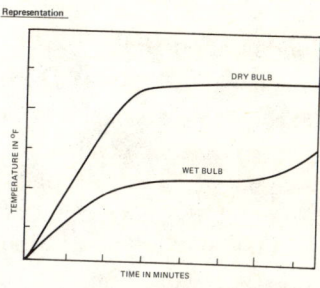

Fig. 13.

an equilibrium value, and then rise again towards the dry bulb temperature once the wick becomes dried. The inflection point where the temperature reaches equilibrium is normally considered to be the wet bulb temperature, as shown in Figure 13. However, for some sources the cloth will dry so rapidly that the inflection point is not discernible because of the high temperature of the flue gas.

A second method for determining wet bulb temperatures involves withdrawal of the stack gas through a ¾ in pipe at 2.0 cfm to cool it to approximately 212°F, and take the wet and dry bulb readings at the reduced temperature [8] [9]. To obtain accurate measurements, it may be necessary to insulate the pipe. It is usually best to use a water aspirator as a prime mover because a pump may become damaged by water and particles, and a small venturi meter may be useful for flow measurement. It is important to measure the static pressure at the point of temperature measurement. See Figure 14.

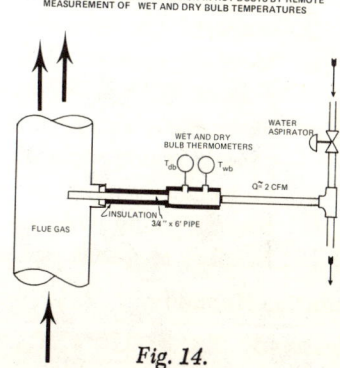

Fig. 14.

Only limited success has been obtained by the authors in the field in using these modifications to the wet and dry bulb method. However, the method has proved successful for gas temperatures in the range between 212 and 270°F, particularly if the gas stream is at or near saturation.

2. Condensation Method

The condensation method makes use of moisture content determination by the equivalent vapor volume of condensed water from a given volume of flue gas, plus the saturated water vapor at the conditions of the last gas-liquid contact phase. The condensation method involves withdrawing stack gas through a condenser or impinger train immersed in an ice water bath, and measuring the amount of water condensed from a known volume of gas. The condensed water is considered to be a vapor at meter conditions for purposes of calculations. The gas stream passing through the meter is considered saturated at the liquid temperature of the last condensing stage, and a thermometer is normally maintained in this impinger. The moisture content of the saturated gas stream can then be calculated from Dalton's law and the vapor pressure data. The total volume of moisture is then the equivalent gaseous volume at meter conditions of the condensed water plus the water passing through the meter. The total gas volume is the observed volume at the meter plus the equivalent volume of condensed water. The moisture content is then computed as the ratio of total moisture volume to total gas volume.

First the equivalent volume of water vapor condensed is determined by equating the weight of condensed water to weight of equivalent water vapor from rearrangement of the ideal gas law.

$$P_m \cdot V_c = \left[\frac{m_w}{MW_w}\right] \cdot RT_m \quad (43)$$

$$m_w = \rho_w \cdot L_c = \frac{MW_w}{R} \cdot \frac{P_m}{T_m} \cdot V_c \quad (44)$$

$$V_c = \left[\frac{R}{MW_w}\right] \cdot \frac{T_m}{P_m} \cdot \rho_m \cdot L_c$$

$$= \left[\frac{1}{454} \cdot \frac{21.85}{18.00} \cdot 1.0\right] \frac{T_m}{P_m} \cdot L_c \quad (45)$$

$$V_c = 2.67 \times 10^{-3} \cdot \frac{T_m}{P_m} \cdot L_c \quad (46)$$

where:

- P_m = Absolute pressure at meter conditions in inches of mercury
- T_m = Absolute temperature at meter conditions in °R
- V_c = Equivalent gaseous volume of condensed water in ft³
- L_c = Liquid volume of condensed water in milliliters
- ρ_w = Density of water is 1.0 grams per liter
- MW_w = Molecular weight of water is 18.00 lb per lb-mole
- R = Universal gas constant is 21.85 ft³-in. Hg per lb-mole—°R
- m_w = Weight of condensed water in grams

The volume of water vapor at saturated conditions, in the last gas-liquid condensing stage is then determined from Dalton's law as follows:

$$V_v = \frac{(P_w)_m}{P_m} \cdot V_m \quad (47)$$

$$MC_m = \frac{(P_w)_m}{P_m} \cdot 100 \quad (48)$$

where:

- V_v = Volume of saturated water vapor at meter in ft³
- $(P_w)_m$ = Vapor pressure of water at meter in inches of mercury
- P_m = Absolute static pressure at meter in inches of mercury
- V_m = Indicated gas volume at meter in ft³
- MC_m = Moisture content of saturated gas stream at meter in percent by volume

The total water volume is the sum of the condensed water plus the saturated vapor.

$$V_w = V_c + V_v \quad (49)$$

where:

- V_w = Total water vapor volume at meter conditions in ft³

The total gas volume is the sum of the condensed water at meter conditions plus the observed meter volume.

$$V_T = V_c + V_m \quad (50)$$

$$MC_s = \frac{V_w}{V_T} \cdot 100 = \frac{V_c + V_v}{V_c + V_m} \cdot 100 \quad (51)$$

where:

V_T = Total volume of gas plus water at meter in ft³

MC_s = Moisture content of flue gas in percent by volume

A sample calculation for a specific problem is shown below:

Sample Problem No. 3 — Moisture Content Determination — Condensation Method

A volume of 350.0 milliliters of water was condensed from 50.0 cubic feet of flue gas (indicated at meter conditions) drawn from a duct through a condensation train. The gas temperature at both the meter and last condensing stage was 70°F while the meter vacuum was 5.0 inches of mercury. The barometric pressure was 29.92 inches of mercury. Compute the moisture content of the flue gas.

Observed: $T_m = 70 + 460 = 530°R$
$P_m = 29.92 - 5.00$
$= 24.92$ in. Hg
$(P_w)_m = 0.739$ inches of mercury at 70°F

Condensed Water Volume:

$$V_c = 2.67 \times 10^{-3} \frac{T_m}{P_m} (L_c)$$

$$= 2.67 \times 10^{-3} \frac{(530)}{(24.92)} (350)$$

$V_c = 19.85$ ft³

Water Vapor Volume:

$$V_v = \frac{(P_w)_m}{P_m} (V_m) = \frac{0.739}{24.92} (50.0)$$

$V_v = 1.48$ ft³

Total Water Volume:

$V_w = V_c + V_v = 19.85 + 1.48$
$V_w = 21.33$ ft³

Total Volume:

$V_T = V_c + V_m = 19.85 + 50.00$
$V_T = 69.85$ ft³

Moisture Content:

$$MC = \frac{V_w}{V_T} \cdot 100 = \frac{21.33}{69.85} \cdot 100$$

$MC = 30.50\%$ by volume

The condensation method is very useful for flue gases at high temperatures. However it has limitations for sources at or near saturated conditions, such as following scrubbers, because water droplets may show up as condensed moisture, thus indicating supersaturation conditions. For exploratory tests, it is advisable to draw in a substantial gas volume, approximately 50 cubic feet, for an accurate determination. The method can also be used for determining moisture during particulate sampling and compared to results obtained with the wet and dry bulb method. The sampling train employed is shown in Figure 15.

3. Adsorption Method

A known volume of source gas is withdrawn from a flue through a drying agent such as silica gel or other desiccants. The water collected is determined by the difference in weight of the solid adsorbent before and after sampling. The method is simple and potentially useful, but cer-

MOISTURE CONTENT BY THE CONDENSATION METHOD

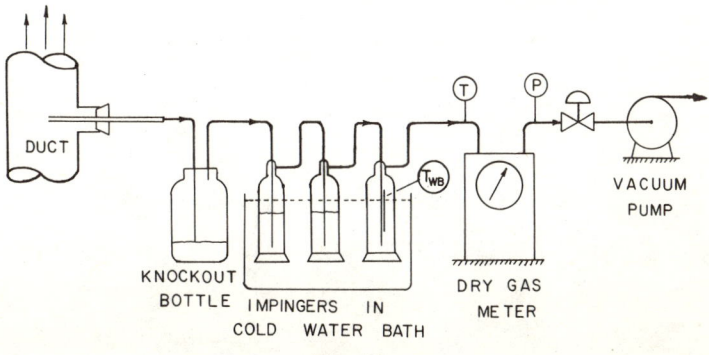

Fig. 15.

tain gases may be adsorbed if present in large quantities. It is also necessary to remove particulates by filtration upstream of the adsorbent, which may require a heated sampling line.

4. Evacuated Flasks

For gases which are more than sixty-percent steam, a method has been devised by withdrawing the gas into an evacuated flask, cooling, and measuring the amount of condensate and the final gas temperature and pressure [5].

D. Gas Composition and Density

It is necessary to determine the composition and density of flue gases in calculating velocities, gas concentrations in terms of specified conditions, and emission rates in weight of pollutant per unit weight of flue gas. Gas composition in combustion and certain other process exit gas streams may be determined with an Orsat flue gas analyzer, which is used to determine the quantities of carbon dioxide, oxygen, and carbon monoxide present on a dry volumetric basis. Each of these constituents is selectively removed by passing through a series of scrubbing solutions (an application of Amagat's law), after which the residual is usually presumed to be nitrogen. Corrections for moisture content are then made in converting gas concentration levels from a dry to wet basis. It is also possible to make use of continuous carbon dioxide and oxygen flue gas monitors based on thermal conductivity or infrared, and paramagnetic detection principles, respectively, to analyze for these gases [11]. Razbegaeva [12] designed an eccentric cam-actuated system for automating Orsat absorption stage analyses using a conventional unit. The Orsat analyzer is illustrated in Figure 16.

The average molecular weight of the flue gas is computed from the relative amounts and respective molecular weights for each of the constituent gases present. The gas density may then be calculated from the ideal gas law relationship in terms of the average molecular weight at the observed conditions of temperature and pressure. Molecular weights and densities for gas streams

ORSAT ANALYZER FOR MEASURING CO_2, O_2, AND CO CONTENTS OF FLUE GAS STREAMS

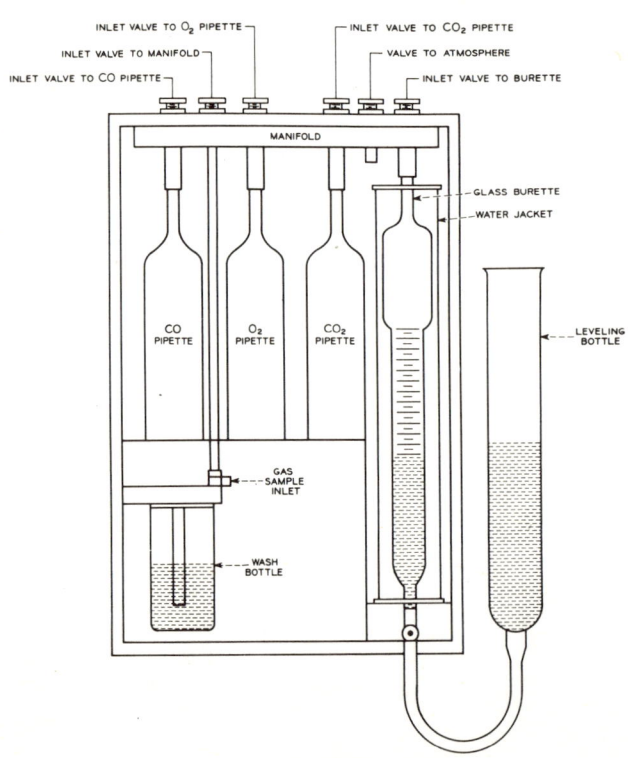

Fig. 16.

may be computed on both dry and wet gas bases to facilitate concentration calculations at both standard and source conditions, as follows:

1. Dry gas:

$$(\%N_2) = 100.0 - [(\%CO_2) + (\%O_2) + (\%CO)] \quad (52)$$

$\%CO_2, \%O_2, \%CO$ = Percent by volume of each gas on dry basis (from Orsat analyses).

All gas volume concentrations are in percent by volume on a dry basis.

$$MW_o = \frac{\Sigma_i (\%Vol_i)_o (MW_i)}{100.0} = lb/lb\text{-}mole \quad (53)$$

$$\rho_o = \frac{M}{V} = \left[\frac{MW_o}{R} \cdot \frac{P_s}{T_s}\right] = lb/ft^3 \quad (54)$$

where:

ρ_o = Density of dry flue gas in lb per lb-mole

MW_i = Molecular weight of a constituent gas in lb per lb-mole

$(\%Vol_i)_o$ = Percent by volume of a constituent gas on a dry basis

MW_o = Molecular weight of dry flue gas in lb per lb-mole

2. Wet Gas:

$$(\%Vol_i)_s = \frac{(100 - \%H_2O)}{100} (\%Vol_i)_o \quad (55)$$

$$MW_s = \frac{\Sigma_i [(\%Vol_i)_s (MW_i)]}{100.0} \quad (56)$$

$$\rho_s = \frac{MW_s}{R} \cdot \frac{P_s}{T_s} \quad (57)$$

where:

$(\%Vol_i)_s$ = Percent by volume of a constituent gas at stack conditions

MW_s = Molecular weight of flue gas at stack conditions in lb per lb-mole

ρ_s = Density of flue gas at stack conditions in lb per ft³

The format for computing flue gas density and molecular weight is shown in Table 7.

A sample problem follows, to show the method of calculations:

Sample Problem No. 4—Gas Composition and Density—Orsat Analyzer

An Orsat analysis of a flue gas showed 15.0 percent carbon dioxide, 3.0 percent oxygen, and 0.0 percent carbon monoxide, where all are expressed in percent by volume on a dry basis. A subsequent moisture content analysis of the flue gas showed 30.0 percent by volume at a temperature of 275°F, and an absolute static pressure of 29.70 inches of mercury. Calculate the molecular weight and density of the flue gas on a dry and wet basis.

Measured:

T_s = 275 + 460 = 735°R; P_s = 29.70 in. Hg

CO_2 = 15.0%; O_2 = 3.0%; CO = 0.0%

N_2 = 100.0 - ($CO_2 + O_2 + CO$)
    = 100.0 - 18.0 = 82.0%

$F_{H_2O} = \frac{100 - MC}{100} = \frac{100 - 30}{100} = \frac{70}{100} = 0.70$

		Dry Gas Basis		Wet Gas Basis	
Gas	MW	(% Vol)$_o$	(% Vol)$_o$(MW)	(% Vol)$_s$	(% Vol)$_s$(MW)
CO_2	44	15.0	660.0	10.5	462.0
O_2	32	3.0	96.0	2.1	67.2
CO	28	0.0	0.0	0.0	0.0
N_2	28	82.0	2296.0	57.4	1607.2
H_2O	18	—	—	30.0	540.0
			$\Sigma_o = 3{,}052.0$		$\Sigma_s = 2{,}676.4$

Dry Gas:

$$MW_o = \frac{\Sigma_i[(\%\ Vol)_o(MW)]}{100.0}$$

$$= \frac{3,052.0}{100.0} = 30.52\ \text{lb/lb-mole}$$

Stack Conditions:

$$\rho_o = \frac{(MW_o)}{R}\frac{P_s}{T_s} = \frac{(30.52)}{(21.85)}\frac{(29.70)}{(735)}$$

$$= 0.0564\ \text{lb/ft}^3$$

Standard Conditions:

$$\rho_o = \frac{MW_o}{R}\frac{P_o}{T_o} = \frac{(30.52)}{(21.85)}\frac{(29.92)}{(520)}$$

$$= 0.0804\ \text{lb/ft}^3$$

Stack Gas:

$$MW_s = \frac{\Sigma_i[(\%\ Vol)_s(MW)]}{100.0}$$

$$= \frac{2,676.4}{100.0} = 26.76\ \text{lb/lb-mole}$$

$$\rho_s = \frac{MW_s}{R}\frac{P_s}{T_s} = \frac{(26.76)}{(21.85)}\frac{(29.70)}{(735)}$$

$$= 0.0496\ \text{lb/ft}^3$$

Sample Problem No. 5 — Gas Volume Corrections

Basic Relationships:

1. Meter to Standard Conditions:

$$V_o = \frac{P_m}{P_o}\frac{T_o}{T_m}\frac{(100 - MC_m)}{(100 - MC_o)}V_m$$

2. Meter to Stack Conditions:

$$V_s = \frac{P_m}{P_s}\frac{T_s}{T_m}\frac{(100 - MC_m)}{(100 - MC_s)}V_m$$

Terminology:

V_m = Indicated gas volume passing through meter during sampling in cubic feet

V_o = Total gas volume sampled at standard conditions in cubic feet

V_s = Total gas volume sampled at stack conditions in cubic feet

MC_m = Moisture content of gas stream at meter conditions in percent by volume (includes only moisture of saturation in gas stream)

MC_o = Moisture content at standard conditions (0.0 percent by volume)

MC_s = Moisture content at stack conditions in percent by volume

T_m = Absolute temperature at meter conditions in °R

T_o = Absolute temperature at standard conditions (60°F or 520°R)

T_s = Absolute temperature at stack conditions in °R

P_m = Absolute pressure at meter conditions in inches of mercury

P_o = Absolute pressure at standard conditions (29.92 in. Hg)

P_s = Absolute pressure at stack conditions in inches of mercury.

Note: Standard conditions are defined as 60°F, 1.0 atm or 29.92 in. Hg at sea-level, for dry gas.

Sample Problem:

Fifty cubic feet of gas at meter conditions were withdrawn from a source. Meter conditions were 70°F, 5.0 inches of mercury vacuum, and 3.0 percent water by volume. Correct the gas volume sampled to both standard and stack conditions, where the observed stack conditions were 275°F, 29.70 inches of mercury, and 30.0 percent water by volume.

Problem Solution:

P_m = 24.92 in. Hg
T_m = 530°F
MC_m = 3.0% by vol.

P_s = 29.70 in. Hg
T_s = 735°R
MC_s = 30.0% by vol.

P_o = 29.92 in. Hg
T_o = 520°R
MC_o = 0.0% by vol.

1. Standard Conditions:

$$V_o = \frac{P_m}{P_o} \frac{T_o}{T_m} \frac{(100 - MC_m)}{(100 - MC_o)} V_m$$

$$= \frac{24.92}{29.92} \frac{520}{530} \frac{97.0}{100.0} (50.0)$$

$$V_o = (0.795)(50.0) = 39.70 \text{ ft}^3$$

2. Stack Conditions

$$V_s = \frac{P_m}{P_s} \frac{T_s}{T_m} \frac{(100 - MC_m)}{(100 - MC_s)}$$

$$= V_m \frac{24.92}{29.70} \frac{735}{530} \frac{(100-3)}{(100-30)} (50.0)$$

$$V_s = (1.62)(50.0) = 80.80 \text{ ft}^3$$

III. VELOCITY DETERMINATION

A. Velocity Greater than 600 fpm

Velocity is determined from the kinetic energy head in height of the fluid flowing, and is normally measured as the equivalent pressure differential between impact (dynamic + static) and static pressures at a point in a flowing gas stream. However, the differential is normally expressed in equivalent inches of water, because direct measurement in height of the fluid flowing is extremely cumbersome. Converting the velocity relationship to dry air at 29.92 in. Hg static pressure, with velocity head expressed in inches of water, it becomes the following expression:

$$\frac{U^2}{2g} = \left(\frac{U^2}{2g} + P_s\right) - P_s = \Delta P \quad (58)$$

$$U_s = 60 \cdot (2.90) \sqrt{T_s \cdot \Delta P} = 174.0 \sqrt{T_s \cdot \Delta P} \quad (59)$$

where:

U_s = Stack velocity in feet per minute.
T_s = Stack temperature in °R.
ΔP = Pressure differential in inches of H$_2$O.

1. Standard Pitot Tube

Direct determination of pressure differential as a measure of velocity is made with a standard pitot tube by inserting it into the duct with the open tip pointed directly into the flow stream. It is a double concentric tube, open at the tip of the inside tube, with static taps on the outer tube. The tube has a 90 degree bend to facilitate its insertion and measurement of velocities in stacks. Standard pitot tubes have the disadvantage of becoming easily plugged in ducts heavily laden with particles because of the small size opening in the tip. They may be used in relatively clean ducts and are sufficiently accurate for calibrating other flow-measuring devices. However, their lower limit of accuracy is normally about 500 feet per minute (8 ft/sec) [10], unless devices such as micromanometers are used for more sensitive pressure measurement [13]. A diagram of a standard pitot tube is shown in Figure 17.

2. Stausscheibe (S-type) Pitot Tube

The Stausscheibe, or S-type, pitot tube is used in most source testing applications because of the susceptibility of standard pitot tubes to plugging. The S-type unit consists of two parallel tubes with openings in opposing directions at the end inserted into the duct, and is shown in Figure 18. It is placed perpendicular to the gas stream so that one opening is pointed directly into the flow stream. A correction must be made here for the deviation from actual static pressure in the duct caused by a suction effect on the downstream tube. At a constant pressure differential, as measured by respective S-type and standard pitot tubes, the following correction factor applies:

$$F_{PT} = \frac{U_{s\text{-Type Pitot Tube}}}{U_{\text{std Pitot Tube}}} \quad (60)$$

For dry air at 29.92 in. Hg:

$$U_s = 174.0 (F_{PT}) \sqrt{T_s \cdot \Delta P} \quad (61)$$

For gases other than air at pressures other than atmospheric, the following equation may be used.

$$U_s = (174.0)(F_{PT}) \sqrt{\frac{28.84}{MW_s} \frac{29.92}{P_s}}$$

$$\cdot \sqrt{(T_s)(\Delta P)} \quad (62)$$

MW_s = Molecular weight of the stack gas in lb per lb/mole
P_s = Static pressure in the duct in inches of Hg

Stausscheibe-type pitot tubes are normally used in most gas velocity measurements, and may be used in large ducts by adding extension sections with only a negligible pressure loss [14]. The large openings of the S-type pitot tubes considerably reduce the susceptibility to plugging in heavily particulate-laden streams, but

STANDARD PITOT TUBE

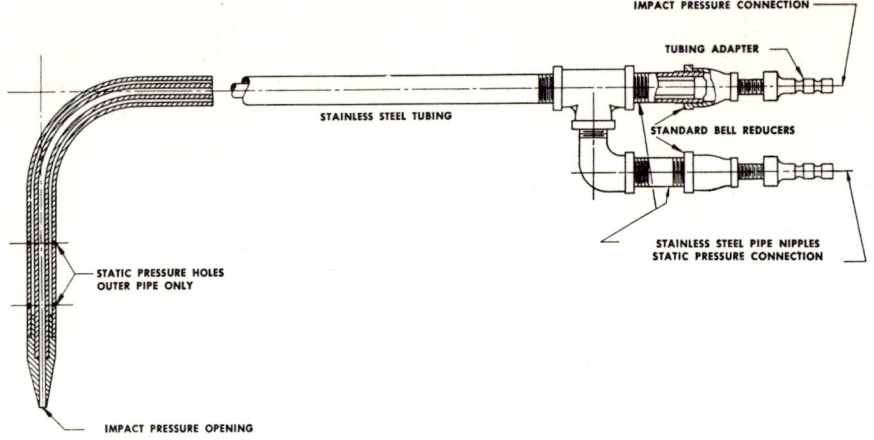

Courtesy of Western Precipitation Corp., Los Angeles, California. [8]

Fig. 17.

STAUSSCHEIBE (S-TYPE) PITOT TUBE

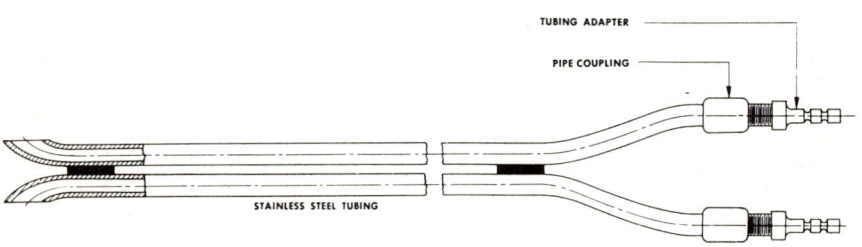

LENGTH IS APPROXIMATELY 5 FEET.
EXTENSION SECTIONS ADDED FOR LONG DUCTS

Courtesy of Western Precipitation Corp., Los Angeles, California. [8]

Fig. 18.

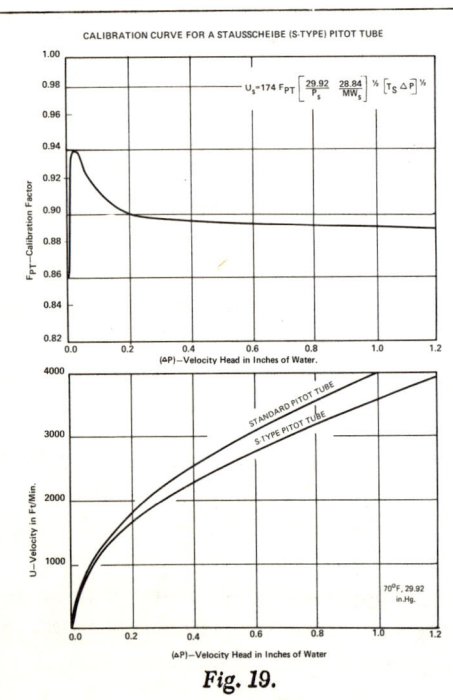

Fig. 19.

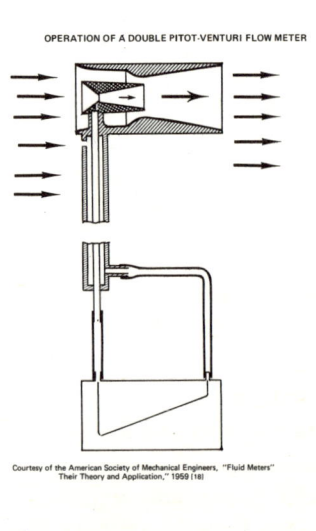

Fig. 20.

also suffer from flow eddies and non-uniformities at low velocities because of the larger cross-sectional area [15]. Their limit of sensitivity is approximately 600 feet per minute or ten feet per second [10] [14]. The correction factor for velocity measured by an S-type pitot tube as compared to a standard pitot tube varies with pressure differential for a given pitot tube, as shown by a typical calibration curve in Figure 19 [5]. Variations in the correction factor of from 0.02 to 0.08 have been noted for individual tubes in tests run on twenty different instruments [17]. Average calibration factors for twenty different pitot tubes have ranged from 0.62 to 0.99, with average values from 0.84 to 0.91 observed [16]. The mean for all tests on all tubes tested was approximately 0.87, with a maximum deviation of 0.05 for most commercially fabricated S-type pitot tubes. These values compare with values of 0.835, 0.85, and 0.855 published in the literature [2] [3] [8] [9] [10] [15]. Significant deviations from these values can occur, particularly with so-called "home-made" S-type pitot tubes. It is advisable to calibrate individual S-type pitot tubes for the velocity range of interest prior to making actual flow measurements to assure accurate results. A procedure which utilizes a constant velocity orifice at the exit of a wind tunnel to calibrate an S-type pitot tube by placing a standard pitot tube adjacent to it is described by Hama [17].

3. Venturi Pitot Tube

Hama [13] describes the use of a venturi type pitot tube to increase the pressure differential at low gas flow rates, which has proved useful for measuring velocities down to 200 feet per minute. It consists of two concentric venturi elements, where the low pressure tap is connected to the throat of the inner venturi and into the throat of the outer venturi [5] [17], as shown in Figure 20.

4. Spherical Pitometer

Pengelly [19] [20] describes the use of a water-cooled, multipoint pitot tube with a spherical sensing head to measure gas velocities in flue gases at high temperatures. The principle of operation is that there is a maximum pressure differential between the point on a sphere facing directly into the flow (impact pressure) and a point approximately 40° off center from this point, because there is no net suction or pressure at this point. Rotation of the spherical head allows determination of the speed and direction of gas flow by determination of pressure differentials at the different points [19]. However, the application is limited to relatively clean stacks and published material describing its performance is limited. Its principle of operation is illustrated in Figure 21.

OPERATION OF A DIFFERENTIAL PITOT TUBE
PRESSURE DISTRIBUTION AROUND A CYLINDER NORMAL TO THE DIRECTION OF FLOW OF THE GAS STREAM.

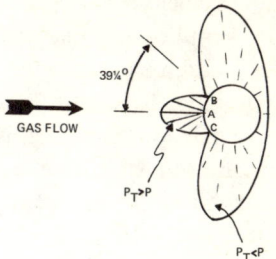

SECTIONAL VIEW OF A DIFFERENTIAL PITOMETER

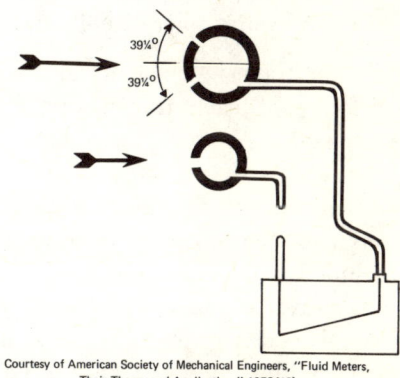

Courtesy of American Society of Mechanical Engineers, "Fluid Meters, Their Theory and Application," 1959 [18]

Fig. 21.

B. Velocity Tables

Velocity tables [21] [22] simplify the procedure for computing velocity because they provide its direct reading from knowledge of stack temperature and velocity pressure differential. However, the tables are based on measurement of dry air velocity with a standard pitot tube at 29.92 in. Hg, and should be corrected for varying densities and static pressures as follows.

$$U_s = 60 \cdot F_{PT} \cdot \sqrt{\frac{28.84}{MW_s} \frac{29.92}{P_s}} \cdot U_{\text{Tables}}$$
(63)

C. Number of Zones

It is necessary to divide a duct into zones of equal area in order to make velocity determinations and to take samples from flue gases. The number of zones required is dependent on the total area of the duct, and increases as the cross-sectional area increases, as shown in Table 8 [1] [9] [10].

D. Traverse Procedure

The common procedure for making a velocity traverse is to divide the duct into zones of equal area, determining the average velocity within each zone, then taking the average of all the zones as the average velocity. The purpose of this procedure is to simplify calculations. The velocity at the center of each zone is determined, and used for determining isokinetic sampling rates. The average velocity in the duct is then determined from velocities in each zone and used to calculate the total flow rate. The average velocity is calculated as follows.

$$U_s = \frac{\Sigma_i(U_i)}{N} \quad \begin{array}{l} U_s = \text{Average Velocity.} \\ U_i = \text{Velocity at one point.} \\ N = \text{Number of points.} \end{array} \quad (64)$$

The pattern for traverses in ducts involves testing the velocity at each traverse point while inserting into and in removing the pitot tube from the duct to give two readings. The traverses should be made perpendicular to each other in round ducts with the center included as a check but not included in the calculations. The location of individual traverse points in circular ducts for zones of equal area is given in Table 9, and has been previously illustrated in Figure 6. The midpoint (50%) should be as a check only and not included in the calculations.

Sample Problem No. 6—Sample Traverse Location and Velocity Measurement.

1. Traverse Location
Principle:

Make use of the previously listed tables. These make use of the percent of the distance across a round duct to a given traverse point.
Problem:
Calculate the necessary distances from the edge of the duct to the traverse point in a round duct of 10.0 feet diameter when dividing the duct into four, five, and six zones of equal area, respectively.

$$D = (12.0)(10.0) = 120.0 \text{ inches}$$

$$\text{Dist} = \frac{(\%)}{100}(D) = \frac{(\%)}{100}(120.0)$$

Solution:

Point	Number of Zones					
	4		5		6	
	%	Dist.[1]	%	Dist.[1]	%	Dist.[1]
1	3.3	4.0	2.2	2.6	2.0	2.4
2	10.5	12.6	8.2	9.7	6.7	8.0
3	19.5	23.4	14.5	17.4	11.8	14.1
4	32.1	38.5	22.7	27.2	17.7	21.2
5	50.0	60.0	34.4	41.2	25.0	30.0
6	67.9	81.5	50.0	60.0	35.4	42.5
7	80.5	96.6	65.6	78.8	50.0	60.0
8	89.5	107.4	77.3	92.8	64.6	77.5
9	96.7	116.0	85.5	102.6	75.0	90.0
10	—	—	91.8	110.3	82.3	98.8
11	—	—	97.8	117.4	88.2	105.9
12	—	—	—	—	93.3	112.0
13	—	—	—	—	98.0	117.8

[1] Distances are in inches

2. Velocity Measurement

Basic Expression:

1. Formula

$$U_s = 174.0 \, (F_{PT}) \sqrt{\frac{28.84}{MW_s} \frac{29.92}{P_s}} \sqrt{T_s \, (\Delta P)}$$

2. Table

$$U_s = 60.0 \, (F_{PT}) \sqrt{\frac{28.84}{MW_s} \frac{29.92}{P_s}} \, (U_T)$$

Terminology:

U_s = Velocity in feet per minute
F_{PT} = Calibration factor for pitot tube (dimensionless)
MW_s = Molecular weight of stack gas in lb per lb-mole
P_s = Static pressure of duct in inches of mercury
T_s = Stack temperature in °R
ΔP = Measured pressure differential in inches of water gauge
U_T = Equivalent velocity from tables at a given velocity head differential in feet per second. (Dry air at 29.92 inches of mercury as measured with a standard pitot tube.)

Sample Problem:

The velocity head differential read from the draft gauge for one point in a duct with an S-type pitot tube is 0.32 inches of water, where the calibration factor is 0.87. The stack conditions were measured as follows:

$T_s = 275°F$; $P_s = 29.70$ in. Hg; $MW_s = 27.5$ lb per lb-mole. Calculate the velocity from the above information and also from the velocity tables.

Solution:

$$T_s = 275 + 460 = 735°R$$

a. Formula:

$$U_s = 174.0 \, F_{PT} \sqrt{\frac{28.84}{MW_s} \frac{29.92}{P_s}} \sqrt{T_s} \sqrt{\Delta P}$$

$$U_s = 174.0 \, (0.87) \sqrt{\frac{28.84}{27.50} \frac{29.92}{29.70}} \sqrt{735} \sqrt{(0.32)}$$

$$U_s = (151.5)(1.03)(27.15)(0.566)$$
$$= (4,237.)(0.566)$$

$$U_s = 2,400. \text{ ft/min.}$$

b. Velocity Tables:

$U_T = 44.5$ ft/sec

$$U_s = 60 \, (F_{PT}) \sqrt{\frac{28.84}{27.50} \frac{29.92}{29.70}} (U_T)$$

$$U_s = (60)(0.87)(1.03)(44.5)$$
$$= (53.8)(44.5)$$

$$U_s = 2,400. \text{ ft/min}$$

The pattern for traverses in rectangular ducts has been previously illustrated in Figure 7. The positions in rectangular ducts at the midpoints of zones of equal area are shown in Table 10, where two or more ports are presumed to be in parallel to each other in the same plane.

An alternative procedure is applicable when there are two ports located perpendicular to each other at the center points of adjacent sides in a rectangular duct. The procedure is analogous to that for round ducts, where the flue is divided into zones of equal area in the form of quadrants, as shown in Figure 22. Traverse positions for these zones are listed in Table 11.

However, the procedure is subject to limitations in potential accuracy in gas streams with nonuniform velocity profiles because it does not take measurements in the respective corners of the duct. This is of considerable importance because experience of the authors indicates that velocity profiles are more likely to be nonuniform in rectangular than in round ducts. This is particularly true following fans, bends or constrictions, where nearly all the flow is often on one side of the duct.

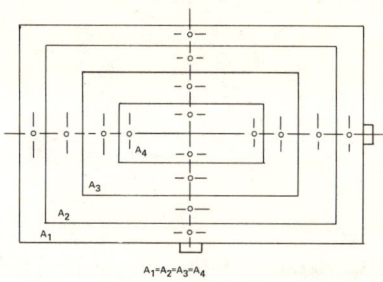

Fig. 22.

IV. TOTAL GAS FLOW RATE

The total gas flow rate is determined from the principle of conservation of mass, where the volumetric rate of flow is the product of the velocity and the cross-sectional area. The mass rate of flow is the gas density at the given conditions times the volumetric rate of flow, where all the values are at stack conditions.

Volume: $\quad Q_s = A \cdot U_s \quad$ (65)

Mass: $\quad M_s = 60 \cdot \rho_s \cdot A \cdot U_s \quad$ (66)

where:
- Q_s = Volumetric rate of flow in ft³ per min.
- U_s = Velocity in feet per minute.
- A = Cross-sectional area of duct in ft².
- M_s = Mass rate of flow in lbs per hour.
- ρ_s = Gas density in lbs per ft³ (wet basis) at stack conditions.

Sample Problem No. 7—Gas Flow Rate

Gas is flowing in a duct of 6.0 ft diameter. First, compute the distance to the traverse points across the duct by dividing it into five zones of equal area. Velocity head measurements for the individual traverse points were observed as listed below. Second, calculate the velocity at each traverse point, and the average velocity in the duct. Third, compute the total gas flow rate on both a volumetric and mass basis.

$T_s = 275°F = 735°R$
$P_s = 29.80$ in. Hg
$MW_s = 27.60$ lb/lb-mole
$F_{PT} = 0.90$

$D_s = 6.0$ ft = 72 in.
$A_s = 0.785(6)^2 = 0.785(36) =$
$A_s = 28.32$ ft²

Point	% Diameter	Distance, Inches	ΔP in. H₂O	$\sqrt{\Delta P}$	Velocity ft/min
1	2.2	1.6	0.09	0.300	1,305
2	8.2	5.9	0.14	0.374	1,630
3	14.5	10.5	0.26	0.510	2,220
4	22.7	16.4	0.35	0.592	2,580
5	34.4	24.8	0.32	0.566	2,460
6	50.0	36.0	0.28	0.528	2,300*
7	65.6	47.2	0.30	0.547	2,380
8	77.3	55.6	0.36	0.600	2,615
9	85.5	61.5	0.34	0.582	2,535
10	91.8	66.1	0.27	0.519	2,260
11	97.8	70.4	0.16	0.400	1,740

* Exclude from calculations.

$\Sigma_i U_i = 24,025$
$N = 11 - 1 = 10$

1. Calculation of Distances:

$$\text{Distance} = \frac{(\% \text{ of Diameter})(\text{Diameter})}{100}$$

For Pt. 5: Dist. $= \frac{34.4}{100.0}(72) = (0.344)(72)$
$= 24.8$ inches

2. Calculation of Velocity:

$$U_s = 174(F_{PT})\sqrt{\frac{29.92}{P_s} \cdot \frac{28.84}{MW_s}} \sqrt{T_s \cdot \Delta P}$$

$$U_s = 174(0.90)\sqrt{\frac{29.92}{29.80} \cdot \frac{28.84}{27.60}} \sqrt{735} \sqrt{\Delta P}$$

$U_s = (156.7)(1.023)(27.1)\sqrt{\Delta P}$
$U_s = 4,350 \sqrt{\Delta P}$

For Pt. 5: $U_s = 4,350 \sqrt{0.32}$
$= 4,350 (0.566) = 2,460$ ft/min

3. Average Velocity

$$U_s = \frac{\Sigma_i U_i}{N} = \frac{24,025}{10} = 2,400 \text{ ft/min}$$

4. Volumetric Flow Rate

$Q_s = A_s U_s = (28.32 \text{ ft}^2)(2,400 \text{ ft/min})$
$Q_s = 68,000 \text{ ft}^3/\text{min}$

5. Mass Flow Rate

$$\rho = \frac{MW \, P_s}{R \, T_s} = \frac{(27.60)(29.80)}{(21.85)(735)}$$
$= 0.0512$ lb/ft³

$M_s = 60\rho Q_s$
$= (60)(0.0512 \text{ lb/ft}^3)\left(68,000 \frac{\text{ft}^3}{\text{min}}\right)$

$M_s = 209,000$ lb/hr

V. LOW VELOCITY MEASUREMENT

The following discussion is related to techniques for measurement of velocities of less than 600 feet per minute (10 feet per second) and gas flow rates generally less than 200 cfm in flue gases. Methodology is based on one of the following principles: 1) temperature differential 2) mechanical displacement 3) distance traversed by an indicator per unit time 4) cross-sectional area adjustment 5) use of volumetric flow rate meters and 6) dilution of an indicator material.

There are three different methods for gas flow measurement based on temperature differential.

A. Thermo-anemometer

Thermo-anemometers are often referred to as "hot-wire" anemometers and have several modifications, all of which operate on the principle that the degree of heat loss to a gas stream for a resistance wire (at a given current level and stack temperature) is proportional to gas velocity passing the sensing element [8]. Gas flow rate is determined either from temperature change in the resistance wire, or by the amount of heating of an air mass passing the heating wire. The latter principle employs two temperature-sensing elements, one heated and the other unheated, where these elements may be either thermocouples or thermometers. It is necessary to know the temperature in the duct to use these instruments, they must be calibrated frequently, and their accuracy becomes severely limited in particle-laden flue gases if the wire becomes coated [6]. However, when properly used and calibrated such as in relatively clean ducts and air streams, they are accurate for velocities down to 100 feet per minute [2].

B. Thermistor Anemometer

The thermistor anemometer is similar to the thermoanemometer except that thermistors, or special heated resistors, are used in place of the resistance wires as heating and sensing elements. Source gas is caused to pass through a narrow hole, heated by the first resistor and the increase in temperature determined by the second resistor. At a given current and stack temperature the velocity is proportional to the square root of temperature change between the two resistors. The device has been found to be sensitive to velocities of less than 20 feet per minute, and the elements are not as easily affected by particles as the resistance wires [24]. However, the device is relatively costly, is still somewhat subject to inaccuracy because of coating by particles in dirty stacks, and must be calibrated frequently.

C. Wall Temperature Difference

The third method involves placing germanium-tipped temperature-sensing elements in contact with the side of the wall of a duct and recording the temperature continuously. A volume of water is injected into the flue gas at a point a given distance upstream from the sensing element. The time after injection when the temperature decrease because of the cooling effect of evaporating water is noted, and the velocity then computed as the distance traveled per given time. The method has proved successful in measuring gas flows in heat exchanger applications, but may be subject to serious errors in large diameter ducts with thick linings, and in flue gases with high water vapor contents. The authors are not aware of other experience with this technique.

The following devices operate on the principle that mechanical rotation or displacement due to impact pressure of the flue gas on the sensing element is proportional to velocity.

D. Vane Anemometer

The vane anemometer consists of a radial series of slotted, diagonal vanes which are caused to rotate when a gas stream is directed toward the unit. The integrating type device provides that the number of rotations per unit time is proportional to the velocity [10], and a remotely located counter and long handle allows use in measuring flows from large flues. However, the device is not accurate and becomes severely corroded in wet or heavily particle- or gas-laden flue gases, and does not stand up at temperatures above 300°F. It is primarily used in measuring air flows in large vents or grilles for air conditioning or ventilation systems in buildings. However, it also may be used for measuring low gas velocities in relatively clean and dry low temperature flue gases. For gases at greater than 250°F, it should not be exposed for longer than 10 to 15 seconds at a time [10].

E. Swinging Vane Anemometer

The swinging vane anemometer, or velometer, is a device where the gas stream is directed against a metal strip vane connected at one end to a sensor. The degree of deflection is proportional to the impact pressure resulting from velocity, and a direct readout for velocity or velocity head is usually provided as an integral part of the instrument. It is useful for air vents and relatively clean flue gases but normally should not be used at temperatures above 150°F or in dirty stacks. Frequent calibration is required at such conditions, and the device may become corroded or damaged [10].

The following methods for velocity measurement are based on injecting a tracer material at a given distance upstream from a sensing device. The time required to traverse the given distance is determined with a stopwatch and the velocity then calculated.

F. Balloons

Walther [25] successfully measured the flow in the vent gases from a Kraft mill continuous digester by adding several balloons to the duct and measuring the time required to flow to a second port approximately 200 feet downstream. However, the gases were at relatively low temperatures, and it was possible to watch the balloons passing the downstream port by placing a transparent plastic window over it. The method could not be used in hot gas streams because the balloons would break.

G. Colored Smoke

Sullivan [26] described injecting colored powders into a flue gas stream at a given point via a long tube squeeze bulb and determining the time required to travel a given distance to the top of the stack. Pararosaniline (red) and phthalocyanine (blue) have brilliant colors which can be seen in most flue gas streams by the human eye, while titanium dioxide (white) is also useful in nonwhite plumes which are not overly dense. The method is relatively simple and is useful in most plumes which are not extremely dense or dark, provided the color can be seen by the observer.

H. Chemical Addition

Moody [27] injected a given volume of 0.25 N ammonium hydroxide liquid into a sulfite pulp mill digester vent stack with a syringe. The ammonia reacted with the sulfur dioxide present to form a ammonium bisulfite-sulfite aerosol. The time required for the white cloud to travel a given distance and appear at the top of the stack was determined with a stopwatch, and the net velocity computed.

An alternative procedure is useful when there is an insufficient concentration of sulfur dioxide present in the flue gas. Syringes of 1.0 N hydrochloric acid and 0.25 N ammonium hydroxide are added simultaneously to the flue gas as shown in Figure 23. The time required for the white cloud produced by generation of the ammonium chloride aerosol to travel the given distance and appear at the top of the stack is measured, and the velocity is then computed. However, these methods cannot be used in saturated stacks because the cloud produced cannot be easily seen.

I. Radioactive Tracers

Roth [28] described the use of radioactive isotopes for gas flow measurement and determining the time required to travel a given distance downstream, using both the peak and total count methods for analysis, by use of a Geiger counter [23]. Radioactive krypton, iodide compounds, carbon-14, and tritium (which requires a scintillation unit) have been used. The method is potentially the most versatile and useful, but requires complicated and expensive equipment, highly trained personnel, only small amounts of tracers must be used, and sufficient safety precautions must be taken.

J. Cross Sectional Area Adjustment

For certain flue gases with discharges at low elevations, it may be possible to construct a projection, either temporary or permanent, to reduce the cross sectional area of a duct [9]. This in-

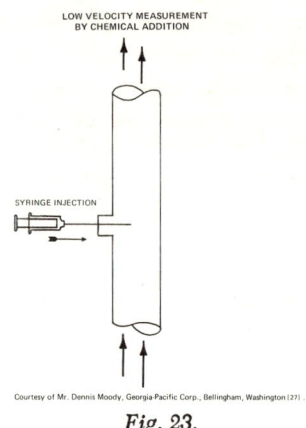

Fig. 23.

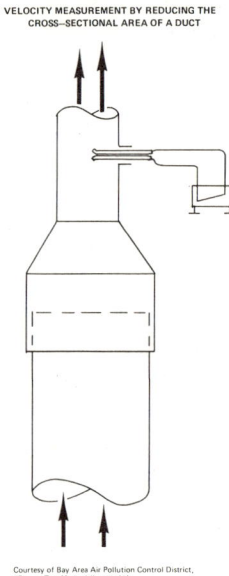

Fig. 24.

volves construction of a tapered convergent nozzle on top of the stack as shown in Figure 24, to increase the velocity to above 60 fpm into the range where it can be measured with a pitot tube. However, physical limitations imposed by height, access, and duct size may limit this method.

VI. TOTAL VOLUME MEASUREMENT

The equation for volumetric rate of flow is determined from the continuity-of-mass principle. For a duct where the average velocity has been determined by a traverse with a pitot tube or other means, the rate of flow in cfm is the product of the cross sectional area of the duct and the average velocity in fpm.

$$Q = A \cdot U$$

The basic equation for gas velocity through flow measuring devices is determined from the relation for conservation of energy, where the velocity is proportional to the pressure drop across the device.

$$U = 60\,C\sqrt{2gh} = 60\,C\sqrt{2g\frac{\Delta P}{\rho}} \quad (67)$$

where:

C = Net Coefficient of discharge (dimensionless).
$\Delta P = P_{In} - P_{Out}$ in lb/ft².
ρ = Density of flue gas in lb per ft³.

The coefficient of discharge (C) is the ratio of actual to theoretical flow through a flow measuring device, which is always less than unity because of stream contraction and frictional effects. Values for this constant for different types of flow meters are described and listed in several other works [5] [6] [18] [29]. Values for these constants have been obtained empirically by experiment. The expansion factor (Y) accounts for the change in gas density as the gas expands adiabatically from the upstream to downstream pressure in going through the flow measuring device. The value of the specific heat ratio for air and most flue gases is approximately 1.4 [5].

The general expression for flow through the restrictions then becomes [6]:

$$Q = 862\,CY\frac{(D_2)^2}{\sqrt{1-\left(\frac{D_2}{D_1}\right)^4}}\sqrt{\frac{\Delta P}{\rho_1}} \quad (68)$$

where:

Y = Expansion factor
Q = Rate of flow in ft³ per minute
D_1 = Upstream diameter in feet
D_2 = Constriction diameter of narrowest point in feet
ΔP = Pressure drop in inches of H$_2$O
ρ_1 = Gas density in lb per ft³

A. Venturi Meters

Venturi meters consist of conically-shaped converging and diverging nozzles in a pipe which are connected by a narrow pipe collar of about one-third the pipe diameter. Pressure taps are placed just upstream of the convergent section and at the point of narrowest diameter to measure the pressure drop as a function of flow rate. These are normally used for measurement of gas volumes in pipes and ducts of 20 to 200 cfm, and normally involve pressure drops of one to six inches of water. However, they are not normally used for smaller gas flows.

B. Convergent Nozzles

These are devices which have a smooth rounded nozzle placed in a pipe with a diameter approximately one-half of the pipe [5]. They have pressure drop characteristics similar to venturi meters, but are not normally used in gas flow measurement applications.

C. Orifice Meters

An orifice meter consists of a plate placed in a pipe, with a small, sharp-edged, round opening in the center. The flow rate is proportional to the pressure drop across the restriction. Flow rates may be measured over a wide range by use of different sized openings, and these devices are used for both flue gas and sample flow measurements. They may involve pressure drops ranging from 0.5 to over 15.0 inches of mercury, and flow rates from less than one to more than 300 cfm, depending on the orifice.

VII. SPECIAL PROBLEMS

A. Pulsating Sources

Ower [15] and Perry [5] state that in measuring velocities with a pitot tube, pulsations of greater than ± 20 percent from the average readings should be damped to prevent results from being more than one percent too high. It is possible to obtain more representative velocities from a pitot traverse on highly pulsating sources, such as high pressure drop venturi scrubbers, by use of the surge bottle technique shown in Figure 25. Placing surge bottles in the lines from the pitot tube to the draft gauge tends to dampen the large oscillations in pressure differential

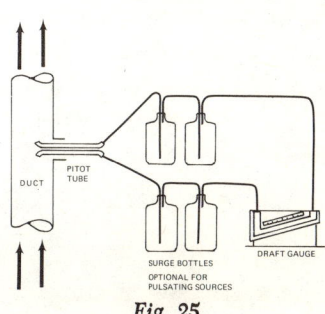

Fig. 25.

where the number of damping stages required depends on the degree of oscillation.

B. Variable Flow Measurement

A technique involving use of a time-sequence camera to record pressure differentials across flow-measuring devices was used to determine unsteady state flow rates [30]. A multiple pitot-tube system was used for a large duct when little back pressure could be tolerated. The duct was calibrated in terms of time-sequence photographs across several pitot tubes in a duct so that a single point reference could then be used to determine duct flow. The same technique was applied to measure the pressure drop across a convergent nozzle in a small duct. Knowledge of gas flow conditions allowed flow calculations to be made in both cases. It is possible to substitute a continuous recorder for the time sequence camera as a means of measuring flow in a duct.

C. Tracer Material Addition

It is possible to measure volumetric flow rates in ducts by addition of a metered amount of an indicator material to a gas stream at a point and measuring the concentration at a point downstream. The method is particularly useful in ducts where it is difficult to make measurements with a pitot tube or other devices. These include, upstream or downstream from induced-draft fans, bends, tangential inlets to round ducts, or in stacks following cyclone separators or collectors, where a swirling flow pattern often exists. The method is useful because turbulent mixing conditions normally exist in most ducts within five diameters downstream for gases.

Magill [32] presented a discussion of the techniques for adding chemicals to a flue gas to measure flow rates. The general equation for a gaseous tracer material is as follows, where it is usually preferable to use a material which is not already present in the stack gas. It is first necessary to correct the metering gas flow to equivalent source conditions to make a material balance using volumetric flows and concentration alone. A material balance may then be written as follows:

$$C_m Q_m + (0) Q_s = C_s (Q_s + Q_m) \quad (69)$$

where:

Q_s = Stack gas flow in cfm
Q_m = Metering gas flow in cfm at equivalent source conditions
C_m = Concentration of tracer in metering gas stream in ppm by vol.
C_s = Final concentration of tracer in stack gas stream in ppm by vol.

If $Q_s \gg Q_m$:

$$C_m Q_m = C_s Q_s$$

$$Q_s = Q_m \frac{C_m}{C_s} \quad (70)$$

A number of materials may be metered into the gas stream at a known rate for a given period so as to reach steady state conditions for resultant concentration determination. Roth [28] added radioactive tracers at a known rate to measure gas flows in heat exchangers. Nieuwenhuizen [32] added methane at a known rate from a cylinder to the inlet of an induced draft fan and measured the exit concentration with a continuous flame ionization hydrocarbon analyzer to measure the gas flow rate. A diagram of the system is shown in Figure 26. This system may be used on other sources as well.

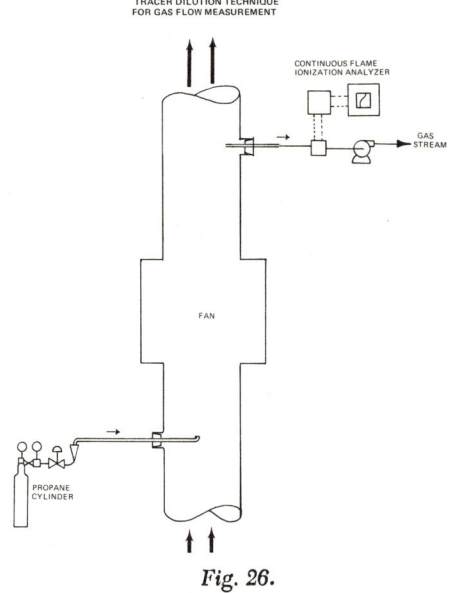

Fig. 26.

VIII. REFERENCES

1. Hawksley, P. G. W., Badzioch, S., and Blackett, J. H., *Measurement of Solids in Flue Gases*, British Coal Utilization Research Association, Leatherhead, Surrey, England.
2. Clayton, G. D., ed., "Stack Sampling," Ch. 9 in *Air Pollution Manual, Part 1—Evaluation*, American Industrial Hygiene Association, Detroit, Michigan, 1961.

3. "Test Methods for Gas Volume and Dust Sampling Determinations," Research-Cottrell, Inc., Bound Brook, New Jersey, 1968.
4. Himmelblau, D. M., *Basic Principles and Calculations of Chemical Engineering*, 2nd ed., Prentice Hall, Inc., Englewood Cliffs, New Jersey, 1967.
5. Perry, J. H., *Chemical Engineers' Handbook*, McGraw-Hill Book Co., New York, New York, 1963.
6. Jorgensen, R., ed., *Fan Engineering*, Buffalo Forge Co., Buffalo, New York, 1961.
7. *Steam, Its Generation and Use*, 37th edition, Babcock and Wilcox Co., New York, New York, 1963.
8. Haaland, H. H., ed., "Methods for Determination of Velocity, Volume, Dust and Mist Content of Gases," Western Precipitation Corp., Los Angeles, California, 1968.
9. Wolfe, E. A., ed., "Source Test Methods," Bay Area Air Pollution Control District, San Francisco, California, 1961.
10. Devorkin, H., Chass, R. L., Fudurich, A., and Kanter, C. V., Holmes, R. G., ed., "Source Testing Manual," Los Angeles County Air Pollution Control District, Los, Angeles, California, 1965.
11. Rose, A. H., Stenburg, R. L., Corn, M., Horsley, R., Allen, D., and Kolp, P., "Air Pollution Effects of Incinerator Firing Practices and Combustion Air Distribution," *Journal of the APCA*, Vol. 8, No. 4, pp. 297-309, February 1959.
12. Razbegaeva, A. P., "Mechanization of the Orsat Gas Analyzer, "*Coke and Chemistry*, (Moscow, USSR), No. 4, pp. 53-54, 1958, trans. in Levine, B. S., "USSR Literature on Air Pollution and Related Occupational Diseases," Vol. 7, U. S. Public Health Service, Washington, D. C., 1962.
13. Hama, G. M., and Curley, L. C., "Instrumentation for Measuring Low Velocity with a Pitot Tube," *Air Engineering*, Vol. 9, No. 7, pp. 28-33, July 1967.
14. Kennedy, E. D., and Wilson, L. D., "Calibration of the Alnor Duct Jet, Double Pitot Tube with Sectioned Extensions for Stack Sampling," *American Industrial Hygiene Association Journal*, Vol. 25, No. 6, pp. 587-588, Nov.-Dec. 1964.
15. Ower, E., *The Measurement of Air Flow*, Chapman and Hall, Ltd., London, England, 1933.
16. Personal communication with Prof. Arthur D. Hughes, Dept. of Mechanical Engineering, Oregon State University, Corvallis, Oregon, 1968.
17. Hama, G. M., "A Calibration Wind Tunnel for Air Measuring Instruments," *Air Engineering*, Vol. 9, No. 12, pp. 18-20, December 1967.
18. *Fluid Meters–Their Theory and Application*, American Society of Mechanical Engineers, New York, New York, 1959.
19. Pengelly, A. E. S., "New Equipment for Flame and Furnace Research," *Journal of the Institute of Fuel*, Vol. 35, No. 256, pp. 210-219, May 1962.
20. Pengelly, A. E. S., "Apparatus for the Measurement of Gas Velocity in Furnaces and Models," *Journal of Scientific Instruments*, Vol. 37, No. 9, pp. 343-346, September 1960.
21. "Velocity Tables," Los Angeles County Air Pollution Control District, Los Angeles, California, January 1960.
22. "Velocity Tables," Research–Cottrell, Inc., Bound Brook, New Jersey, 1957.
23. Jacobs, M. B., *The Chemical Analysis of Air Pollutants*, Interscience Publishers, Inc., New York, New York, 1960.
24. Murphy, D. E., and Sparks, R. E., "A Thermistor Anemometer for Measurement of Low Fluid Flows," *Industrial and Engineering Chemistry Fundamentals*, Vol. 7, No. 4, pp. 642-645, November 1968.
25. Personal communication with Mr. James S. Walther, Crown-Zellerbach Corp., Camas, Washington, 1968.
26. Personal communication with Mr. F. Sullivan, Standard Oil Corp., San Francisco, California, 1968.
27. Personal communication with Mr. Dennis Moody, Georgia Pacific Corp., Bellingham, Washington, 1968.
28. Roth, H., "A Thermal Method of Determining the Rate of Flow in Pipes," *Combustion*, Vol. 35, No. 3, pp. 33-36, September 1963.
29. Streeter, V. L., *Fluid Mechanics*, 3rd ed., McGraw-Hill Book Co., New York, New York, 1962.
30. Harding, C. I., Hendrickson, E. R., and Sholtes, R. S., "Measuring Non-Steady Flow in Industrial Vents," *Journal of the APCA*, Vol. 16, No. 1, pp. 12-14, January 1966.
31. Magill, P. L., Holden, F. R., and Ackley, C., *Air Pollution Handbook*, McGraw-Hill Book Co., New York, New York, 1956.
32. Nieuwenhuizen, J. K., and Posthumus, H., "An Accurate and Simple Method for Air-Flow Measurement in Field Tests on Air-Cooled Heat Transfer Equipment," *Journal of the Institute of Fuel*, Vol. 40, No. 313, pp. 45-47, January 1967.

Table 7. Determination of Flue Gas Density

Gas	Molecular Weight	Dry Gas Basis		Wet Gas Basis	
		(% by vol)$_o$	(% by vol)$_o \cdot$ (MW)	(% by vol)$_s$	(% by vol)$_s \cdot$ (MW)
CO_2	44.0				
O_2	32.0				
CO	28.0				
N_2	28.0				
H_2O	18.0				
			$\Sigma_o =$		$\Sigma_s =$

$$MW_o = \frac{\Sigma_i \, (\% \, Vol_i)_o \, (MW_i)}{100.0} \quad \text{(53—Repeated)}$$

$$\rho_o = \frac{MW_o}{R} \frac{P_s}{T_s} \quad \text{(54—Repeated)}$$

$$MW_s = \frac{\Sigma_i \, (\% \, Vol_i)_s \, (MW_i)}{100.0} \quad \text{(56—Repeated)}$$

$$\rho_s = \frac{MW_s}{R} \frac{P_s}{T_s} \quad \text{(57—Repeated)}$$

Table 8. Number of Traverse Points in Ducts

Round		Rectangular	
Diameter (inches)	Number	Area ft^2	Number
8-10	2	0-2	4
12-20	3	2-9	9
22-8	4	10-17	16
30-42	5	18-25	20
43+	6	25+	21+

Table 9. Equal Area Zones for Velocity Traverses in Round Ducts

Percent of Flue Diameter from Port to Test Point

Point No.	Number of Areas				
	2	3	4	5	6
1	6.2	4.4	3.3	2.2	2.0
2	25.0	14.7	10.5	8.2	6.7
3	50.0*	29.4	19.5	14.5	11.8
4	75.0	50.0*	32.1	22.7	17.7
5	93.8	70.6	50.0*	34.4	25.0
6	—	85.3	67.9	50.0*	35.4
7	—	95.6	80.5	65.6	50.0*
8	—	—	89.5	77.3	64.6
9	—	—	96.7	85.5	75.0
10	—	—	—	91.8	82.3
11	—	—	—	97.8	88.2
12	—	—	—	—	93.3
13	—	—	—	—	98.0

* Do not include in calculations.

Table 10. Equal Area Zones for Velocity Traverses in Rectangular Ducts

Percent of Distance Across Duct

Point	Number of Test Points per Port						
	2	3	4	5	6	8	10
1	25.0	16.7	12.5	10.0	8.7	6.3	5.0
2	75.0	50.0	37.5	30.0	25.0	18.8	15.0
3	—	66.7	62.5	50.0	42.3	31.3	25.0
4	—	—	77.5	70.0	59.7	43.8	35.0
5	—	—	—	90.0	75.0	56.2	45.0
6	—	—	—	—	91.3	68.7	55.0
7	—	—	—	—	—	81.2	65.0
8	—	—	—	—	—	93.7	75.0
9	—	—	—	—	—	—	85.0
10	—	—	—	—	—	—	95.0

Courtesy of Western Precipitation Corp., Bulletin WP-50, 1968, [8].

Table 11. Velocity Traverse Points in Rectangular Ducts with Perpendicular Ports

Percent of Distance Across Duct

Point	Number of Zones Across Duct				
	2	3	4	5	6
1	7.3	4.6	3.4	2.6	2.2
2	32.3	15.2	10.7	8.3	6.8
3	50.0*	35.6	19.8	14.8	12.0
4	82.3	50.0*	36.0	23.1	17.8
5	92.7	64.4	50.0*	38.8	25.4
6	—	84.8	64.0	50.0*	39.8
7	—	95.4	80.2	61.2	50.0*
8	—	—	89.3	76.9	60.2
9	—	—	96.6	85.2	74.6
10	—	—	—	91.7	82.2
11	—	—	—	97.4	88.0
12	—	—	—	—	94.2
13	—	—	—	—	97.8

* Do not include in calculations. Use as a check only.
Courtesy of Western Precipitation Corp., Bulletin WP-50, 1968, [8].

Table 12. Dimensionless Constants for Gas Flow Measuring Devices [5]

Flow Meter	Discharge Coefficient C	Expansion Value	Factor-Y Equation
Orifice	0.60	0.90-1.00	$1 - \dfrac{1-r}{K}(0.41 + 0.35 B^4)$
Venturi	0.98	0.83-1.00	$\left[r^{2/K} \left(\dfrac{K}{K-1} \right) \left(\dfrac{1 - r^{K-1/K}}{1-r} \right) \left(\dfrac{1-B^4}{1-B^4 r^{2/K}} \right) \right]^{1/2}$
Conventional Nozzle	0.95		

where: $r = \dfrac{P_2}{P_1}$; $K = \dfrac{C_p}{C_v} \cong 1.4$; $B = \dfrac{D_2}{D_1}$

and: Subscript p = pressure
Subscript v = volume

CHAPTER 5

SAMPLING TRAINS

I. INTRODUCTION

A. Elements

Gas is withdrawn from a source at a known rate through a sampling train to facilitate collection of one or more materials for subsequent analysis. Normally, a sample is taken properly if it is representative of the flue gas stream, is free of interfering contaminant materials, and has not undergone chemical or physical change, except in a predictable manner [1].

Sampling trains are usually arranged in series, and consist of the following sections:
1) Sample removal; 2) Pretreatment; 3) Collection equipment; 4) Flow measurement and control; 5) Prime mover. They are also normally arranged in this order, and are shown in Figure 27.

B. Requirements

It is necessary to consider both the materials to be collected and the characteristics of the collecting equipment and other necessary components in designing a sampling train. The following properties of substances to be collected should be considered. 1) the physical and chemical properties of the material. These include whether it is in the liquid, solid, or gaseous state, its reactivity, and for particles, the size, shape, surface properties, etc.; 2) the concentration of the materials present; 3) the collection properties, including solubility, electrical charge, and others; 4) properties of the material following collection, such as reactivity with air, container walls, or the collection medium; 5) the possible interference of other materials collected; and 6) safety considerations in sampling and analysis of materials.

The following criteria apply in the design, construction, and operation of sampling trains for collection of particulate and gaseous materials: 1) the temperature and pressure characteristics of the flue gas. This may require pressure reduction equipment, thick wall tubing, cooling devices, or electric heating sections. 2) the type and number of collection stages required, where this depends on the types and amounts of materials being measured. 3) the rate of flow through the sampling train, and the necessary require-

ELEMENTS OF A SOURCE SAMPLING TRAIN

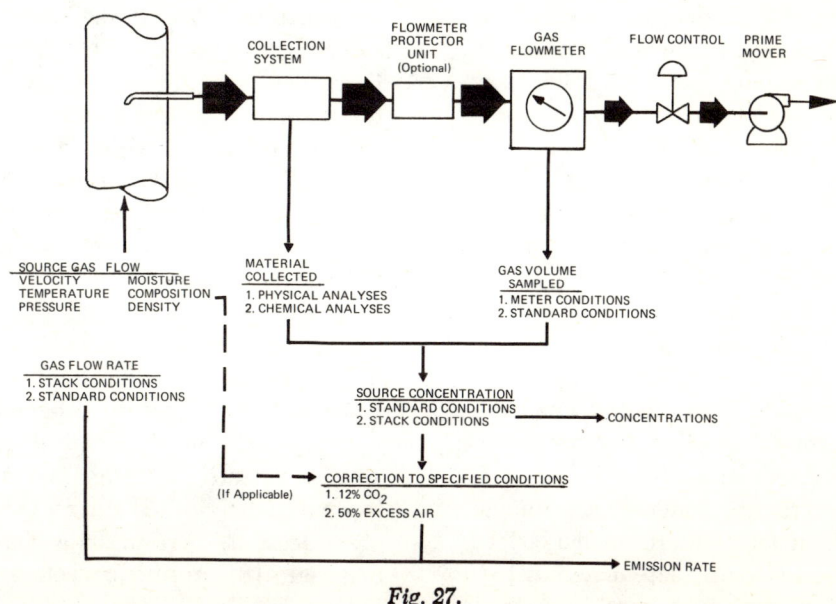

Fig. 27.

ments for the prime mover. 4) the pressure drop across the sampling train at the maximum flow rate expected. 5) electric power, water, ice, and steam required for the sampling train for pumps, heating or cooling probes, and so forth. 6) the relative simplicity of the sampling train. The less need for complex and heavy equipment, the better. 7) the ease of removal and cleaning collection equipment and sample probes. 8) safety and equipment breakage considerations. Equipment should be kept in light, easily carried, and protected containers. Electrical equipment should be kept fully enclosed, especially in wet weather. 9) the cost of the equipment.

II. COMPONENTS

A. Sample Removal

The sample removal section consists of a probe, nozzle, and heating or cooling lines. Probes are normally long tubes with 90° bends at one end to be pointed either into the direction of flow for particulate sampling, or away from the direction of flow in gaseous sampling. The nozzle at the end of the probe is normally sharp-edged, pointing inward from the outside edge. The probe can be made of glass, stainless steel, teflon, or titanium, depending on the conditions to which it is exposed. It is important that the material used be unreactive with the substances of interest being withdrawn. Stainless steel is a relatively rugged material which is useful for many applications. However, its deep pore structure may cause small particles to become trapped, and thorough cleaning of the probe may then be required after sampling. It may be necessary to add a glass or teflon liner, which has much smaller pores than stainless steel [2]. However, the full significance of this effect has not been evaluated. While glass is relatively unreactive, it has the problem of possible breakage if not handled properly. Titanium may be necessary in certain applications but is expensive and difficult to machine. Bends in the probe should be as gentle as possible to minimize particle deposition on the probe walls by deviation from gas flow streamlines. The distance from the nozzle to the collection device should be as short as possible to avoid losses from deposition by impaction, sedimentation, or thermophoresis.

It is necessary to use water-cooled sample probes for sources at temperatures above 800°F, to avoid combustion of particulate materials [3], to prevent exceeding maximum temperatures of filtration materials, and boiling temperatures of liquids. Such probes consist of a concentric two-tube system where the inner tube is used for the flue gas and the outer tube contains the circulating coolant, usually water [4] [5]. It is usually best to add the water at the hot end of the outer tube to avoid boiling.

Simultaneous sampling at more than one point normally involves a considerable amount of equipment and personnel. Chojnowski [6] designed a water-cooled sampling probe for simultaneous sampling at three locations in a duct. It consists of three concentric tubes with a probe emerging from each one.

B. Pretreatment Section

This element is an optional item which is only added for certain applications. Several examples are, filters added in gas sampling lines to remove particulates upstream from absorbers, condensers to cause formation of sulfuric acid mist when sulfur trioxide is present, and glass sections to cause hydrogen fluoride to form silicon tetrafluoride, which can pass through a filter into gas adsorption bottles. Each special application will be discussed in the appropriate section of the following chapter.

C. Collection Equipment

A number of different devices are used for collecting particulate and gaseous samples prior to analysis. Gaseous samples may be collected in evacuated bottles, purged through small containers until a representative mixture is obtained, absorbed in a liquid medium, adsorbed on a solid packing, or collected in freeze-out traps. Particulate samples may be collected by filtration, wet or dry impingement, impaction, electrostatic or thermal precipitation, or centrifugal action. These may be used singly or in series in a number of different combinations, depending on the types and amounts of materials present for a given source. Sometimes the filtration unit is included as an integral part of the sample probe assembly to maintain it at a heated condition, as an alundum thimble assembly [7]. Bloomfield [1] presents a summary of the uses and properties of collection devices used for gas and particulate sampling. The ASME standards for particulate collection requires an efficiency of at least 99 percent by weight for incoming materials [8]. If the collection efficiency of a device (or series of devices) in a sampling train is lower than this figure, it should be known or predictable.

D. Flow Measurement and Control

The flow measurement and control section usually consists of three elements. First, the filtration and drying assembly to remove particles, moisture, and corrosive gases upstream of the meter. This is frequently accomplished by using an impinger containing silica gel or drierite followed by a packed tube containing glass wool. Second, the flow measuring assembly may consist of a device to measure the rate of gas flow, such as a rotameter or orifice meter, and a device to measure the total volume sampled, such as a dry gas meter. Third, the gas flow is regulated, usually by a valve-bleed arrangment which provides for a sensitive and relatively stable flow arrangement. The flow measuring section is normally placed after the collection device, so as to have the gas stream cooled to near ambient conditions and minimize contamination of the flow meters. The flow meter is normally placed upstream of the prime mover because of the tendency for vacuum pumps to leak, and thus cause inaccurate total volume measurements. A subsequent section will be devoted to sample flow measuring devices.

E. Prime Mover

The prime mover is the source of vacuum necessary to cause gas to flow out of the source into the sampling train. The prime mover is necessary for all sources except those under sufficient positive pressure as to require no vacuum source. Vacuum pumps are used for relatively low flow sources from zero to five cfm with significant pressure drop from one to more than ten inches of mercury. Their size is normally excessive for greater flow rates, however. Centrifugal blowers, vacuum cleaner and high-volume sampler motors are useful for high gas flow rates from five to 50 cfm, but are not able to withstand large pressure drops greater than one or two inches of mercury. Water aspirators are useful as constant vacuum sources over a wide range of gas flows, depending on the water flow rate. However, a suitable drain facility must be available nearby, and the line must not be subject to wide flow surges. Aspirators are useful in sampling applications where vacuum pumps may be rendered inoperative by dirt, moisture, or corrosive gases, such as the modified moisture content determination apparatus for cooling flue gas below 212°F [7]. Air-actuated aspirators have not been found satisfactory in many applications because of variable source pressure, particularly in industrial plant locations.

III. SAMPLE FLOW MEASUREMENT

Stack gas is caused to flow through a measuring unit after being withdrawn from the stack and through the collection device. The sampling flow measuring devices must be calibrated on a regular basis to assure their continued accuracy. Flow measuring and calibration devices operate on the basis of either rate of flow or total volume.

A. Orifice Flow Meter

Gas is drawn from the stack through a sampling train followed by a restricted opening in the line. Here the rate of flow is proportional to the pressure drop across the restriction, which limits the flow to a certain maximum value. This is useful because it limits the flow rate to a relatively constant value when there is a variable downstream vacuum source, such as a vacuum pump or plant aspiration system. Then a large change in pressure produces only a small change in flow rate. However, a suitable vacuum source must be provided when using orifice flow meters.

Orifices should be made of stainless steel to minimize corrosion, and must be calibrated prior to use. They should be located downstream of sampling trains to avoid contamination, and preceded by a drying agent and glass wool filter to minimize the chance for plugging. They should also be cleaned by washing after each use. Different sized orifices may be used for different flow ranges. The general relationship for an orifice flow meter is as follows [9]:

$$Q_o = (60) \, CA \sqrt{\frac{\Delta P_o}{\rho}} \qquad (71)$$

where:

Q_o = Rate of flow at orifice in ft³/min
C = Constant for orifice
A = Cross sectional area in ft²
ΔP_o = Pressure drop across orifice in inches of Hg.
ρ = Gas density in lb/ft³

Values for the orifice constant "C" may be determined for each orifice. However, the normal procedure is to set up a calibration curve by drawing air through the orifice with the sampling train in place exactly as used in the field. The variation in flow rate with pressure drop is then determined across the orifice. A typical calibration train and curve are shown in Figure 28. Correction of flow rates from calibration conditions (1) to field conditions (2) is shown as follows:

Substitution of the ideal gas law relationships for density allows corrections for flow rates in terms of pressure and temperature changes.

$$Q_1 = \sqrt{\frac{MW_2}{MW_1} \frac{P_2}{P_1} \frac{T_1}{T_2}} (Q_2) \qquad (72)$$

where:

Q = Volumetric rate of flow
MW = Molecular weight of gas
P = Pressure
T = Temperature

all for conditions 1 and 2 respectively.

Orifice flow meters find extensive usage in both gas and particulate sampling because of their accuracy and stability.

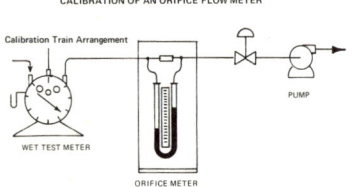

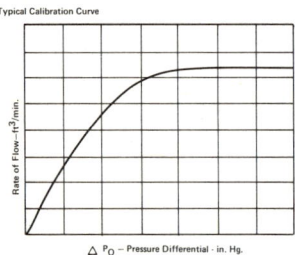

Fig. 28.

B. Rotameter

The rate of flow through a rotameter is proportional to the height of rise of a metering float in a transparent, vertical tube. Rotameters are devices used for low, and essentially constant, pressure drops. They can be used in both gas and particulate sampling. Rotameters have several limitations in use for gas sampling flow measurements. They suffer from float oscillation with variable vacuum sources. A surge bottle added downstream of the rotameter helps to minimize this problem. They should be operated at near to atmospheric pressure, or else calibrated for use at conditions substantially different from atmospheric. They are subject to condensation of moisture, which will render flow measurements useless. This problem may be eliminated by placing a drying agent upstream of the rotameter or heating the rotameter. The first method is better because it is less complicated.

The general equation for rotameters is the same as for orifice meters, and the flow correction in terms of pressure and temperature change is as follows [10] [11]:

$$Q_1 = \sqrt{\frac{MW_2}{MW_1} \frac{P_2}{P_1} \frac{T_1}{T_2}} Q_2 \qquad (73)$$

C. Cyclone Meters

The pressure drop across a cyclone collector is measured as a function of flow rate, following calibration of the unit. Hawksley [12] presents a complete discussion of the methods used, where the expression is the same as for an orifice meter. This device is normally used as a combined particle collection-flow measuring device in high rate particulate sampling.

D. Dry Gas Meter

The dry gas meter is a total-volume, positive-displacement device which uses a gas stream moving a diaphragm as a means of flow measurement where the rate of flow is the volume displaced per unit time [13]. The meters are high-flow-rate devices (0.1-2.0 cfm) which find extensive use in particulate sampling. They are not usually accurate for lower flow rates except when small meters are used. They should be calibrated frequently to maintain accuracy; a standard orifice (of plus-or-minus two percent error) and calibration curve are supplied by the manufacturer. They should not be operated at negative pressures exceeding ten inches of mercury.

E. Bubble Tube Meter

Gas is drawn from the stack through the sampling train and through a cylindrical tube of known volume. A soap solution in the bottom of the tube is caused to form bubbles by passage of gas through the liquid The rate of rise per unit time provides a direct measure of volumetric flow rate. Bubble tubes are usually small flow devices of less than one liter per minute capacity because high flows and large diameter bubbles cause surface tension effects which tend to promote bubble breakage. They are extensively used with gas chromatographs and coulometric titrators. The use of a bubble tube meter as a calibration device is shown in Figure 29.

IV. FLOW CALIBRATION DEVICES

Gas flow measuring units must be calibrated on a regular basis to assure their continued accuracy, especially if used for sampling in moist, dirty, or corrosive atmospheres.

A. Standard Orifice

Standard orifices are the same as the orifice flow meters previously described, and are used to calibrate dry gas meters, rotameters, and high-volume suspended particulate samplers used in ambient air monitoring. The principle used is to measure the pressure drop across an orifice of known characteristics as shown in Figure 30. The equation for calibration is as follows [14]:

$$Q = K \sqrt{\frac{T_m}{P_b} (\Delta P)} \qquad (74)$$

where:

T_m = Ambient temperature in °R.
P_b = Barometric pressure in inches Hg.
K = Orifice constant.
Q = Rate of flow in cfm.
ΔP = Pressure drop across orifice, in inches of Hg.

B. Wet Test Meter

The wet test meter is composed of a case containing submerged rotor vanes divided into compartments of equal volume [13], where gas is passed consecutively into each compartment, thereby causing the rotors to turn. They are total volume devices where the rate of flow is the volume of water displaced per unit time. They are specifically intended for laboratory calibration of flow measuring devices *only* and should *not* be used for field sampling. They should be checked periodically against a standard orifice or spirometer for the sake of accuracy.

C. Spirometer

The spirometer is an absolute calibration device which consists of two vertical cylinders. One is closed at the top and fits inside the other,

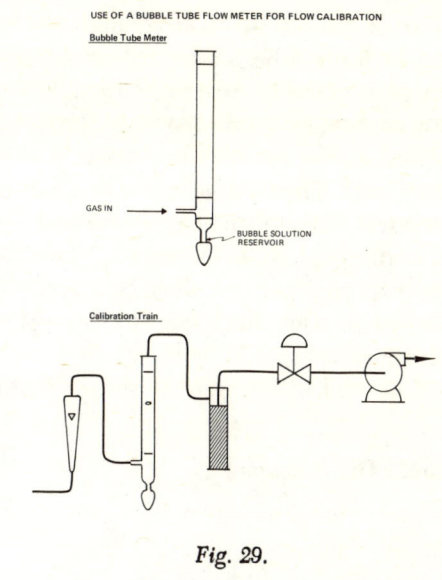

Fig. 29.

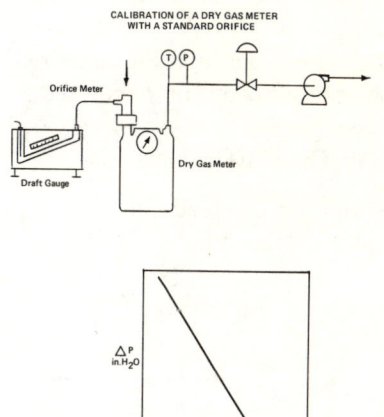

Fig. 30.

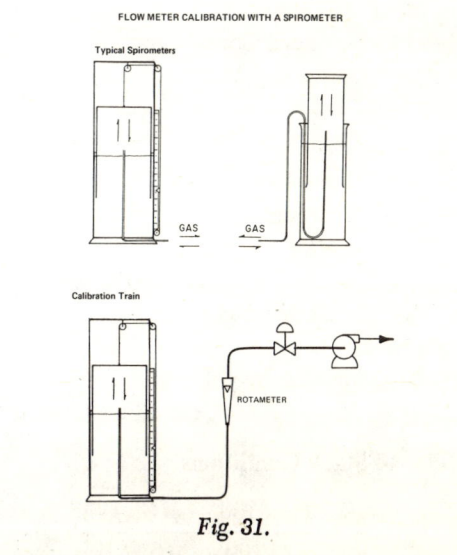

Fig. 31.

which is closed at the bottom and filled with water. The inner tube is attached to pulleys, and is counterweighted. It is caused to slide up and down a calibrated measuring bar. Gas is added to the floating inner tube at a point just above the fluid level. The volume is the product of the cross-sectional area of the inner tube and the distance displaced, while the rate of flow is the volume displaced per unit time. Spirometers provide an absolute measurement of volume, but are cumbersome and should be used only for laboratory calibration. A diagram is shown in Figure 31.

D. Liquid Displacement

It is possible to generate a gas volume under positive pressure at a known rate, as shown in Figure 32. Water is added to the upper container at a constant rate, causing the container to act as a constant head tank. Water then enters the calibrated lower bottle at the same rate, and the flow meters may thus be calibrated.

V. FLOW CORRECTIONS

The volume sampled for a flow measuring device at meter conditions is expressed as follows:

$$V_m = (Q_m)(t) \quad (75)$$

where
- V_m = Volume at meter conditions in ft³
- Q_m = Flow at meter conditions in ft³/min
- t = Time in minutes

Corrections can then be made as follows by use of the ideal gas laws:

A. Meter to Standard Conditions

$$V_o = \frac{P_m}{P_o} \frac{T_o}{T_m} \frac{(100 - MC_m)}{(100 - MC_o)} (V_m) \quad (76)$$

where: MC = Moisture content of gas, in percent

and: T_o = 60°F = 520°R;
$\qquad\qquad\qquad\qquad$ o = Std Conditions
P_o = 29.92 in. Hg;
$\qquad\qquad\qquad\qquad$ m = Meter Conditions
MC_o = 0.0% by vol.;
$\qquad\qquad\qquad\qquad$ s = Stack Conditions

B. Meter to Stack Conditions

$$V_s = \frac{P_m}{P_s} \frac{T_s}{T_m} \frac{(100 - MC_m)}{(100 - MC_s)} (V_m) \quad (77)$$

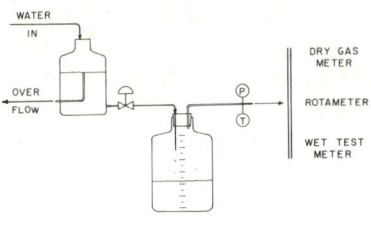

Fig. 32.

VI. REFERENCES

1. Bloomfield, B. D., "Source Testing," Ch. 28 in Stern, A. C., ed., *Air Pollution*, Vol. 2, pp. 487-536, Academic Press, Inc., New York, New York, 1968.
2. Razbegaeva, A. P., "Mechanization of the Gas Analyzer Orsat, *"Coke and Chemistry,* (Moscow, USSR), No. 4, pp. 53-54, 1958, trans. in Levine, B.S., "USSR Literature of Air Pollution and Related Occupational Diseases," Vol. 7, U. S. Public Health Service, Washington, D. C., 1962.
3. Rehm, F. R., "Test Methods for Determining Emission Characteristics of Incinerators," *Journal of the APCA*, Vol. 15, No. 3, pp. 131-135, March 1965.
4. Evans, D. G., "A Water-Cooled Probe for Sampling Gases from Shaft Furnaces," *Journal of the Institute of Fuel*, Vol. 37, No. 278, pp. 108-112, March 1964.
5. Smith, J. F., Hultz, J. A., and Orning, A. A., "Sampling and Analysis of Flue Gas for Oxides of Sulfur and Nitrogen," U. S. Bureau of Mines Report of Investigations 7108, 1968.
6. Chojnowski, B., "A Multipoint Solid and Gas Sampling Probe,"*Journal of the Institute of Fuel*, Vol. 37, No. 277, pp. 79-81, February 1964.
7. Haaland, H. H., ed., "Methods for Determination of Velocity, Volume, Dust, and Mist Content of Gases," Bulletin WP-50, Western Precipitation Corp., Los Angeles, California, 1968.
8. "Determining Dust Concentration in a Gas Stream," Performance Test Code 27-1957, American Society of Mechanical Engineers, New York, New York, 1957.
9. "Sampling and Analysis of Waste Gases and Particulate Matter," *Manual on the Disposal of Refinery Wastes,* Vol. 5, American Petroleum Institute, New York, New York, 1954.
10. Hudson, R. G., *The Engineers' Manual,* John Wiley and Sons, Inc., New York, New York, 1963.
11. Perry, J. H., *Chemical Engineers' Handbook,* 4th ed., McGraw-Hill Book Co., New York, New York, 1963.
12. Hawksley, P. G. W., Badzioch, S., and Blackett, J. H., *Measurement of Solids in Flue Gases,* British Coal Utilization Research Association, Leatherhead, Surrey, England, 1961.
13. Jacobs, M. B., *The Chemical Analysis of Air Pollutants,* Interscience Publishers, Inc., New York, New York, 1960.
14. Orifice Calibrator Chart provided by Western Precipitation Corp., Los Angeles, California, 1965.

CHAPTER 6
PRINCIPLES OF PARTICULATE SAMPLING

I. INTRODUCTION

Particulate sampling is done on sources where solid and liquid particles are being emitted, and is done for several reasons. These include determining the total particle emission levels for compliance with air pollution regulations, total volatile and nonvolatile fractions for combustion control, chemical compositions as indices of material balances and process control, and particle size distribution for potential effects and the feasibility of operation of different types of control devices. Particulate sampling involves the added complication of continuous velocity determination because of inertial effects of many particles. Numerous standards and manuals have been published previously describing various techniques for many different sources [1] [2] [3] [4] [5] [6] [7] [8] [9] [10] [11] [12] [13] [14] [15]. These should be consulted to provide both background and specific information.

II. ISOKINETIC SAMPLING

A. Theory

It is necessary to withdraw particulate samples into the sample probe at the same velocity as the average velocity of the gas stream at that point in order to obtain a representative sample. This condition is known as isokinetic sampling. Sampling at other than isokinetic velocities induces systematic errors for two reasons. 1) First, sampling at greater or less than the isokinetic rate tends to cause a respectively larger or smaller volume to be withdrawn from the flue than accounted for by the cross-sectional area of the probe. 2) Second, particles of greater than three to five microns in size have sufficient inertia so that particle motion may deviate significantly from the gas flow streamline pattern. In that case, particles are selectively drawn into the probe in a size distribution different from that existing in the duct. Fahrenbach [16] observed that if the sampling velocity is greater than the isokinetic rate, the sample will have a lower mass concentration of particulate material than the main stream, because of the greater percentage of the fine particles. However, if the sampling velocity is less than the isokinetic rate the particulate sample has a higher mass concentration than actually present, with a lower concentration of fine particles. The effect is shown in Figure 33 for sampling at less than, at, and greater than the isokinetic velocity condition.

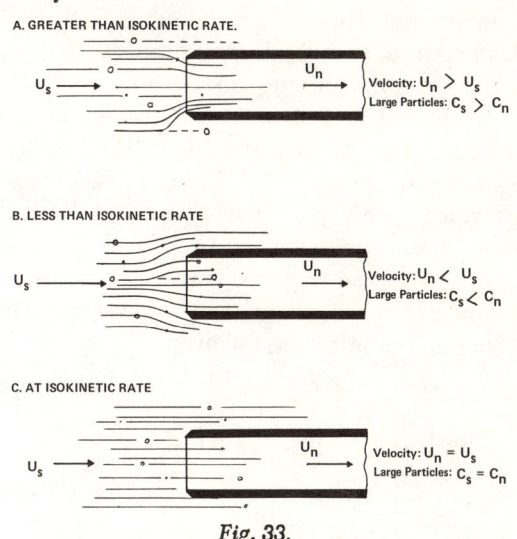

Fig. 33.

Parker [17] found that it was necessary to sample isokinetically and in regions of minimum flow disturbance in order to obtain a representative sample. Particles of less than four microns in size were found to closely follow the gas flow streamlines because their inertial effects became negligible, while particles greater than 50 microns had sufficient inertia to move relatively independently of gas flow streamlines. An inertia parameter K_I was used to describe the relative behavior of particles between four and fifty microns in size, where the larger its value the greater the degree of inertia [17].

$$K_I = \frac{C \cdot \rho_p \cdot D_p^2 U_s}{18 \mu_g \cdot D_n} \qquad (78)$$

where:

C = Dimensionless correction factor
ρ_p = Particle density in lb per ft^3
D_p = Particle diameter in feet.
U_s = Gas velocity in feet per second
μ_g = Gas viscosity in lb per ft-second
D_n = Sampling nozzle diameter in ft.

If the value of the inertia parameter was less than 0.05, inertial effects were negligible; if the value was greater than 50.0, the particles moved nearly independently of gas flow streamlines.

B. Effect of Sampling Rate

Several studies have been made to determine the effects of sampling at other than isokinetic velocities on observed particulate concentrations as compared to actual concentrations, and also in terms of particle size. Hemeon and Haines [18] found that the ratio of measured to actual concentration increased as the ratio of sampling velocity to actual velocity decreased in tests run at velocities of 1,000 and 2,000 fpm on test aerosols in the five to 25 micron size range [5]. These results are in agreement with predictions. Watson [19] developed an expression to describe the relationship of measured to actual particle concentration in terms of deviation of sampling velocity from isokinetic conditions, as follows:

$$\frac{C_m}{C_s} = \left(\frac{U_s}{U_n}\right)\left\{1 + f_{(p)}\left[\left(\frac{U_n}{U_s}\right)^{1/2} - 1\right]\right\}^2 \quad (79)$$

where:

- C_m = Measured particle concentration in grains per cubic foot
- C_s = Actual particle concentration in gas stream in grains per cubic foot
- U_s = Actual velocity of gas stream at point of sampling in feet per minute
- U_n = Mean gas velocity at sampling nozzle in feet per minute
- D_n = Inside diameter of sampling nozzle tip in feet
- ρ_p = Density of particles in pounds per cubic foot
- D_p = Diameter of particle in feet
- μ_g = Viscosity of gas stream in pounds per foot-second

$$p = \frac{\rho D_p^2 U_s}{18 \mu_g D_n} \quad (80)$$

The corrections for concentrations in terms of particle size were derived from wind tunnel tests at a velocity of 900 feet per minute using spherical spore particles of 4 and 32 micron diameter, respectively. This is expressed in the relationship between p and $f_{(p)}$, as shown in Figure 34.

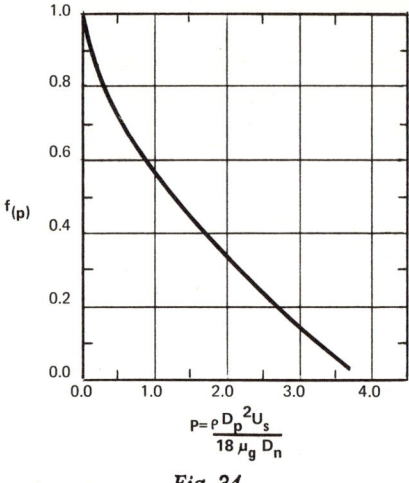

Fig. 34.

It is necessary to have knowledge of the particle size distribution of the gas stream being sampled in making corrections, however.

Watson [19] also determined the effect of an isokinetic sampling conditions on the ratio of measured to actual concentration in terms of particle size. The ratio of measured to actual concentration is plotted as a function of the ratio of actual duct velocity to probe velocity in Figure 35. Results showed an error of less than ten percent in measured concentration for particles of four micron diameter where the probe velocity was less than the isokinetic rate. Significantly larger errors were involved with increase in particle size as the deviation from isokinetic sampling conditions increased. The theoretical limit in measuring errors for very

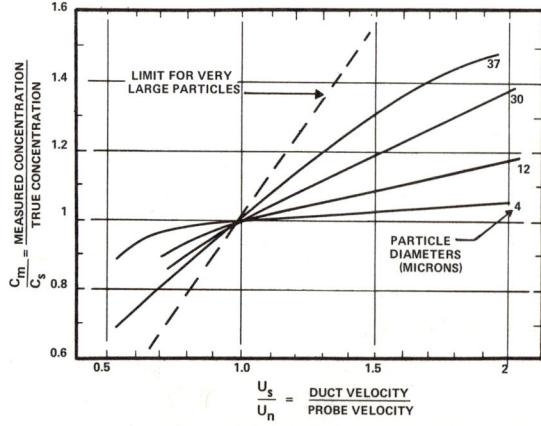

Fig. 35.

Fig. 34 & 35—Courtesy of *Amer. Industrial Hygiene Assoc. Journal*, Article by H. H. Watson, Jan. 1954.

large particles of sufficient inertia so as not to be affected appreciably by changes in sampling velocity relative to duct velocity were also computed.

Badzioch [20] studied the ratio of measured to actual particulate concentrations over a range of velocities, from 25 to 200 percent of an isokinetic sampling velocity of 1,500 feet per minute (25 feet per second). The degrees of deflection of four different particle sizes were studied using a one inch sampling nozzle. For the large particles, only those in the isokinetic flow stream of the nozzle will enter it independent of deflection by changes in flow stream direction. This was shown by the relatively small negative error at velocities greater than isokinetic, but a rapid increase in positive error at less than isokinetic rates. The opposite patterns were displayed by the small particles, which were subject to deflection by changes in flow stream direction caused by anisokinetic sampling condition. It is noted that for particles smaller than five microns in size, errors are relatively small at anisokinetic conditions, either positive or negative, because they tend to follow the gas flow streamlines.

Badzioch formulated a mathematical relationship to describe changes in observed particle concentrations compared to actual levels with deviations from isokinetic sampling velocities [5] [10] [20]:

$$\frac{C_m}{C_s} = \alpha \left(\frac{U_s}{U_m}\right) + (1 - \alpha) \qquad (81)$$

where:

C_m = Measured concentration
C_s = Actual concentration
U_m = Average velocity at sampling nozzle
U_s = Actual gas stream velocity
α = Inertia parameter

The inertia parameter "α" represents the fraction of particles present in the gas volume which are deflected because of the anisokinetic conditions. For fine particles less than five microns in size the value of "α" tends to zero; it tends to unity for large particles. Typical values are from 0.3 to 0.7, with an average of 0.5 for coal-fired boiler flue gases and nearly zero for particulate matter in oil-fired boiler flue gases because of their small sizes.

Bonnet [21] also studied the errors associated with sampling at other than isokinetic sampling rates for several particle sizes. He constructed a chart to facilitate corrections for sampling errors in terms of sampling velocity relative to isokinetic velocity and particle size.

Vitols [22] made use of a computer to predict theoretical errors in particulate sampling by computing the trajectories of spherical particles in a frictionless fluid so that inertial effects of particles were presumed to predominate. Results showed good agreement with those obtained by Badzioch [20] and represent the maximum error limits in particulate sampling.

Bloomfield [5] recently presented a table describing the effect of departure from isokinetic sampling conditions on sample concentrations, as shown in Table 13.

Sehmel [23] made an experimental study of the errors resulting from subisokinetic sampling of zinc sulfide particles using a membrane filter for collection. Sampling velocities from 0.008 to 1.00 of the isokinetic rate were investigated for air velocities from 238 to 2,640 feet per minute (3.96 to 44 feet per second). A mathematical model was used to calculate the ratio of measured to actual concentration based on the efficiency of particle impaction. It was presumed that an air cushion in the probe caused by the differences in velocity between the gas stream and the probe would preferentially allow the larger particles to enter the probe based on their greater inertia. Results shown in Figure 36 indicated that small particles of less than five microns in size were relatively unaffected by reducing the sampling rate below the iso-

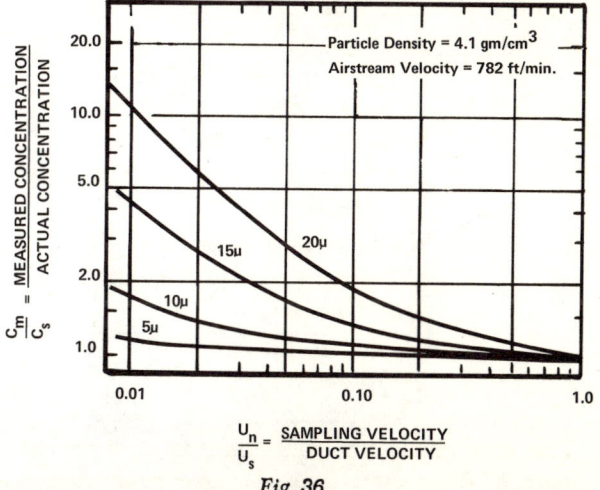

Fig. 36.

Courtesy of: *American Industrial Hygiene Association Journal*, Article by G. A. Sehmel, May-June 1967 (23).

kinetic velocity. However, for particles of 10 to 30 microns in size, the apparent concentration could be from twice to ten times the actual concentration at subisokinetic conditions. A method was also included for correcting measured to actual particulate concentrations, but its use would require a knowledge of particle size distribution of the gas stream.

C. Effect of Probe Alignment

It is necessary to align the probe in the direction of the incoming gas stream in order to obtain a representative sample of the particulate materials present. Misalignment of the probe causes selective withdrawal of smaller particles without as many of the larger particles as are actually present; hence, an error in measurement. Watson [19] studied the effect of changing the angle of probe alignment with the gas stream upon the ratio of measured to actual concentrations for different sized particles while maintaining the probe velocity the same as that in the duct. Results shown in Figure 37 indicate that the measured concentration was less than the actual concentration in all cases studied. The difference between measured and actual concentrations increased as the angle of misalignment increased, and as the particle size increased. The effect was especially pronounced at angles of greater than thirty degrees, and for particles larger than four microns in size. Additional work by the National Council for Air and Stream Improvement [24] recommends permitting a maximum error of plus or minus five degrees in probe alignment with the direction of the gas stream.

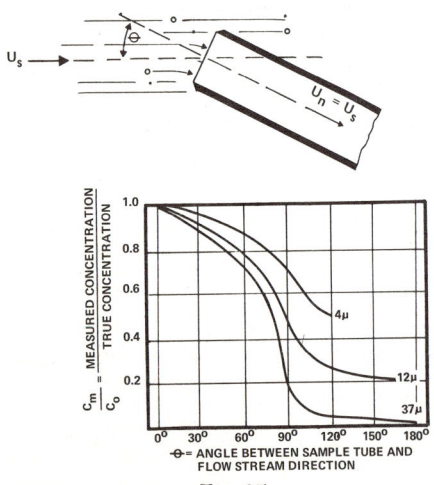

Fig. 37.

Courtesy of: *American Industrial Hygiene Association Journal*, Article by H. H. Watson, January 1954 (19).

An important aspect of probe alignment occurs in ducts where the direction of gas flow is not the same as the alignment of the duct. The problem is particularly common following bends or obstructions where there is a nonuniform flow pattern, or following cyclonic collectors or scrubbers where there is a swirling flow pattern. When an S-type pitot tube is placed adjacent to the sample probe, the probe nozzle should be aligned in the direction of the maximum velocity head differential across the pitot tube. This method, in contrast to sampling the velocity component in the direction of the duct, is an attempt to assure obtaining a representative sample.

D. Effect of Nozzle Shape

Nozzle configuration has an effect on the accuracy of particulate sampling by affecting the pattern of gas flow streamlines. Parker [17] observed that knife-edge probes tapered from the outside inward to a sharp edge produce the minimum disruption of gas flow streamlines. Hawksley [10] also observed that square-edged probes tend to selectively allow larger particles to bounce into the probe, and that the probe should be slanted from the outward edge inward. Badzioch [20] observed, however, referring to the previous discussion, that the use of blunt-edged probes had only a small effect on the value of the inertia parameter "α" and that blunt-edge probes may not yield significantly different results than sharp-edged ones.

However, subsequent work by Whiteley and Reed [25] evaluated several different types of probes by simultaneous sampling in a dust chamber using coal flyash from a precipitator, and found blunt-edge probes produced significant errors. They found that the sharp-edged probe (15° angle) produces similar results to a short probe with a 120° angle. The side port tube probe, which is capped at one end with a hole facing in one direction near the tip, was slightly less accurate than the two above probes at velocities greater than the isokinetic rate. However, it was found sufficiently accurate at velocities less than the isokinetic rate, as substantiated in subsequent work by Lundgren and Calvert [26]. It is noted that deviation of measured from actual concentrations did not exceed plus-or-minus ten percent for the sharp-edged probe over a velocity range of 25 to 200 percent of the isokinetic rate, as

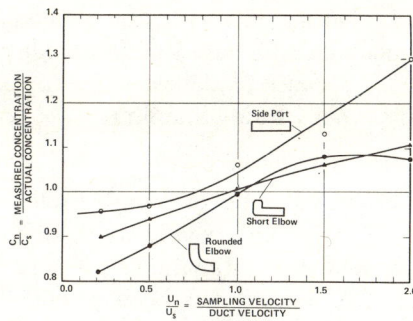

Fig. 38.
Courtesy of: *Journal of the Institute of Fuel* (London), Article by A. B. Whiteley and L. E. Reed, July 1959 (25).

shown in Figure 38 [25]. Dennis and Samples [27] investigated the flow characteristics of five different probe shapes over a velocity range of 1,000 to 6,000 fpm and found similar results.

E. Effect of Nozzle Size

There is a lower limit to the diameter of sampling nozzles used in particulate determination, for two reasons. First, systematic error is induced by selectively excluding large particles because of inadequate cross-sectional area. Second, the volume of source gas sampled per unit time decreases by the 1.5 power as the nozzle cross-sectional area decreases, thus tending to magnify any random errors. The ASME [9] and Los Angeles County [1] recommend a minimum sampling nozzle size of ¼ inch, which particularly applies to coal and oil-fuel combustion sources and many manufacturing sources as well. Rehm [28] recommended a minimum nozzle size of ¾ inch for sampling refuse incinerator flue gases because of the presence of large flake and char particles.

Two studies have shown the potential effects of nozzle size on the measured particulate concentrations. Hemeon and Haines [18] made a study on the effect of three nozzle diameters (⅛″, ¼″, ⅜″) on particle concentration for particles in the size range of 5 to 25 microns, and also in the range from 400 to 500 microns. For the small size particles there was no significant difference between the three nozzle sizes. However, for the large particles (not a typical sampling condition for many sources) use of the ⅛ inch nozzle resulted in collection of a significantly smaller quantity of particulate material from the same gas stream than with the larger sized nozzles. Results of the study are shown in Table 14.

Recent studies by Miller and Abrams [29] were made by taking consecutive samples using 3/16, 1/4, and 3/8 inch sampling nozzles, respectively, on the flue gases from a waste wood-fired power boiler at constant firing conditions. Results showed the particulate concentrations measured using the ¼ and ⅜ inch nozzles were approximately the same, but the value measured when using the 3/16 inch nozzle was substantially below that of the other two sizes.

Parker [18] presents a statistical expression for minimum nozzle size, which states that nozzle size becomes important only when the quantity of material present is small and is being sampled at low velocities, and for short time intervals. However, it does not account for systematic errors caused by large particles, except through the term for fractional error "F" after a given sampling time.

$$D_n^2 = \frac{1.274\, T^2}{t\, U_s C_s F^2} \qquad (82)$$

where:

- D_n = Nozzle diameter in meters
- t = Sampling time in minutes
- U_s = Isokinetic velocity in meters per minute
- C_s = Actual particle concentration in number/m³
- T = Confidence level
- F = Fractional error after sampling time t

F. Nozzle Selection

The sampling nozzle used must be large enough so as to obtain a representative sample including the large particles, and to obtain a large enough sample volume to collect a sufficient amount of material. It also must be of small enough diameter to maintain isokinetic velocities at the nozzle within the flow capacity of the prime mover and the pressure drop characteristics of the sampling train. However, it is usually best to use sampling nozzles greater than ¼ inch in diameter to avoid systematic errors [1] [9]. The normal range of sampling nozzle sizes is from ¼ to 1½ inches in diameter, depending on the source and the equipment used.

The exact nozzle sizes for given velocity ranges depend on the characteristics of the sampling train in use [5] [10] [14]. Sampling trains designed to operate using thimbles, impingers, and filtration systems, and within the gas flow range of 0.5 to 2.0 cfm at pressure drops ranging from 1.0 to 10.0 inches of mercury, can make use of the nozzle selection chart in Table 16A. The above considerations apply to methods specified by ASME [8] [9] Los Angeles County [1], Dade County [7] and Bay Area [6] manuals, Western Precipitation [2], Research-Cottrell [11], National Council [24], Public Health Service [30], and others. Values are valid within plus- or minus-twenty percent over the temperature range from 150°F to 1000°F for most flue gases. A value of approximately 450°F at ten percent moisture and twelve percent carbon dioxide power boiler flue gas is used as the mean for the basis of calculations, and applies to most flue gas conditions. Values listed in Table 16A have been derived from actual observations from combination thimble-impinger sampling trains for gas flows from 0.5 to 2.0 cfm.

A similar set of values for nozzle size has been constructed by the British Standards Institution for a higher range of flow rates. It uses an internally-located cyclone-glass fiber filter particulate collection system over a flow range of 4 to 18 cfm at a maximum pressure drop of 36 inches of water, which is approximately three inches of mercury [13]. Calculations are again based on power boiler flue gas at 450°F and results are shown in Table 16B.

An expression was developed by Toynbee and Parkes [31] for the sampling nozzle diameter in terms of Reynolds' number when using a pressure null balance sampling system:

$$D_n = [0.33 \log (N_{Re}) - 0.491]^{0.25} \quad (83)$$

$$N_{Re} = \frac{\rho U_n D_n}{\mu} \quad (84)$$

where:

N_{Re} = Reynolds Number (dimensionless)
ρ = Density of flue gas stream in pounds per cubic foot
U_n = Nozzle velocity in feet per minute
D_n = Nozzle diameter in feet
μ = Viscosity of flue gas in pounds per foot-minute

Bloomfield [5] presents an additional method employing a relatively constant sampling flow rate, and varying nozzle size with changes in velocity. However, the method could present difficulties for rapid and variable velocity changes, because of the necessity for changing sample probes during sampling.

G. Sampling Rate Estimation
1. Pitot Tube Parallel to Sample Probe

It is necessary to provide a rapid and accurate method for estimating isokinetic sampling rates. A method has been developed for estimating sampling rates in terms of flow characteristics of the sampling train with varying probe sizes and flue gas characteristics to achieve a considerable time-saving in running tests. The equation for isokinetic sampling rate shows the flow rate to be directly proportional to the velocity in the duct and inversely proportional to the absolute meter pressure where all other quantities may be measured:

$$Q_m = \left[\frac{T_m \cdot P_s}{T_s} \cdot \frac{(100 - MC_s)}{(100 - MC_m)} \cdot A_n \right] \cdot \frac{U_s}{P_m} \quad (85)$$

$$Q_m = K \cdot \frac{U_s}{P_m}; \quad Q_m \cdot P_m = K \cdot U_s \quad (86)$$

$$U_s = \frac{1}{K} \cdot Q_m \cdot P_m \quad (87)$$

where:

K = Dimensionless constant
Q_m = Sample flow rate at meter in cubic feet per minute
T_m = Temperature of gas at meter in °R
P_m = Absolute static pressure at meter in inches of mercury
T_s = Absolute temperature of flue gas in °R
P_s = Absolute static pressure of flue gas in inches mercury
MC_m = Moisture content of gas at meter in percent by volume
MC_s = Moisture content of flue gas in percent by volume
U_s = Velocity of flue gas at probe tip in feet per minute
A_n = Cross sectional area of probe tip in square feet

The sampling rates and meter pressures are recorded at a series of conditions with the complete sampling train in line for each size probe. It is then possible to select the proper sample probe size, and estimate the isokinetic sampling rate from simply a direct velocity measurement. The procedure used to set up a curve of isokinetic sampling rate versus stack velocity for a given sample probe size with the complete sampling train in line is shown in the following table for a 5/16 inch diameter nozzle.

where: Q_R = Indicated Rotameter flow rate in liters per minute.

A family of curves observed for a given sampling train used to sample a power boiler is shown in Figure 39. Once the chart has been constructed for a given source, direct pressure differential readings on the draft gauge may be substituted for the velocity readings, as shown in Figure 40. Correction of actual conditions to those used in computing the charts can be made by changes in the proportionality constant, as follows:

$$K_t = \frac{(T_m)_o}{(T_m)_t} \cdot \frac{(P_s)_o}{(P_s)_t} \cdot \frac{(T_s)_t}{(T_s)_o} \cdot K_o \quad (88)$$

where:

K_o = Constant for equation 86 at initial conditions (dimensionless)
K_t = Constant for equation 86 during test, at time t
$(T_m)_o$ = Absolute meter temperature in °R at initial conditions
$(T_m)_t$ = Absolute meter temperature in °R during test at time t
$(T_s)_o$ = Absolute stack temperature in °R at initial conditions
$(T_s)_t$ = Absolute stack temperature in °R during test at time t
$(P_s)_o$ = Absolute static pressure of flue gas in inches of mercury at initial conditions
$(P_s)_t$ = Absolute static pressure of flue gas in inches of mercury during test at time t

There are often changes in gas velocity during sampling which require changes in sampling

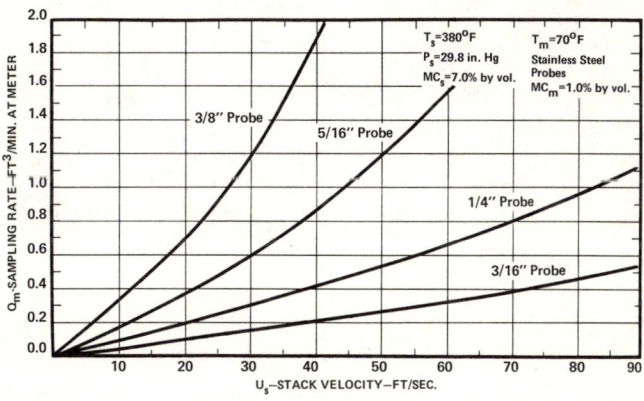

Fig. 39.

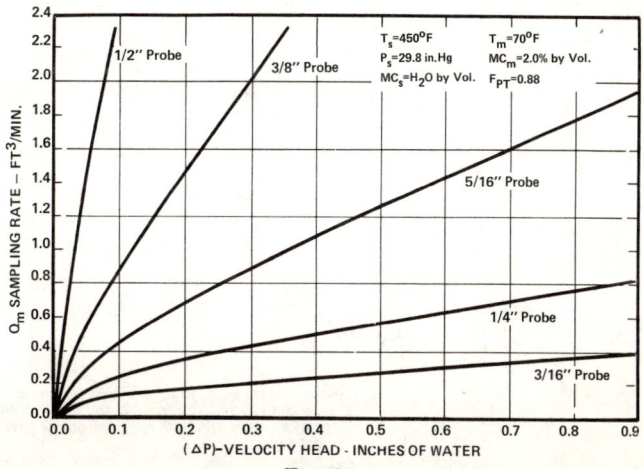

Fig. 40.

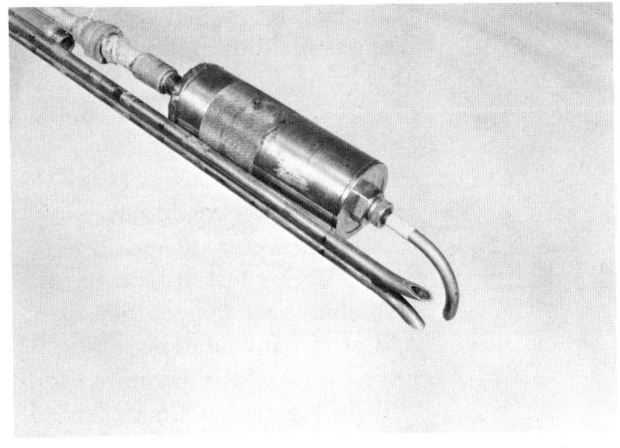

Fig. 41.

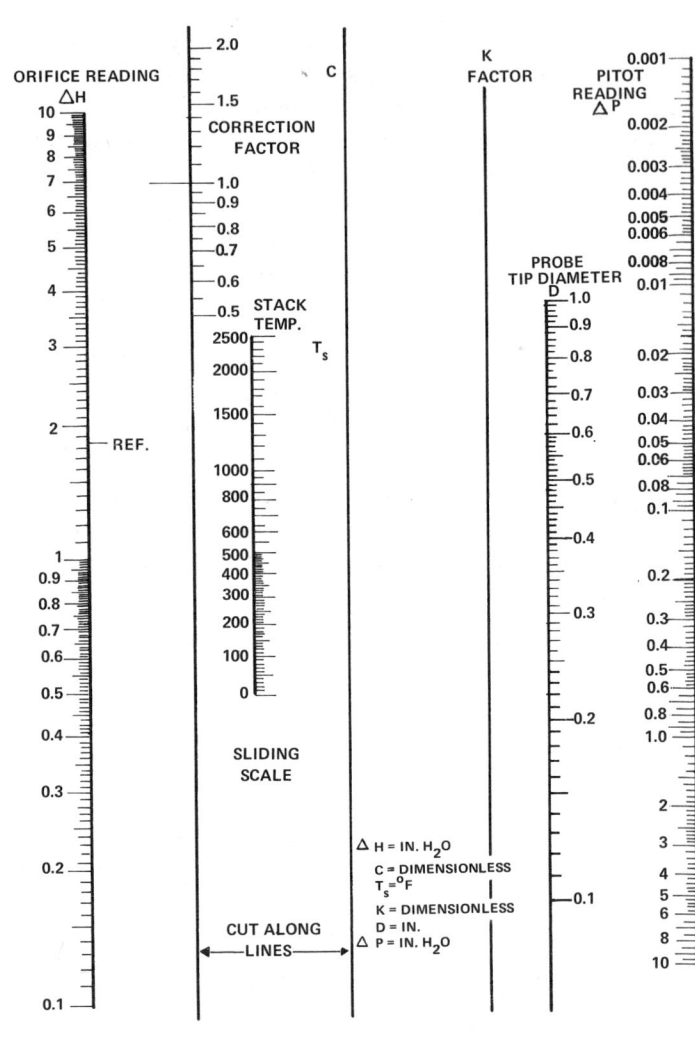

Fig. 42.

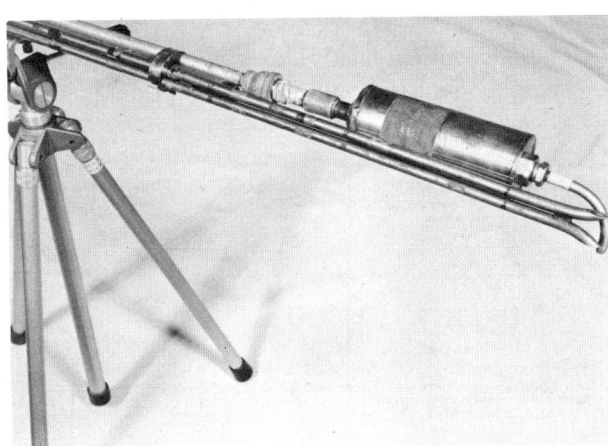

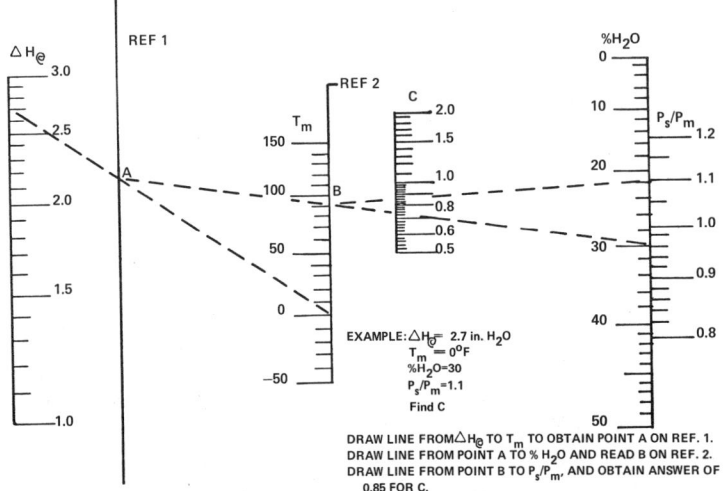

EXAMPLE: $\Delta H_@ = 2.7$ in. H_2O
$T_m = 0°F$
$\%H_2O = 30$
$P_s/P_m = 1.1$
Find C

DRAW LINE FROM $\Delta H_@$ TO T_m TO OBTAIN POINT A ON REF. 1.
DRAW LINE FROM POINT A TO $\% H_2O$ AND READ B ON REF. 2.
DRAW LINE FROM POINT B TO P_s/P_m, AND OBTAIN ANSWER OF 0.85 FOR C.

Fig. 43.

Courtesy of: Mr. Walter S. Smith, National Air Pollution Control Administration, Durham, North Carolina, 1968 (30).

rate to maintain isokinetic conditions. A pitot tube is normally installed adjacent to the sampling probe at a distance of 1 to 1-½ inches, to continuously measure velocity as shown in Figure 41. [10] [14] [30] [32]. The pitot tube should be close enough to the probe for accurate velocity measurement, but not so close as to interfere with the gas flow streamlines to the sampling nozzle. This is often done by clamping the pitot tube and sample probe assembly on opposite sides of a 1 or 1-½ inch pipe, with both lines axially in the same direction.

It is also necessary to continually check the sampling rate during testing to maintain isokinetic sampling conditions, especially if filtration techniques are used, because of the added resistance to flow resulting from particles being deposited on the filter media and the sampling lines. To maintain the same isokinetic velocity, it is then necessary to frequently increase the flow rate during the sampling period to compensate for the increased resistance to flow. The relationship used to make the corrections in flow rate, presuming constant isokinetic velocity and source conditions is the ratio of the meter pressure at the start of the test to the meter pressure at time "t" during the test. Corrections for changes in sampling rate caused by flow resistance increases for different initial meter pressures are listed in Table 18.

$$(Q_m)_t = \frac{(P_m)_o}{(P_m)_t} (Q_m)_o = \frac{(P_m)_o}{(P_m)_o - \Delta P_m} (Q_m)_o \tag{89}$$

where:

$(Q_m)_o$ = Sampling rate at start of test in cubic feet per minute
$(Q_m)_t$ = Sampling rate during test at time t in cubic feet per minute
$(P_m)_o$ = Absolute static pressure at meter at start of test in inches of mercury
$(P_m)_t$ = Absolute static pressure at meter during test in inches of mercury
ΔP_m = Change in absolute meter pressure during test between start of test and time t, in inches of mercury

However, this correction for changes in flow characteristics is not normally greater than ten percent for most sampling trains, except when thimbles or fine filters are used.

The preceding treatment is based upon maintaining the flow measuring section upstream of the prime mover, so as to avoid possible leakage problems in vacuum pumps. Either an orifice flow meter or a rotameter is added to indicate the gas flow rate, and is placed in series with the dry gas meter used to measure the total volume sampled. The accuracy of dry gas meters for accurate volume measurement is subject to severe limitations at greater than ten inches of mercury vacuum at the meter; this level should not normally be exceeded. It is also necessary to shield the gas meter in some applications where it may be exposed to sunlight or to excessive temperatures, to prevent erroneous meter temperature measurements. Maintenance of constant meter temperature simplifies sampling rate-calculations considerably.

Smith [30] uses a similar technique for estimating isokinetic sampling velocity by continuously measuring gas velocity with a pitot tube. He uses a nomograph to make corrections in the sampling rate to correspond to velocity changes; the nomograph is illustrated in Figures 42 and 43. The flow measurement and control section is placed downstream of the prime mover, which allows it to remain at an essentially constant meter inlet pressure which is relatively independent of changes in flow resistance upstream of the pump. Leakage through the oil-lubricated vacuum pump was observed to be insignificant during special flow tests [33]. Therefore, nomograph can be constructed for an essentially constant meter pressure, which considerably simplifies the calculations for sampling rate. However, caution should be used in applying this sampling train arrangement because many vacuum pumps have considerable air leakage through the shift seals, which may introduce significant error in flow measurement. This is particularly true of many oil-less carbon vane vacuum pumps; the effect tends to increase with wear and age of the pump. The nomograph must also be modified if a water aspirator is used as the prime mover because it is not possible to place the meter downstream of the aspirator. However, the general technique is very useful in most source test applications, particularly for nomograph construction. Supplementary information on nomography is available in Perry and other references [34] (35).

However, both of the previously described methods are used for sampling flow rates from 0.5 to 1.0 or 2.0 cfm. The cyclone-filter system described by the BCURA [10] and British Standards Institution [13] is a high flow rate [4.0 to 18.0 cfm], where pressure drop across an orifice meter or the cyclone collector is used as the basis of sampling rate determination. However, pressure drops through the sampling train are considerably lower than with the others, seldom exceeding three inches of mercury. A sample calibration curve is shown in Figure 44.

The procedures for constructing calibration curves for orifice flow meters are described by Bloomfield [5] from work done by Gallaer [36]. Pitot tube readings and stack temperatures are converted to equivalent orifice pressure differentials to maintain isokinetic sampling velocities, and are presented in a graphical form to simplify usage in the field. The equation for pressure drop through a sampling orifice corresponding to isokinetic conditions is taken from ASME Performance Test Code 27 [9].

$$\Delta P_O = \left(\frac{A_n}{C_O A_n} \right)^2 \cdot \frac{P_s}{P_O} \frac{T_O}{T_s} \cdot (\Delta P)_{PT} \quad (90)$$

where:

A_n = Area of the sampling nozzle in ft²
A_O = Area through the orifice in ft²
C_O = Orifice coefficient—dimensionless
P_s = Static pressure in the duct in inches of Hg
P_O = Pressure at the orifice outlet in inches of Hg
T_s = Temperature of the flue gas in °R
T_O = Temperature of the gas at orifice outlet in °R
ΔP_O = Pressure drop across the orifice in inches of water

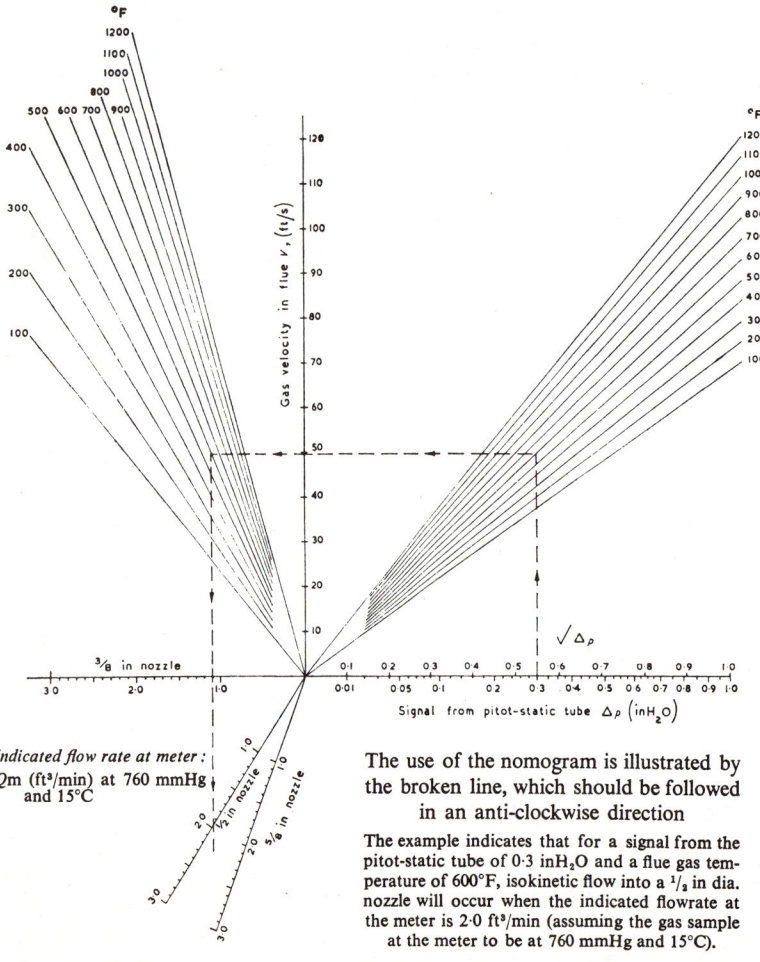

The use of the nomogram is illustrated by the broken line, which should be followed in an anti-clockwise direction

The example indicates that for a signal from the pitot-static tube of 0·3 inH₂O and a flue gas temperature of 600°F, isokinetic flow into a ½ in dia. nozzle will occur when the indicated flowrate at the meter is 2·0 ft³/min (assuming the gas sample at the meter to be at 760 mmHg and 15°C).

Fig. 44.

Courtesy of: British Standards Institution, London, England, British Standard 3405, 1961 (13).

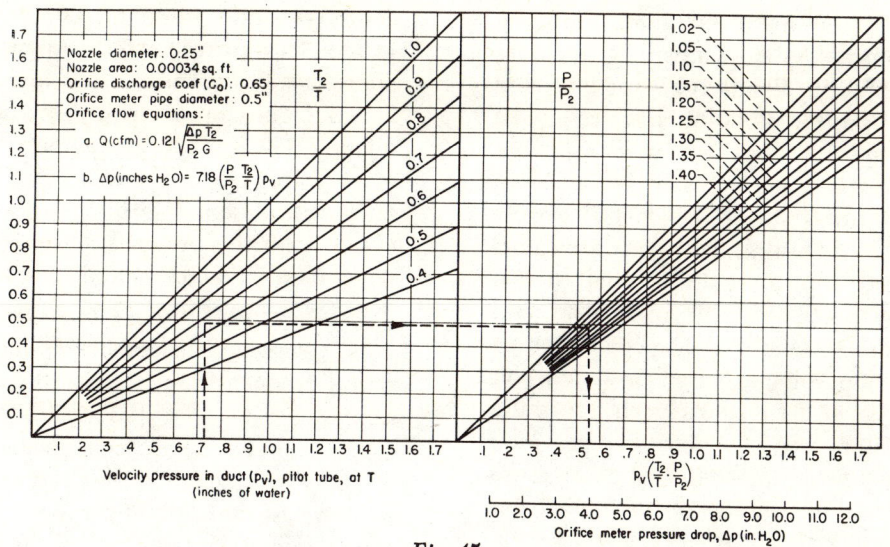

Fig. 45.
Courtesy of: *Power* Magazine, McGraw-Hill Publications Company, Article by C. A. Gallaer, January 1957 (36).

ΔP_{PT} = Velocity pressure differential at draft gauge in inches of H_2O

Note: $C_O = 0.60$ to 0.65

By letting C_O equal 0.65 [36], and substituting for cross-sectional areas, the equation becomes:

$$\Delta P_O = 2.37 \left(\frac{D_n}{D_O}\right)^4 \cdot \left(\frac{P_s}{P_O}\right) \cdot \left(\frac{T_O}{T_s}\right) \cdot \Delta P_{PT} \quad (91)$$

A nomograph may then be constructed to facilitate calculations, as shown in Figure 45 [36].

2. Continuous Automatic Sampling Rate Adjustment

Moore [37] designed a system for continuous adjustment of the sampling rate to maintain isokinetic conditions from pitot tube measurements in making particulate measurements in the exhaust of a coal-fired gas turbine. The sample flow rate was automatically maintained at the isokinetic conditions by an air-operated valve actuated by pressure differential cells connected to the pitot tube through pneumatic impulse relays. Flow rate was measured by the pressure differential across an orifice, also connected to the impulse relays. It also had a system to automatically move the entire system from one sampling point to several others on a cyclical basis. The device made use of particulate collection by filtration through a thimble, and it allowed taking particulate samples without the need for an operator being present during the test. It proved successful during approximately 500 hours of continuous operation. A diagram of the system is illustrated in Figure 46.

Subsequent work by Moore and Fasching

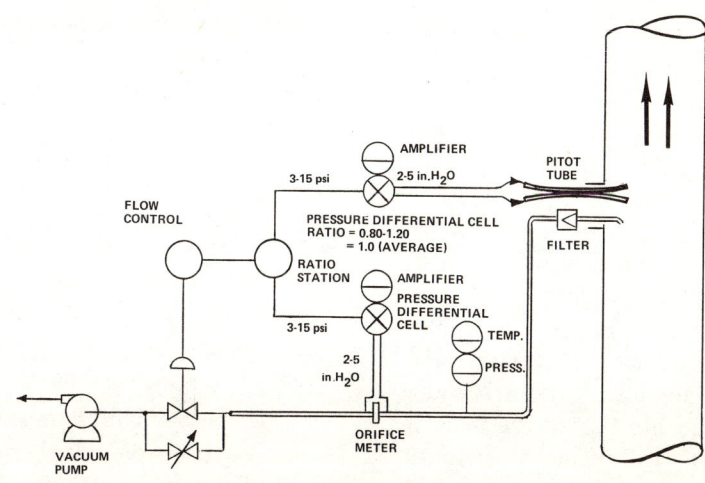

Fig. 46.

[38] resulted in an improved automatic sampling rate control system employing pressure activated transducers. A pitot tube was placed in parallel to the sample probe directly facing the gas stream. Flow rate was regulated by means of a motor driven valve actuated by a servo-mechanism control system. The control system consisted of two identical analog compensators for the respective pressure differentials across the orifice flow meter and the pitot tube, and inverter and a servo amplifier. The compensators used were analog voltages based on a ten volt dynamic range, with the transducers for static and differential pressures, which also provided an output of ten volts for the full range pressure inputs. Ten volts output signal was also available from the thermocouple amplifier based on absolute temperature input. Isokinetic conditions were maintained by the servo control system by preserving the following electronic equality, where all variables were absolute and converted to voltages:

$$\frac{P_m \Delta P_O}{T_m} = \frac{P_s \Delta P_{PT}}{T_s} \qquad (92)$$

where:

- P_m = Absolute static pressure at the orifice
- ΔP_O = Pressure differential across the orifice meter
- T_m = Absolute temperature at the meter
- P_s = Absolute static pressure in the duct
- ΔP_{PT} = Pressure differential across the pitot tube
- T_s = Absolute temperature in the duct

Physical design features of the revised system were similar to the earlier model, and have been illustrated in Figure 47.

Flue gas was withdrawn through a probe into a thimble where the particles were filtered out, and then through an orifice flow meter. Flow was controlled by a bleed valve arrangement, and a vacuum pump was used as a prime mover. The orifice was a low pressure drop device employing a full scale output of 0.50 psia at sample line flow at meter conditions of 200°F, and 40.7 inches of mercury absolute. This produced a nozzle inlet velocity equal to that in the duct at a temperature of 700°F, 40.7 inches of mercury absolute pressure, and a pressure differential across the pitot tube of 1.61 inches of water gauge. This corresponded to a stack velocity of approximately 6,000 feet per minute at a meter flow rate of approximately five cubic feet per minute. The system has potential usefulness in other source test applications, and was produced at a cost of about $6,000. It proved more accurate than the previous system, but at the expense of additional cost and complexity.

3. Pressure Null Balance Adjustment

Pressure null balance sampling systems have found frequent usage for estimating isokinetic sampling rates [9] [28] [39] [40], and two typical nozzles have been illustrated in Figure 48. Their principle of operation is that static pressure in the sampling tube equals the static pressure in a duct only when the velocity in the tube is the same as that in the duct at a given point [9]. The system is relatively simple in its operation and does not require sophisticated equipment. However, numerous problems have been observed in attempting to accurately maintain true isokinetic sampling conditions because the existence of equal pressures at outer and inner probe walls does not necessarily mean that equal velocities exist at both points. Differences in frictional flow losses between inner and outer surfaces caused by turbulence and surface nonuniformities, progressive coating and possible plugging of the inner static tap by particles, and possible differences of static tap location may all produce these conditions. Parker [17] found that null balance systems had limited usage for large probes greater than ¾ inch diameter. Toynbee and Parkes [31] postulated that by a slight expansion of the rear section of the probe the inner frictional losses could be reduced inside the nozzle, and the system could be used over the velocity range from 600 to 2,500 fpm. However, subsequent comments by Nonhebel [31] in the same issue stated that the plugging problems associated with the inner static taps could not be overcome. Work by Dennis [27] and Hemeon and Haines [18] indicated that it was not always possible to assure isokinetic sampling conditions, and found the errors at different velocities for two nozzle sizes when departing from nozzle conditions [5]. These results are as shown in Figure 49.

4. Thermal Null Balance Adjustment

The principle of thermal null balance has only recently been applied toward maintaining

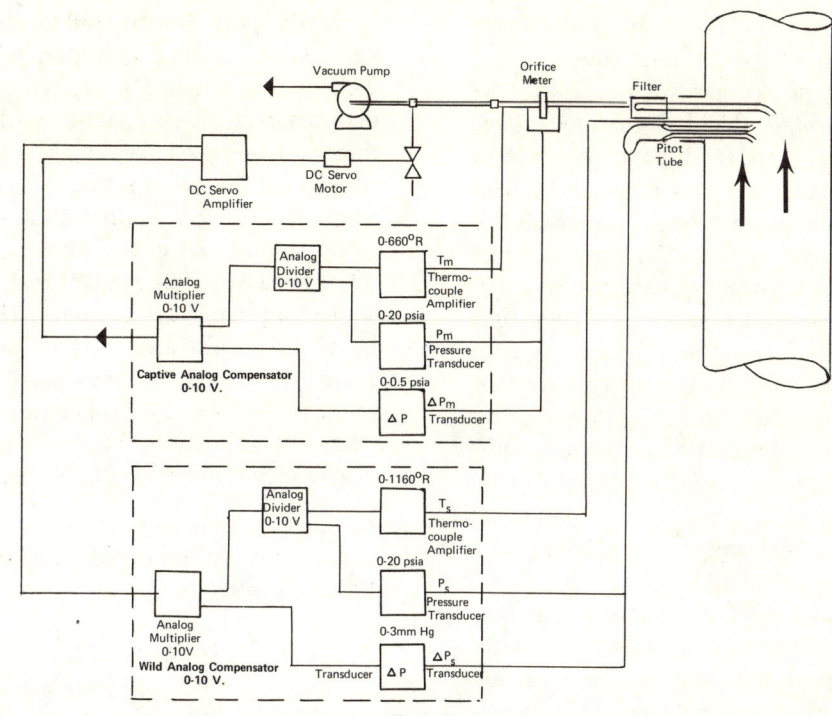

Fig. 47.
Courtesy of: Mr. Albert S. Moore, U.S. Bureau of Mines, Morgantown, West Virginia, 1969 (38).

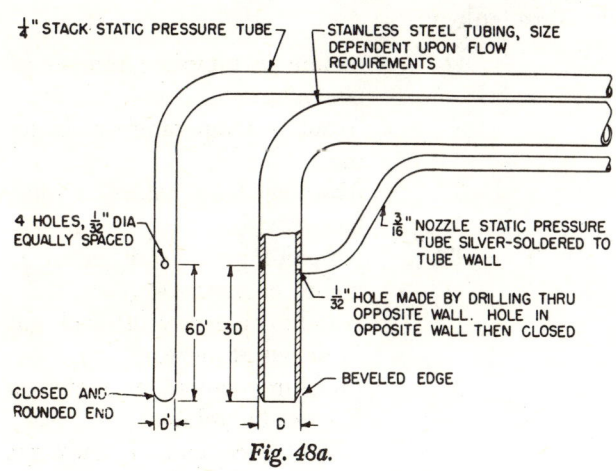

Fig. 48a.

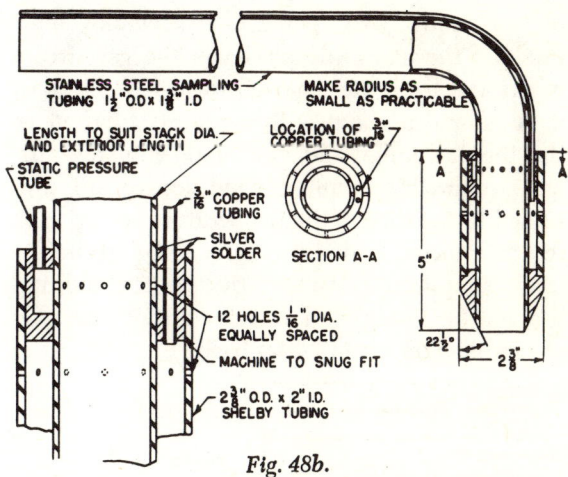

Fig. 48b.

Courtesy of: *Journal of the Air Pollution Control Association*, Article by W. C. Hemeon and G. F. Haines, November 1954 (18).

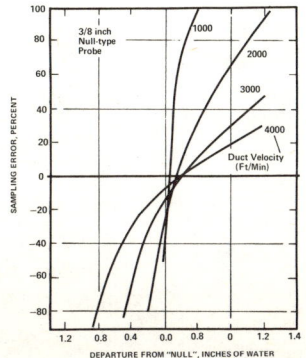

Fig. 49.
Courtesy of: *Industrial and Engineering Chemistry*, Article by R. Dennis, W. Samples, et al., February 1957 (27).

isokinetic conditions in flue gases. The device operates by means of two parallel tubes placed into the direction of the gas stream, both containing axially located shielded thermocouples. One of the tubes is opened at both ends to allow the gas stream to pass through it. The other tube has a 90 degree elbow into which the sample is withdrawn. When the temperatures recorded at each thermocouple are the same, it indicates that the velocities of the two gas streams are also the same. When the temperature in the sampling nozzle tube is greater than that of the other tube, the sampling flow rate must be increased to obtain isokinetic sampling conditions, by providing additional cooling. If the flow rate is greater than isokinetic, it must be reduced to lessen the degree of cooling and return to isokinetic sampling.

The thermal null balance system has potential usefulness for maintaining isokinetic conditions, particularly at low gas velocities of 600 feet per minute or less. Boothroyd [41] developed a thermal null balance system which was found to maintain isokinetic conditions within plus-or-minus two percent. It employed a shielded thermocouple placed in the gas stream where the fluid was allowed to pass freely past it. A sample probe was then placed adjacent to the thermocouple, and gas pulled into the nozzle. Isokinetic sampling conditions were presumed to exist when the temperature recorded at the thermocouple was the same as before beginning sampling. However, the device did not provide for changes in velocity during the time of sampling on a continuous basis.

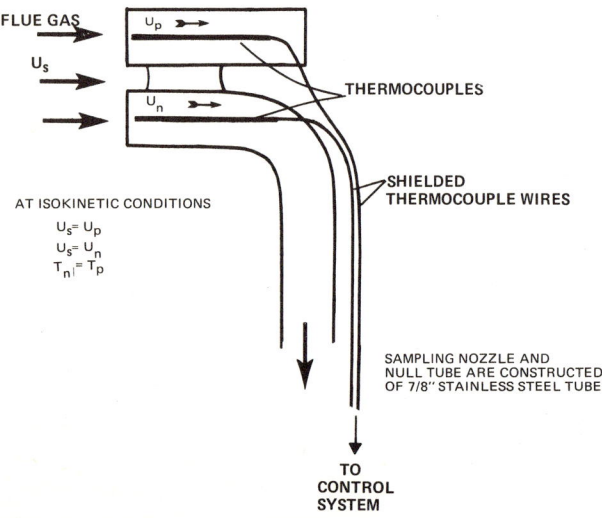

Fig. 50.
Courtesy of: Mr. Paul Hyde, Oregon State University, Corvallis, Oregon, 1968 (42).

Hyde [42] constructed a thermal null balance system using two parallel tubes as previously described. The device provided for the maintenance of isokinetic sapling on a continuous basis by balancing the temperatures recorded by the two probes. The device has been successfully used for monitoring wigwam waste wood burner exit gases, where gas velocities are often between 300 and 600 feet per minute and subject to rapid fluctuations. The device could be electronically modified to provide for automatic, continuous compensation of sampling rates to provide for isokinetic conditions, even when the velocity is subject to rapid fluctuation. The device is illustrated in Figure 50.

Sample Problem #8;
Calculation of Isokinetic Sampling Rate
Basic Expression:

$$Q_m = (0.545 \times 10^{-2}) \frac{T_m}{T_s} \frac{P_s}{P_m} \frac{(100 - MC_s)}{(100 - MC_m)} (D_n)^2 (U_s) \quad (93)$$

Terminology:

Q_m = Isokinetic sampling rate in ft³ per minute
T_m = Absolute temperature at meter in °R
T_s = Absolute temperature of flue gas in °R
P_m = Absolute pressure at meter in inches of mercury
P_s = Absolute pressure of stack gas in inches of mercury
MC_m = Moisture content at meter in percent by volume
MC_s = Moisture content of stack gas in percent by volume
D_n = Internal diameter of sampling nozzle in inches
U_s = Velocity at point of sampling in feet per minute

Sample Problem

Determine the isokinetic sampling rate in a duct where the velocity was found to be 2,400 feet per minute at stack conditions of 275°F, 29.70 inches of Hg, and 30.0 percent moisture by volume. The probe used has an internal diameter of 5/16 inch at the tip, and the meter conditions are estimated as 70°F, 5.0 inches of mercury vacuum, and 3.0 percent moisture by volume.

Problem Solution

$$Q_m = (0.545 \times 10^{-2}) \frac{530}{735} \frac{29.70}{24.92}$$

$$\frac{100.0 - 30.0}{100.0 - 3.0} (5/16)^2 (2,400)$$

$$Q_m = (0.545 \times 10^{-2})(0.622)(0.0975)(2400)$$

$$Q_m = 0.80 \text{ ft}^3/\text{min}$$

III. STATISTICAL CONSIDERATIONS

It is necessary to consider fluctuations in both gas flow rate and particle concentration in the duct, when taking samples from a source by relating results to time, location, and process operating conditions. The emission level as determined by a given sample may be different from the actual level because of several random and systematic errors.

A. Sampling Conditions

Achinger and Shigehara [43] have devised a system for classifying source sampling conditions into four categories based on gas flow rate and concentration in terms of time and location. It provides a systematic statistical approach to gaseous and particulate sampling; where 1) the number of measurements necessary, 2) the number and location of sampling points, and 3) the time and duration of each sample measurement must all be considered. Gas flow rates and material concentrations are considered as either steady or unsteady in relation to time; flow and concentration distributions at a given time are either uniform or nonuniform in terms of space at a given location in a duct.

1. Steady Flow—Uniform Concentration

This is the simplest case because there is no variation in gas flow rate or concentration with space or time at a given location. An example of this condition is gaseous sampling in a long flue at constant flow rate where turbulence produces a relatively uniform concentration distribution across the duct. The same conditions may be approximated for particulate sampling, especially if the particles are very small, as is the case for an oil-fired power boiler operating at a steady rate. It is then only necessary to make one measurement in a duct when such conditions exist.

2. Steady Flow—Nonuniform Concentration

The gas flow rate and concentrations remain relatively constant with time, but there is a variation between points across the duct. An example is a coal-fired power boiler operating at a constant rate, but there is a variation in velocity and concentration profiles across the duct. A series of measurements is made at several locations during a single test, to obtain a composite sample, where the more sampling points used the more representative the sample. This condition is specified in several test procedures, where the process unit should be operated at relatively constant conditions during the test [2] [9] [24].

3. Unsteady Flow—Uniform Concentration

The concentration of a material remains essentially constant across a duct because of turbulent mixing at a given time but emission levels vary with time. Examples are sampling for oxides of sulfur or nitrogen in a power boiler flue at variable firing conditions and Kraft mill batch digester blow gases flowing in a small pipe on a cyclical basis. Sampling is then required at only one point at a given time, but a number of samples must be collected over an extended period of time.

4. Unsteady Flow—Nonuniform Concentration

Both concentration and flow rate vary with both position and time, requiring a large number of sampling points at a given time and a number of samples covering an extended time period for one or more points. This is the most complex condition and is frequently observed for offgases from manufacturing process units, smaller power boilers, and refuse incineration units. It is sometimes possible to sample at two or more points simultaneously if sufficient manpower and equipment are available.

Two alternatives exist, depending on the nature and type of the study being made, and whether time or spatial considerations are paramount. Bulletin WP-50 [2] states that when significant time variations in concentration are observed, "it is better to have many observations at a few points than a few observations at each of many points." This applies to both flow rate and concentration measurements. At locations where there is a highly variable flow and concentration profile, such as in a large duct immediately downstream from a bend or obstruction, the only recourse is a large number of sampling points in attempting to obtain a representative sample. This is often necessary in ducts where it is not possible to locate sampling ports so as to meet previously specified criteria. It is possible to use honeycombs, straightening grids and vanes, and turning vanes to minimize non-uniformities, but these are often subject to pressure drop increase, severe plugging and corrosion, and may produce stratification [4] [44].

B. Particle and Gas Distribution

Particle concentrations and gas flow rates in ducts vary with location time, and firing condition, in both random and systematic fashions. An extensive treatment of the theoretical and practical aspects of these variations is presented by Hawksley [10].

1. Location

Particles tend to follow the streamlines of gas flow in ducts, but deviations in particle motion from the gas stream can occur at bends, elbows, junctions, other constructions, and in horizontal ducts. The magnitude of the deviation depends on the sizes of the particles, the flue gas velocity, and the abruptness of the change in direction or length of the horizontal run (10). Particles smaller than three to five microns in size tend to follow the flow streamlines closely while large particles larger than 50 to 100 microns, depending on density, surface, and gas velocity, tend to move independently of the streamlines because they cannot negotiate changes in flow direction as easily as the gas molecules and smaller particles. Particles between these extremes deviate from streamline flow in varying degree. However, the Reynolds' number [or index of turbulent mixing] is normally greater than 200,000, and this condition tends to restore the initial upstream conditions downstream from an obstruction or bend.

The distance downstream from a bend, junction, or fan required for the particulate concentrations to return to relative uniformity depends on the relative number of large particles present and the degree of the obstruction. Hawksley [10] states that the large particles [or grit, specified as particles larger than 76 μ diameter] are selectively drawn to the outer edge of bends because of their relative inability to change flow direction, where the effect increases with sharpness of the bend. Relative uniformity in concentration and velocity profiles is restored within one to three diameters downstream from bends if there are very few large particles present, and in normal cases this occurs within three to five downstream diameters. However, it may be six to eight diameters if a large proportion of the particles are in the "grit" category or greater than 75 to 100 microns.

Induced draft fans tend to force the gas stream to the outer radial edges, normally producing wide nonuniformities in both velocity and concentration profiles. A helical flow pattern often develops in round ducts following cyclone separation stages where solids may or may not be thrown toward the outside of duct. In long horizontal ducts the velocity profile may become relatively uniform, but the large particles tend to settle toward the bottom. Therefore, it is normally better to sample in vertical ducts whenever possible [10].

It is noted that the velocity profile downstream of a bend or obstruction tends to uniformity faster than the particle concentration. This is because the effect of turbulent mixing affects the gas molecules more rapidly and to a greater extent than the particles. This difference increases as the mass and six of the particle increases, because greater energy is then required to produce the necessary mixing. However, there is an insufficient amount of information regarding these phenomena at the present time.

2. Time

Particle and gaseous concentrations and gas flow rates also vary with time, as is the case for sootblowing, precipitator rapping, and cyclical processes such as Kraft mill batch digester blow gases. These may require special sampling equipment because these operations normally last for only a very short period of time—three to ten minutes. A large sample volume in a short period of time is required to obtain a sufficient amount of sample and it may become necessary to sample at two or more points simultaneously because there is normally insufficient time to move the probe between several points. Good organization and extremely rapid work are required in setting up equipment, making flow measurements, adjusting sampling rates, and the sampling interval must coincide exactly with the period of the special condition.

Variations with time may be handled for other situations by making frequent velocity and concentration measurements over an extended time period at one or more points in a duct.

3. Operating Conditions

The emission levels and gas flow rates also

vary with the operating level of a combustion or process unit. Normally, the concentration and flow rate increase as the process loading increases, producing a dramatic and nearly exponential increase in emission rate above a certain level. Hawksley [10] found similar results for particulate emissions from a coal-fired boiler where the emission "E", as percent of the weight rate of fuel added, showed a dramatic increase with velocity. The emission was found to increase with the cube of velocity as shown in Figure 51 and the following equation:

$$E\% = -4.9 \times 10^{-3} + 1.833 \times 10^{-6} U_s^3 \quad (94)$$

where:

$E\%$ = Lb material per pound fuel × 100
U_s = Velocity in feet per second

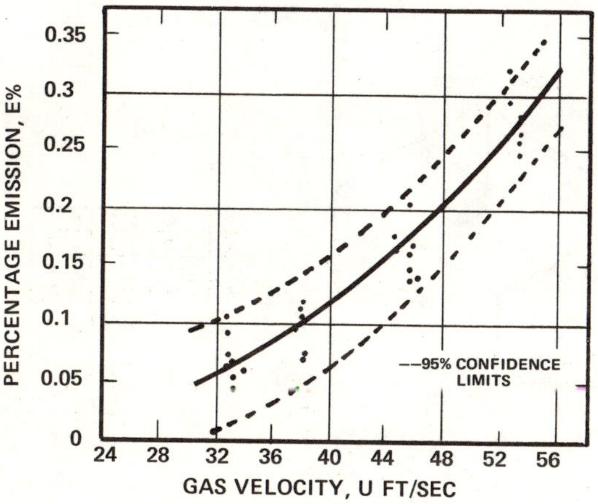

Fig. 51.

Courtesy of: British Coal Utilization Research Association, Book by P. G. Hawksley, S. Badzioch, and J. H. Blackett, *Measurement of Solids in Flue Gases*, Leatherhead, Surrey, England, 1961 (10).

Similar findings have been observed for hydrogen sulfide emissions from Kraft recovery furnaces, where a minor linear increase has been noted up to a certain level and beyond it an exponential increase with loading [45]. The emission appears to rise linearly with loading at low levels where the furnace is able to process the material present. However, when the furnace is pushed to a point sufficiently beyond its rated loading, it is unable to burn or process all material added. Exponential increases in emission levels with small increases in operating rates are then noted, probably because unprocessed or partially processed [or burned] material is then being emitted.

It is then important to correlate operating levels with any samples taken from sources. If possible, a series of operating conditions should be established and allowed to reach equilibrium, and then one or more samples should be taken at each operating level. Emission profiles may then be established for changes in operating levels for process or combustion units.

C. Accuracy

The accuracy in sampling for average concentration and emission rate of materials is dependent upon: 1) systematic errors caused by nonrepresentative sampling methods, and 2) random errors caused by the limitations in number of sampling points and the finite duration of the sampling period [10]. It is often a good procedure to make a detailed introductory study for a source using a number of sampling points for varying conditions, to establish the patterns of emission levels with sample location, time, and operating conditions. Once these patterns are established, it is best to take a number of samples at a few representative points in the duct.

The accuracy of a sample is influenced by the number and location of the sampling points. Sampling points following bends, fans, connections, and other obstructions present particular problems because of the resultant nonuniformities in flow and particulate concentration distribution. Quantitative measurements of certain of these obstructions in ducts which create these nonuniformities have been made by Hawksley [10], and have been shown in Figure 52.

1. Systematic Errors

Systematic errors are induced by shortcomings in the sampling techniques and methods;

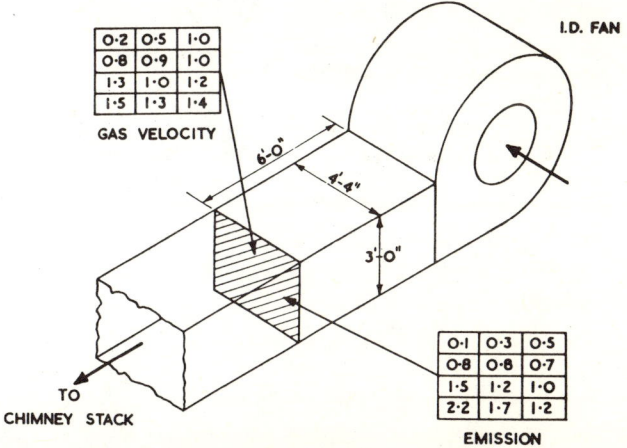

AVERAGE GAS VELOCITY = 43 FT/SEC
AVERAGE EMISSION PER UNIT AREA = 3.0 LB/SQ.FT-HR

Fig. 52.

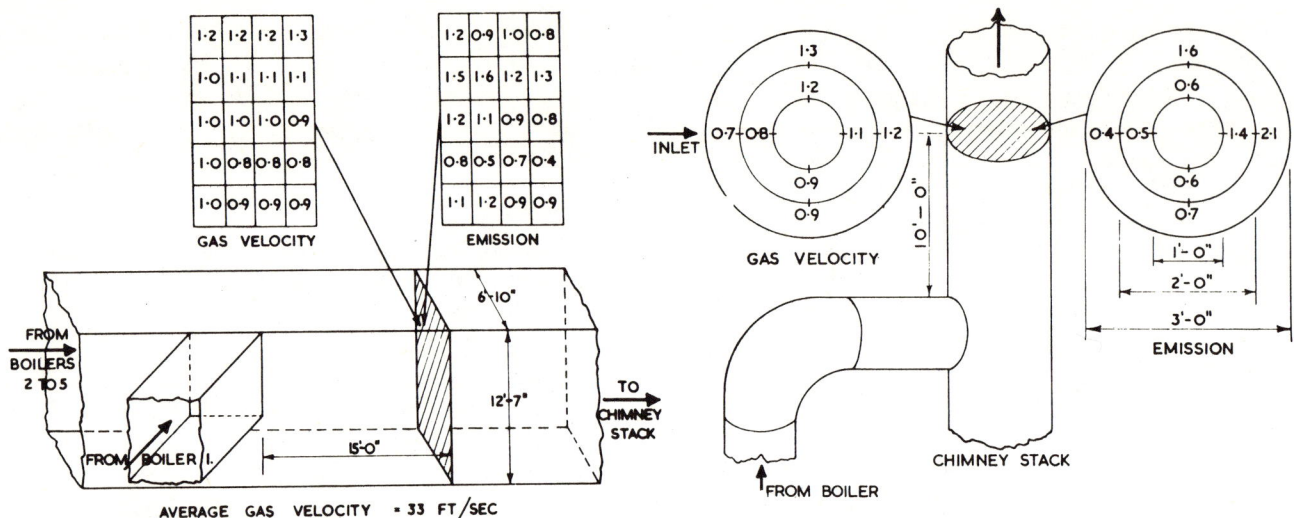

Fig. 52b.

Fig. 52c.

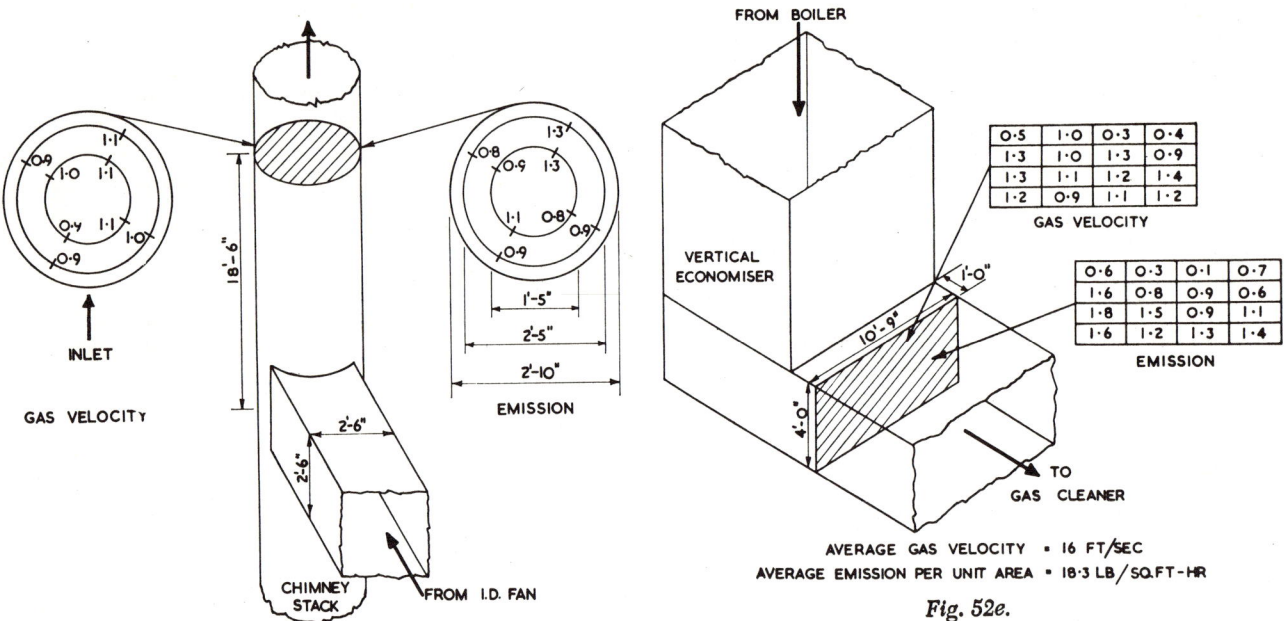

Fig. 52d.

Fig. 52e.

Courtesy of: British Coal Utilization Research Association, Book by P. G. Hawksley, S. Badzioch, and J. H. Blackett, *Measurement of Solids in Flue Gases*, Leatherhead, Surrey, England, 1961 (10).

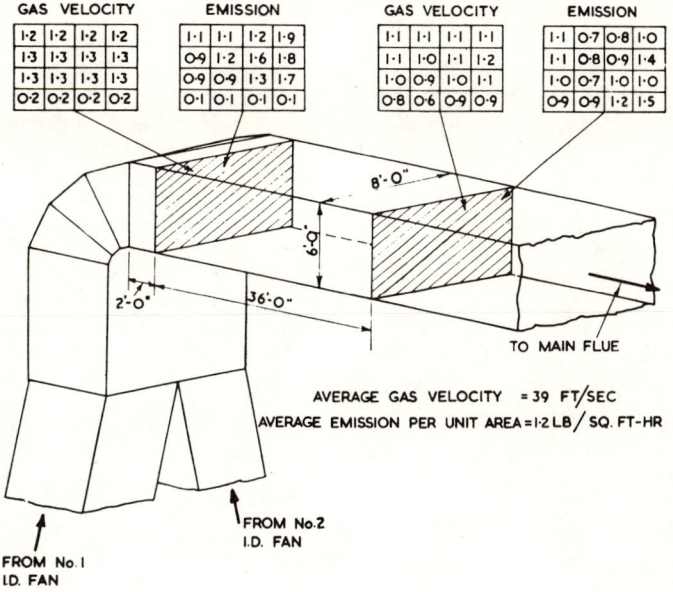

Fig. 52f.

these may or may not be easily corrected. These include errors in calibration of pitot tubes or sample flow measuring instruments, sampling at anisokinetic conditions, and inaccurate measurements of duct area temperature and static pressure. The ratio of weight of material collected to tare weight of the collection device is also important. The ASME [19] recommends that for filter bags this ratio should be at least four to one so that an analytical balance is not then necessary. The use of alundum thimbles or crucibles for evaporating impinger liquid requires accurate weighing techniques using an analytical balance, because it is necessary to take the small difference between two large numbers. A list of the above errors and approximate maximum values is listed in Table 19.

Systematic errors may also result from the finite number of sampling points used and the limited sampling time for a sample [10]. Errors resulting from sampling at a point during a time coinciding with the time the level at that point is higher or lower than its average concentration are minimized by taking several short increments at a point rather than one long interval. A duration of approximately ten minutes per point, and sampling a number of points, is a reasonable compromise between the accurate determination at a single point for an extended period and an extremely large number of points for a very short period. A normal maximum number of points is 16, to reduce the error in mass emission rate determination to plus or minus five percent on a two hour sample.

2. Random Errors

There is a characteristic variation in both concentration and gas flow rate for particulate materials suspended in a gas stream in the turbulent flow region. The magnitude depends on the error in measuring each increment and the total number of increments where a total measurement is actually the mean of a number of individual increments. With increments of five to ten minutes at each sample point the random error is approximately [10]:

$$\% \text{ Error} = \pm 20/\sqrt{N} \qquad (95)$$

where:

N = Number of sample points

This defines the degree of precision of a single sample, and the possible reproducibility of a second sample if the same operating conditions are maintained.

3. Total Error

The total error varies with the total time and number of sampling increments. The error in a one hour sample is approximately ten percent for eight sample points. Minimum error occurs when there is no systematic error in taking the sample, and inaccuracies depend solely on the number of increments in the sample. Maximum error arises when there is a considerable bias in the sample. The average errors have been computed from the root mean square values for random and systematic errors. Hawksley [10] presented a summary of the estimated errors in mass emission rate determinations based on 95 percent confidence limits, as shown in Table 20.

Accuracy does not improve significantly after eight increments per sample. It is dependent on the sampling time, and increases with the number of sampling points (10), as shown in Table 21. Procedures specified by the British Standards Institution [13] are good to plus or minus 25 percent, by the ASME [9] and National Council [24] are good for approximately ten percent error, and by the Public Health Service [46] are good for approximately five percent.

IV. REFERENCES

1. Devorkin, H., Chass, R. L., Fudurich, A., and Kanter, C. V., ed. by Holmes, R. G. "Source Testing Manual," Los Angeles County Air Pollution Control District, Los Angeles, California, 1965.
2. Haaland, H. H., ed., "Methods for Determination of Velocity, Volume, Dust and Mist Content of Gases," Bulletin WP-50, 7th ed., Western Precipitation Corp., Los Angeles, California, 1968.
3. Jacobs, M. B., *The Chemical Analysis of Air Pollutants*, Interscience Publishers, Inc., New York, New York, 1960.
4. Clayton, G. D., ed., "Stack Sampling," Ch. 9 in *Air Pollution Manual, Part 1,–Evaluation*, American Industrial Hygiene Association, Inc., Detroit, Michigan, 1961.
5. Bloomfield, B. D., "Source Testing," Ch. 28 in Stern, A. C., *Air Pollution*, Vol. 20, pp. 487-536, Academic Press, Inc., New York, New York, 1968.
6. Wolfe E. A., ed., "Source Test Methods," Bay Area Air Pollution Control District, San Francisco, California, 1961.
7. Rivera, M. F., "Incinerator Source Testing Manual," Dade County Pollution Control Authority, Miami, Florida, September 1968.
8. "Dust Separating Apparatus," Performance Test Code 21-1941, American Society of Mechanical Engineers, New York, New York, 1941.
9. "Determining Dust Concentration in a Gas Stream," Performance Test Code 27-1941, American Society of Mechanical Engineers, New York, New York, 1941.
10. Hawksley, P.G.W., Badzioch, S., and Blackett, J. H., *Measurement of Solids in Flue Gases*, British Coal Utilization Research Association, Leatherhead, Surrey, England, 1961.
11. "Test Methods for Gas Volume and Dust Sampling Determinations," Research–Cottrell, Inc., Bound Brook, New Jersey, 1968.
12. "Sampling and Analysis of Waste Gases and Particulate Matter," *Manual on the Disposal of Refinery Wastes*, Vol. 5, American Petroleum Institute, Inc., New York, New York, 1954.
13. "Simplified Methods for Measurement of Grit and Dust Emissions from Chimneys," British Standard 3405, British Standards Institution, London, England, 1961.
14. "Flue Dust Sampling Instructions–Cyclone Method," Babcock & Wilcox Co., Alliance, Ohio, 1968.
15. Hardie, P. H., Resume of Methods for Measuring Flue Dust," *Transactions of the American Society of Mechanical Engineers*, Vol. 59, No. 10, pp. 355-358, October 1937.
16. Fahrenbach, W., "The Dynamics of Dust and Its Influence on Dust Measurement," *Forschung auf dem Gebiete des Ingenieurwesens*, Vol. 2, No. 2, pp. 395-407, February 1931.
17. Parker, G. J., "Some Factors Governing the Design of Probes for Sampling in Particle and Drop-Laden Streams," *Atmospheric Environment*, Vol. 2, No. 5, pp. 477-490, September 1968.
18. Hemeon, W. C. L., and Haines, G. F., "The Magnitude of Errors in Stack Dust Sampling," *Air Repair*, Vol. 4, No. 3, pp. 159-164, November 1954.
19. Watson, H. H., "Errors Due to Anisokinetic Sampling of Aerosols," *American Industrial Hygiene Association Quarterly*, Vol. 15, No. 1, pp. 21-25, January 1954.
20. Badzioch, S., "Correction for Anisokinetic Sampling of Gas-Borne Dust Particles," *Journal of the Institute of Fuel*, Vol. 33, No. 230, pp. 106-110, March 1960.
21. Bonnet, M., "Dust Content Determination in Flowing Gases with Suction Probes," *Chemische Ingenieure Technik*, Vol. 39, No. 16, pp. 972-977, August 1967.
22. Vitols, V., "Theoretical Limits Due to Anisokinetic Sampling of Particulate Matter, "*Journal of the APCA*, Vol. 16, No. 2, pp. 79-84, February 1966.
23. Sehmel, G. A., "Estimation of Air Stream Concentrations of Particulates from Subisokinetically Obtained Filter Samples," *American Industrial Hygiene Association Journal*, Vol. 28, No. 3, pp. 243-251, May–June 1967.
24. "Manual for the Sampling and Analysis of Kraft Mill Recovery Stack Gases," National Council for Air and Stream Improvement Atmospheric Pollution Technical Bulletin No. 14, New York, New York, October 20, 1960.
25. Whiteley, A. B., and Reed, L. E., "The Effect of Probe Shape on the Accuracy of Sampling Flue Gases for Dust Content," *Journal of the Institute of Fuel*, Vol. 32, No. 222, pp. 316-319, July 1959.
26. Lundgren, D., and Calvert, S., "Aerosol Sampling with a Side Port Probe, "*American Industrial Hygiene Association Journal*, Vol. 28, No. 3, pp. 208-215, May–June 1967.
27. Dennis, R., Samples, W., Anderson, D., and Silverman, L., "Isonkinetic Sampling Probes," *Industrial and Engineering Chemistry*, Vol. 49, No. 2, pp. 294-302, February 1957.
28. Rehm, F. R., "Incinerator Testing and Test Results," *Journal of the APCA*, Vol. 6, No. 4, pp. 199-204, February 1957.
29. Miller, A. M., Brown, J., and Abrams, R., "Applied Techniques of Analyses of Stack Emissions," Presented at the West Coast Regional Meeting of the National Council for Air and Stream Improvement, Portland, Oregon, October 2, 1968.
30. Smith, W. S., Martin, D. M., Durst, D., and Hyland, R., "Stack Gas Sampling Improved and Simplified with New Equipment," Paper No. 67-119, Presented at the 60th Annual Meeting of the APCA, Cleveland, Ohio, June 13, 1967.

31. Toynbee, P. A., and Parkes, W. J. S., "Isokinetic Sampler for Dust Laden Gases," *International Journal of Air and Water Pollution*, Vol. 6, No. 2, pp. 113-120, March–April 1962.
32. Gerstle, R., Cuffe, S., Orning, A., and Schwartz, C., "Air Pollutants from Coal-Fired Power Plants," *Journal of the APCA*, Vol. 15, No. 2, pp. 59-64, February 1965.
33. Personal communication with Mr. Walter S. Smith of U. S. Public Health Service, Durham, North Carolina, 1968.
34. Perry, J. H., *Chemical Engineers' Handbook*, 4th edition, McGraw-Hill Book Co., New York, New York, 1963.
35. Davis, D. S., *Empirical Equations and Nomography*, McGraw-Hill Book Co., New York, New York, 1943.
36. Gallaer, C. A., "How to Measure Dust in Stacks and Ducts," *Power*, Vol. 101, No. 1, pp. 88-91, January 1957.
37. Moore, A. S., "Sampling Dust in the Bureau of Mines Coal-Fired Gas Turbine," *Combustion*, Vol. 35, No. 4, pp. 28-30, October 1963.
38. Personal correspondence with Mr. Albert S. Moore, U. S. Bureau of Mines, Morgantown Coal Research Center, Morgantown, West Virginia, 1969.
39. Rehm, F. R., "Test Methods for Determining Emission Characteristics of Incinerators," *Journal of the APCA*, Vol. 15, No. 3, pp. 131-135, March 1965.
40. "Incinerator Testing," Bulletin T-6, Incinerator Institute of America, Inc., New York, New York, 1968.
41. Boothroyd, R. G., "An Anemometric Isokinetic Sampling Probe for Aerosols, "*Journal of Scientific Instruments*, Vol. 44, No. 4, pp. 249-253, April 1967.
42. Hyde, Paul, "Particulate Sampling of Wigwam Burners," Forest Research Laboratory, Oregon State University, Corvallis, Oregon, October 1968.
43. Achinger, W. C., and Shigehara, R. T., "A Guide for Selected Sampling Methods for Different Source Conditions," *Journal of the APCA*, Vol. 18 No. 9, pp. 605-609, September 1968.
44. Caplan, K. J., ed., "Performance Testing" Ch. 12 in *Air Pollution Manual–Part 2–Control Equipment*, American Industrial Hygiene Association, Detroit, Michigan, 1968.
45. Harding, C. I., and Landry, J. E., "Future Trends in Air Pollution Control," *Tappi*, Vol. 49, No. 8, pp. 61A-67A, August 1966.
46. "Specifications for Incinerator Testing at Federal Facilities," U. S. Public Health Service, National Air Pollution Control Administration, Durham, North Carolina, October 1967.

Table 13. Effect of Departure from Isokinetic Conditions on Measured Conditions

Velocity Ratio U_s/U_m	Concentration — C_m/C_s	
	Range	Average
0.6	0.75-0.90	0.85
0.8	0.85-0.95	0.90
1.0	—	1.00
1.2	1.05-1.20	1.10
1.4	1.10-1.40	1.20
1.6	1.15-1.60	1.30
1.8	1.20-1.80	1.40

Courtesy of: Academic Press, Inc., New York, New York. "Source Testing," Chapter 24 by Bloomfield, B. D., in Stern, A. C., Air Pollution, Vol. 2, 1968 [5].

Table 14. Effect of Nozzle Size on Efficiency of Particle Collection

Nozzle Size Inches	Concentration Ratio — Fine: 5-25μ	Measured Concentration / Actual Concentration Coarse: 420-500μ
1/8	1.00-1.04	0.78-0.80
1/4	1.00-1.03	0.84-0.90
3/8	1.00	1.00

Courtesy of Journal of the Air Pollution Control Association, Hemeon, W., and Haines, G., "The Magnitude of Errors in Stack Dust Sampling," November 1954 [18].

Table 15. Confidence Levels for Minimum Probe Size Expression

Confidence %	Level T
68.0	1.0
95.0	2.0
99.0	2.6
99.9	3.2

Courtesy of: Parker, G. D., Atmospheric Environment, Pergamon Press, London, September 1968 [17].

Table 16A. Sampling Nozzle Selection for Different Gas Velocities: For Thimble-Impinger Sampling Trains, Where $Q = 0.5\text{-}2.00$ cfm

Nozzle Diameter, Inches	Pressure Differential Range, Inches H$_2$O	Velocity Range	
		ft/min	ft/sec
3/16	0.80-2.50	5,000-12,000	80-180
1/4	0.30-1.50	3,000-7,500	50-120
5/16	0.10-0.50	1,500-3,500	25-60
3/8	0.03-0.15	300-2,000	15-35
1/2	0.02-0.10	600-1,500	10-25
5/8	0.01-0.03	300-900	5-15

Table 16B. Sampling Nozzle Selection for Different Gas Velocities: Cyclone-Filter Sampling Train, Where Q = 4.0-15.0 cfm

Sampling Nozzle		Pressure Differential, inches H_2O	Velocity Range	
Area ft^2	Diameter inches		ft/min	ft/sec
1/100	1.36	0.020-0.105	600-1,500	10-25
1/150	1.12	0.045-0.235	900-2,500	15-40
1/200	0.96	0.080-0.420	1,500-3,500	25-55
1/300	0.78	0.180-0.940	2,500-5,000	40-85
1/400	0.67	0.320-1.680	3,000-7,500	50-120

Courtesy of British Standards Institution BS 3405, 1961 [13].

Table 17. Estimation of Isokinetic Sampling Rates

Setting Q_R liters/min	Flow Rate Q_m cfm	Pressure P_m in. Hg	$Q_m P_m$	Velocity U_s ft/min
10	0.30	28.9	8.67	1,020
15	0.52	26.6	13.85	1,630
20	0.68	25.2	17.15	2,020
25	0.87	23.5	20.40	2,400
30	1.06	22.0	23.35	2,750
35	1.24	21.4	26.60	3,130
40	1.42	20.2	28.60	3,370
45	1.61	19.4	31.25	3,680
50	1.80	18.6	33.55	3,950

Table 19. Magnitude of Systematic Errors in Particulate Sampling Measurements

Type of Error	Magnitude
Pitot Tube Calibration	±1.0%
Pitot Tube Measurements	±1.0%
Dry Gas Meter Calibration	±2.0%
Temperature and Pressure	±1.0%
Weighing of Material	±1.0%
Total	±6.0%

Table 20. Estimated Accuracy of Particulate Emission Rate Measurements

Number of Increments	Number of Sampling Points	95% Confidence Limits — % Difference		
		Maximum	Minimum	Average
1	1 (center)	±63	±19	±48
4	1 (4 at center)	±54	±9.5	±45
	4 (1 at each)	±21	±9.5	±15
8	1 (8 at center)	±51	±6.7	±45
	4 (2 at each)	±18	±6.7	±13
	8 (1 at each)	±12	±6.7	±8.7
9	1 (9 at center)	±50	±6.3	±44
	9 (1 at each)	±11	±6.3	±8.0
16	4 (4 at each)	±16	±4.8	±12
	8 (2 at each)	±10	±4.8	±7.3
	16 (1 at each)	±7.6	±4.8	±5.6
24	4 (6 at each)	±15	±3.9	±12
	6 (4 at each)	±11	±3.9	±8.3
	8 (3 at each)	±9.4	±3.9	±6.7
	12 (2 at each)	±7.6	±3.9	±5.4
	24 (1 at each)	±5.7	±3.9	±4.3

Courtesy of British Coal Utilization Research Association, Hawksley, P., Badzioch, S., and Blackett, J., "Measurement of Solids in Flue Gases," Leatherhead, Surrey, England, 1961 [10].

Table 18. Corrections in Sampling Rate with Meter Pressure Changes

$(P_m)_o$ — Initial Meter Pressure Inches of Hg	Flow Rate Correction — $\dfrac{(P_m)_o}{(P_m)_o - \Delta P_m}$ P_m — Change in Meter Pressure in inches of Hg									
	1	2	3	4	5	6	7	8	9	10
29.92	1.04	1.07	1.11	1.15	1.20	1.25	1.30	1.36	1.43	1.50
28.92	1.04	1.07	1.11	1.16	1.21	1.26	1.32	1.38	1.45	1.53
27.92	1.04	1.08	1.12	1.17	1.22	1.28	1.34	1.40	1.48	1.56
26.92	1.04	1.08	1.12	1.18	1.23	1.29	1.35	1.42	1.50	1.59
25.92	1.04	1.08	1.13	1.18	1.24	1.30	1.37	1.44	1.53	1.63
24.92	1.04	1.09	1.14	1.19	1.25	1.32	1.39	1.47	1.56	1.67
23.92	1.04	1.09	1.14	1.20	1.26	1.34	1.41	1.50	1.60	1.72
22.92	1.04	1.10	1.15	1.21	1.28	1.35	1.44	1.54	1.65	1.78
21.92	1.05	1.10	1.16	1.22	1.30	1.38	1.47	1.58	1.70	1.84
20.92	1.05	1.10	1.16	1.23	1.31	1.40	1.50	1.62	1.76	1.92
19.92	1.06	1.11	1.18	1.25	1.34	1.43	1.54	1.68	1.82	2.00

Table 21. Magnitude of Errors in Measuring Particulate Emission Rates

Number of Increments	Sampling Period Minutes	Number of Sampling Points (10 minute increments)	Estimated Error %
1	10	1 at center	±48.0
4	40	4 at center	±45.0
		1 at each of 4	±15.0
8	90	8 at center	±44.5
		2 at each of 4	±13.0
		1 at each of 8	±8.5
16	160	4 at each of 4	±12.0
		2 at each of 8	±7.5
		1 at each of 16	±5.5
24	240	6 at each of 4	±12.0
		4 at each of 6	±8.5
		3 at each of 8	±6.5
		2 at each of 12	±5.5
		1 at each of 24	±4.5
400	67 hours	16 at each of 25	±1.0

Courtesy of: British Coal Utilization Research Association, Hawksley, P., Badzioch, S., and Blackett, J., "Measurement of Solids in Flue Gases," Leatherhead, Surrey, England, 1961 [10].

CHAPTER 7

METHODOLOGY OF PARTICULATE SAMPLING

I. SAMPLING PROCEDURE

Particulate sampling requires determination of duct velocities and flow parameters as previously described, and then selection of the sampling points [1] [2]. The first method is to compute the average velocity, take the point nearest the average, and collect a single sample at that point. It is the simplest but probably least representative method. The second method is to collect an integrated sample from a series of points in the stack (usually 6 or more) by sampling consecutively at each point for about ten minutes and then proceeding to the next. It is necessary to correct the sampling velocity at each point in the stack, but this method gives a composite sample covering several points. The third and perhaps most comprehensive method calls for simultaneously running a series of samples at different points in the duct, but has limitations in number of ports, equipment, manpower, and facilities available.

Points selected for sampling should be as representative as possible of the flow and concentration profile in the duct, and as many points as feasible should be selected. The sampling train and nozzle size used should be able to pull the highest flow rates necessary to maintain isokinetic conditions. It may become necessary to stop the sampling to change nozzle size for points with velocities significantly higher or lower than for the others. If so, all points within the flow range for a given nozzle size should be done at one time to minimize the number of changes required.

The train should be set up, and the approximate sample flow range settings made for the first point by turning on the pump before placing the probe in the stack. The sampling probe should be marked for each sampling position and warmed for five to ten minutes prior to sampling by placing it in the duct with the probe tip pointed downstream so as to reach equilibrium. The probe is then pointed directly into the oncoming stream (at a deviation of less than 5 degrees off center) and sampling commenced at the isokinetic rate. Meter flow rate should be adjusted periodically as meter and stack conditions are changed during the sampling period, as previously described.

Sampling at equal times at each point and adjusting flow rates during sampling eliminate errors introduced by sampling at one point for more than its representative portion of the total flow period. This prevents preferential weighting of one portion of the sample in comparison to the others. Equal volume sampling may be used only if the particulate concentration profile is uniform across the duct, but prior knowledge of this condition does not usually exist. Sample increments are taken at each point chosen until the entire procedure is completed and the gas flow is then shut off and the probe assembly removed from the duct. Particular care should be taken, so as not to spill particles from sampling nozzles, filters, or probes while removing the equipment from the duct by turning the nozzle downward, or bumping the probe against railings or pipes. All probe, nozzle, and tubing areas should be washed carefully with acetone or distilled water (depending on the nature of the particles collected). Filters and liquid washings should be placed in sealed containers for transport to the laboratory.

II. SAMPLING CALCULATIONS

The following is a series of three problems which relate to situations encountered in particulate sampling. The first is a calculation of gas flow rate (this was presented as Sample Problem No. 7, in Chapter 4), the second provides for computing particulate concentration and emission rate; the third provides for these in terms of amount of material emitted per unit weight of flue gas. These problems are typical of those encountered in actual source test applications.

Sample Problem No. 9. Particulate Concentration and Emission Rate.

Basic Expressions:

(1) Concentration at Standard Conditions:

$$C_o = \frac{m_p}{V_o}$$

(2) Concentration at Standard Conditions—Corrected to 12% CO_2:

$$C_{12} = \frac{(12.00)}{(\% \ CO_2)} (C_o)$$

(3) Concentration at Stack Conditions:

$$C_s = \frac{m_p}{V_s}$$

(4) Emission Rate:

$$ER = \frac{60}{7000} \left(\frac{m_p}{V_s}\right)(Q_s)$$
$$= (8.58 \times 10^{-3})(C_s)(Q_s)$$

Terminology

C_o = Particulate concentration at standard conditions, in grains per cubic foot.

C_{12} = Particulate concentration at standard conditions, in grains per cubic foot corrected to 12 percent carbon dioxide on a dry basis.

C_s = Particulate concentration at stack conditions, in grains per cubic foot.

m_p = Amount of particulate material collected during sampling, in grains.

$\% CO_2$ = Percent carbon dioxide by volume on a dry basis in the flue gas.

ER = Emission rate, in pounds per hour.

V_o = Volume sampled at standard conditions, in cubic feet.

V_s = Volume sampled at stack conditions, in cubic feet.

Q_s = Volumetric rate of flow, in cubic feet per minute.

Note: Standard conditions are defined as dry gas at 60°F, 1.0 atm (29.92 in. Hg at sea level).

Sample Problem:

In sample problem No. 7, 77.15 grams of particulate material were collected from the 50.0 ft³ of indicated gas sampled at meter conditions. An Orsat analysis of the flue gas showed it to be 15.0 percent carbon dioxide by volume on a dry basis. Compute the particulate concentration at stack and standard conditions, and when referred to 12.0 percent carbon dioxide by volume. Also compute the emission rate in pounds per hour and in pounds per 1,000 pounds of flue gas on both a dry and wet bases.

d_s = 0.0498 lb/ft³ (wet basis)
Q_s = 70,000 cfm
d_o = 0.0567 lb/ft³ (dry basis)
MC = 30.0% by volume

Problem Solution:

1. Concentration at Stack Conditions:

V_s = 80.80 ft³
m_p = $\frac{77.15}{15.43}$ = 5.0 grains
C_s = $\frac{m_p}{V_s}$ = $\frac{5.00}{80.80}$ = 0.062 gr/ft³

2. Concentration at Standard Conditions:

V_o = 39.70 ft³
C_o = $\frac{m_p}{V_o}$ = $\frac{5.00}{39.70}$ = 0.126 gr/ft³

3. Concentration at Standard Conditions referred to 12% CO_2:

$$C_{12} = \frac{(12.0)}{(\% \ CO_2)}(C_o) = \frac{12.0}{15.0}(0.126)$$
$$= (0.80)(0.126)$$

C_{12} = 0.101 gr/ft³

4. Emission Rate:

ER = $(8.58 \times 10^{-3})(C_s)(Q_s)$
    = $(8.58 \times 10^{-3})(0.062)(70,000)$
ER = 15.10 lb/hr

Sample problem No. 10. Emission per Unit Weight of Flue Gas.

Basic Expressions:

(1.) Gas Flow Rate at Standard Conditions:

$$Q_o = \frac{T_o}{T_s} \frac{P_s}{P_o} \frac{(100 - MC_s)}{(100 - MC_o)} Q_s$$

(2.) Gas Flow Rate at Standard Conditions Referred to 12% CO_2:

$$Q_{12} = = \frac{(\% \ CO_2)}{(12.0)} Q_o$$

(3.) Emission per Unit Weight of Gas at Stack Conditions:

$$ER/M_s = \frac{ER}{60 \ d_s \ Q_s} = \frac{ER}{M_s} \times 10^3$$

(4.) Emission per Unit Weight of Gas at Standard Conditions:

$$ER/M_o = \frac{ER}{60\, d_o\, Q_o} = \frac{ER}{M_o} \times 10^3$$

(5.) Emission per Unit Weight of Gas at Standard Conditions Peferred to 12% CO_2:

$$ER/M_{12} = \frac{ER}{60\, d_{12}\, Q_o} = \frac{ER}{M_{12}} \times 10^3$$

(6.) Molecular Weight and Density of Dry Gas Referred to 12% CO_2:

$$MW_{12} = 5.28 + \frac{(0.88)}{(100 - \%CO_2)}[(\%O_2)(32) + (\%CO)(28) + (\%N_2)(28)]$$

$$d_{12} = \frac{P_o(MW_{12})}{RT_o} = \frac{29.92}{(520)(21.85)} MW_{12}$$

$$d_{12} = 2.63 \times 10^{-3}(MW_{12})$$

Terminology

Q_s = Volumetric rate of flow at stack conditions, in cubic feet per minute.
Q_o = Volumetric rate of flow at standard conditions, in cubic feet per minute.
Q_{12} = Volumetric rate of flow at standard conditions, in cubic feet per minute referred to 12 percent CO_2.
M_s = Mass rate of flow of flue gas at stack conditions, in pounds per hour.
M_o = Mass rate of flow of flue gas at standard conditions, in pounds per hour.
M_{12} = Mass rate of flow of flue gas at standard conditions, in pounds per hour referred to 12 percent CO_2.
MW_{12} = Molecular weight of dry flue gas at standard conditions, in pounds per cubic foot when referred to 12 percent CO_2.
d_s = Density of flue gas at stack conditions, in pounds per cubic foot.
d_o = Density of flue gas at standard conditions, in pounds per cubic foot.
d_{12} = Density of flue gas at standard conditions, in pounds per cubic foot when referred to 12 percent CO_2.
ER = Emission rate of particulate meter, in pounds per hour.
MC_s = Moisture content of flue gas, in percent by volume.
MC_o = Moisture content at standard conditions (0.0% by volume).
P_s = Absolute pressure of flue gas, in inches of mercury.
P_o = Absolute pressure at standard conditions (29.92 in Hg).
T_s = Absolute temperature of flue gas, in °R.
T_o = Absolute temperature at standard conditions (60°F or 520°R).
R = Universal gas constant (21.85 ft³-in. Hg/lb-mole-°R).

Concentrations of CO_2, O_2, N_2, and CO are in percent by volume on a dry basis from Orsat analyses.

Sample Problem:

From information in the previous examples, calculate the gas flow rates at standard conditions and for 12 percent CO_2 for a flow rate of 70,000 cfm at 275°F, 29.70 in. Hg, and 30.0 percent water by volume. Compute the molecular weight and density of the flue gas at standard conditions when referred to 12 percent CO_2, from an Orsat analysis showing 15.0 percent CO_2, 3.0 percent oxygen, and 0.0 percent carbon monoxide. Compute the mass rate of flow for the flue gas at stack and standard conditions and for 12 percent CO_2. Also calculate the emission rate in pounds of particulate per 1,000 pounds of flue gas at stack conditions, standard conditions, and when referred to 12 percent carbon dioxide.

d_s = 0.0500 lb/ft³
ER = 15.1 lb/hr
MW_o = 30.6 lb/lb-mole $N_2 = 100 - (15 + 3 + 0) = 100 - 18 = 82$

(1) Gas Flow Rate at Standard Conditions:

$T_s = 275 + 460 = 735°F$

$$Q_o = \frac{T_o}{T_s}\frac{P_s}{P_o}\frac{(100 - MC_s)}{(100 - MC_o)} Q_s$$

$$= \frac{520}{735}\frac{29.70}{29.92}\frac{(100-30)}{(100)}(70,000)$$

$Q_o = (0.492)(70,000) = 34,400$ ft³/min

(2) Gas Flow Rate for 12 Percent CO_2:

$$Q_{12} = \frac{(\%CO_2)}{(12.0)} Q_o = \frac{(15.0)}{(12.0)}(34,400)$$
$$= (1.25)(34,400)$$

$Q_{12} = 43,000$ ft³/min

(3) Molecular Weight and Density:

$$MW_{12} = 5.28 + \frac{(0.88)}{(100 - 15)}[(\%O_2)(32)$$

$$MW_{12} = 5.28 + \frac{(0.88)}{100-15}\begin{bmatrix}(3.0)(32)\\+(\%CO)(28)+(\%N_2)(28)]\\+(0.0)(28)+(82)(28)\end{bmatrix}$$

$$MW_{12} = 5.28 + \frac{88}{85}[0.96 + 0.00 + 22.95]$$

$$= 5.28 + 1.035(23.91)$$

$MW_{12} = 5.28 + 24.80 = 30.08$

$MW_{12} = 30.1$ lb/lb-mole

$d_{12} = \frac{P_o}{RT_o}(MW_{12}) = 2.63 \times 10^{-3}(30.1)$

$d_{12} = 0.0792$ lb/ft³

$MW_o = 30.6$ lb/lb-mole

$d_o = \frac{P_o}{RT_o}(MW_o) = 2.63 \times 10^{-3}(30.6)$

$d_o = 0.0805$ lb/ft³

(4) Mass Rate of Flow at Stack Conditions:

$M_s = 60\, d_s\, Q_s = (60)(0.0500)(70,0000)$
$M_s = 210,000$ lb/hr

(5) Mass Rate of Flow at Standard Conditions:

$M_o = 60\, d_o\, Q_o = (60)(0.0805)(34,400)$
$M_o = 166,000$ lb/hr

(6) Mass Rate of Flow at 12 Percent CO_2:

$M_{12} = 60\, d_{12}\, Q_{12} = (60)(0.0792)(43,000)$
$M_{12} = 204,000$ lb/hr

(7) Emission Rate at Stack Conditions:

$$ER/M_s = \frac{ER}{M_s} \times 10^3 = \frac{15.1}{210,000} \times 10^3$$

$ER/M_s = 0.0720$ lb part/1000 lb flue gas

(8) Emission Rate at Standard Conditions:

$$ER/M_o = \frac{ER}{M_o} \times 10^3 = \frac{15.1}{166,000} \times 10^3$$

$ER/M_o = 0.0908$ lb part/1,000 lb flue gas

(9) Emission Rate at 12 Percent Carbon Dioxide:

$$ER/M_{12} = \frac{ER}{M_{12}} \times 10^3 = \frac{15.1}{204,000} \times 10^3$$

$ER/M_{12} = 0.0740$ lb part/1,000 lb flue gas

III. SAMPLING EQUIPMENT

A particulate sampling train consists of a sample probe, collection system, flow measuring and control section, and a prime mover. A pitot tube is normally placed parallel to the probe at a distance of one to one and one-half inches away to facilitate continuous velocity measurement. It is also possible to make measurements of temperature, moisture content, and Orsat analyses during the test period.

A. Sample Probes:

Flue gas is drawn through the nozzle at the end of the sample probe and conducted to the collection device. It is sometimes necessary to adjust the temperature of the flue gas passing through the collector, if the efficiency of the collection device is impaired, or if the particles are altered in passing through the probe.

1. Water Cooled Probes:

Water-cooled probes are used when the temperature of flue gas is hot enough to cause combustion of the organic portion of the particles present, fusion of deposited inorganic particles, or even deterioration of the probe material. The minimum stack temperature at which water-cooled probes should be used has been generally estimated as 800 to 1,000°F for refuse incinerators, shaft furnaces, and other sources [3] [4] [5]. However, combustion may well occur at lower temperatures, which would result in significantly low results, especially if the particles were allowed to deposit on the probe walls. This is particularly true for sources where a major portion of the particulate is organic in nature, such as the flue gases from pathological waste incinerators. These problems have not been thoroughly investigated to date.

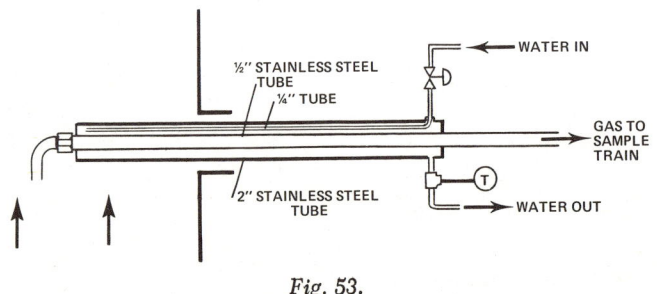

Fig. 53.

Several designs for water-cooled probes are available, where they normally consist of a concentric, two-tube unit. Water should be added by extending a tube to the hot end of the outer tube to prevent vaporization of the water. The leading edge of the sample probe should be tapered similar to an airplane wing so as to provide a minimum interference with the gas flow streamlines. The nozzle should extend approximately one-half to one inch beyond the leading edge of the probe to allow taking representative samples. A diagram of a water-cooled sample probe is shown in Figure 53. It should also be noted that if particle collection devices are used within the stack under these conditions that they also should be waterjacketed to prevent combustion [6]. Normally, it is best to locate the sample collection devices outside of the stack in these extremely hot ducts.

2. Electrically Heated Probes:

It may become necessary to heat the sample probe to prevent condensation in wet or saturated gas streams if dry collection techniques are being employed, as water may render these useless. The gas stream is normally heated to 212°F to prevent condensation. The use of insulated nichrome wire wrapping or heating tapes in combination with a rheostat is helpful. However, it is necessary to enclose the heating section with a waterproof, leakproof, and insulated covering to minimize the danger of electric shocks.

Particle deposition on the inner walls of the probe results in losses if they are not removed at the end of a given sampling period by thorough washing. The problem occurs at the outer edge of the tube following bends and also when tubes become cooled after passing out into the air. Removal of particles by placing the collection device inside the duct minimizes this problems. Smith [7] uses a glass liner inside a stainless steel tube as a sample probe because the walls of glass are smoother and have fewer deep ridges than stainless steel, which results in lower losses from the probe.

B. Collection System:

The collection system involves one or more particle collection devices arranged in series so as to remove the particles for analysis. The devices used, their operating and collection characteristics, as well as limitations, will be covered in a subsequent section.

C. Flow Measurement and Control:

It is necessary to measure the volume of gas withdrawn from a duct, as well as to control the rate of flow. A drying agent is often placed following the final stage of the particle collection system to protect the flow measuring device from moisture and corrosion. This agent may be either silica gel or drierite [7]. However, this adds a substantial increase in pressure drop to the sampling train, and is not normally used at gas flows exceeding two cfm. Also, the capacity of the drying agent to collect moisture is rapidly depleted at these flow ranges. A small glass wool filter placed following the unit containing the drying agent, particularly with silica gel, tends to prevent small drying agent flakes from interfering with the flow measuring device.

The flow measuring system for sample flow rates at meter conditions of 2.0 cfm or less normally consists of a device to indicate the rate of flow, such as an orifice flow meter or a rotameter. It also has a device for indicating the total gas volume sampled, usually a calibrated dry gas meter. The dry gas meter is used because it is able to withstand the vacuum of five to ten inches of mercury normally occurring during sampling, and also the corrosive and moist nature of many flue gases. However, they may become subject to serious inaccuracies if used at vacuum conditions exceeding ten inches of mercury, and should not be used beyond this range [6]. The diaphragms may be replaced and the meter then calibrated prior to reuse, or a new meter substituted in place. However, wet test meters should be held in reserve and used for calibration purposes *only*. They are not designed to operate under several inches of mercury or in corrosive environments, and are normally far more expensive than dry gas meters!

For high flow rate systems where the meter flow is greater than four cfm, the gas flow rate is normally the only quantity measured because dry gas meters become very large and unwieldy for field use. Sampling flow rate is then determined by measurement of pressure drop across a calibrated orifice flow meter [7] [8] [9], or a cyclone collector placed in the sampling train [3] [10] [11]. Drying agents cannot normally be used for these high flow rates, so the moisture must either be condensed prior to the meter or the meter heated above the dew point, particularly for sources exceeding 15.0 percent moisture by volume. It is also necessary to measure the temperature and static pressure at the meter, and the wet bulb temperature of gases entering the meter for moisture content determination. It is necessary to maintain a careful record of gas flow during the sampling period so as to make as accurate a determination of total sample volume as possible.

The flow control assembly is normally placed immediately downstream of the meter. The double needle valve-bleed arrangement shown in Figure 54 has proved useful in controlling sample flow rates over a wide range. It employs a needle valve just downstream of the meter, followed by a tee with another needle valve placed on the attached bleed line. Coarse adjustments in flow are made with the bleed line valve, and fine adjustments to establish the correct flow rate are then made with the sample line valve. Different valves are selected for the various size ranges.

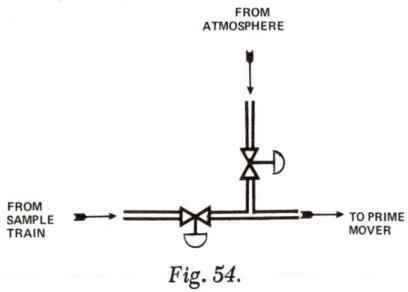

Fig. 54.

D. Prime Mover:

The prime mover causes gas to flow through the sampling train by providing a source of negative pressure, where the maximum flow rate and vacuum possible depend on the characteristics of the individual unit. Vacuum pumps, centrifugal fans, and water aspirators are commonly used as prime movers in source sampling applications. Vacuum pumps of varying sizes can often be used for sample flow rates up to five cfm, but their size and weight may become unwieldy above this range. Serious inflow air leakage has been observed for oil-less carbon-vane vacuum pumps, presumably around the shaft seals. Therefore, it is normally a safe precaution to place the vacuum pump at the downstream end of the sampling train to eliminate any possibility of air leakage through the pump affecting total volume measurements. Smith [7] places the vacuum pump upstream of the meter to maintain the flow measuring unit at constant pressure to simplify isokinetic flow rate calculations when employing a nomograph. Checks of an oil-lubricated vacuum pump indicated negligible (less than two percent) air leakage into the pump [12]. However, extensive wear may result from frequent usage, which will increase the tolerances between moving parts and the housing, thereby heightening the chance of air leakage through the pump. Therefore, the prime mover should be placed *downstream* of the flow measuring device.

Centrifugal blowers have been successfully used as prime movers for high volume sampling systems for flow rates from four to greater than fifty cfm [3] [8] [9] [13]. It is possible to use the motor from a conventional high volume air sampler but the motor internals may become subjected to severe corrosion from the passage of flue gases [10]. Also, their flow characteristics do not normally allow successful operation under vacuum conditions exceeding three to five inches of mercury. It is therefore necessary to restrict usage of collection equipment to those devices which do not have large pressure drops.

Water aspirators can be used as vacuum sources over a wide range of flow rates and are not affected by the corrosive nature of many flue gases [10]. They operate on the principle that a jet of fluid passing a point produces a large vacuum, tending to pull gas molecules in the direction of fluid flow. Adjustments in the range of flow rates are made by changes in the water flow rate. However, it is necessary to have a drain for the water utilized, and the aspirator must be placed at the downstream end of the sampling train.

Air aspirators operate on the same principle as water aspirators, and may be used as sources of vacuum over a wide range by adjustments in the compressed air flow rate [14]. Steam may be used in place of compressed air in many instances. However, their use in industrial establishments has been less successful because of frequent changes in compressed air pressure caused by variations in plant demand. The flow rate has appeared to be more subject to fluctuations with air than water-aspirated systems.

E. Sample Ports:

Sample ports are normally between three and four inches in diameter to allow insertion of the sample probe and other devices into the duct for measurement. Two modifications have been recently described in the Research-Cottrell "Test Methods" manual to overcome problems caused by positive pressure or condensed moisture in flue gases [14].

Positive pressure ducts can present a serious danger to sampling personnel. A sample port assembly consisting of two couplings, a valve, and stopper section allow the sample probe to be inserted into a closed and isolated section of pipe. The sample probe is first inserted into the section of pipe outside of the valve, with the valve closed. The valve is then opened and the sample probe inserted to the given location. The same procedure could be used for the pitot tube when making a traverse by using a different stopper. The system has been previously illustrated in Figure 10.

A water trap assembly can be used in inclined or vertical sample ports to trap condensate and prevent it from washing material out of the sample line. Material collected by washing may then be determined by analysis of the condensate. The system consists of a concentric two-tube pipe where condensate is collected in the outer pipe, which is closed at the bottom. The system has been shown in Figure 11. It is normally a good idea not to have long vertical lines pulling gas upward from a duct to minimize problems of condensate washing. It is highly desirable to drain condensate downstream from the probe to minimize losses caused by absorption of material into the liquid.

IV. SAMPLING TRAINS:

A. Considerations:

The factors involved in selecting a particulate sampling train include the condition of the flue gas stream, the characteristics of the particles to be collected, the type and amount of material collected, and the sampling rate.

1. Flue Gas Conditions:

The flue gas temperature determines whether wet or dry collection devices may be used, or whether special equipment such as electrically heated or water-cooled probes are necessary. The static pressure of a duct can present safety problems with strong positive pressures because hot flue gas may blow in the faces of test personnel. Strong negative static pressures in ducts can cause collapse of flexible tubing in sample lines, thus requiring use of rigid-wall tubing. The presence of sufficient moisture in the flue gas results in condensation when the gas is cooled, which can interfere with filtration and centrifugal particulate collection devices, and interfere with particle size distribution determinations. The use of heated probes, sample lines, and collection devices may become necessary when these problems occur, making for additional complexity in the sampling train, and presenting a possible safety hazard from electrical shocks in rainy weather. The velocity of the stack gas at the point of sampling determines the nozzle size used and the necessary capacity of the prime mover to maintain isokinetic conditions.

2. Particle Concentration:

Characteristics of the material to be collected are also of importance, including both the concentration present and the particle size distribution. The particulate concentration determines the sampling time necessary, where normally one-hour is the minimum sampling time required to obtain a representative sample [6] [10] [15]. The sampling time may be considerably shortened for abnormally high particulate loadings (concentration greater than 3 grains per SDCF) or if a high volume (flow rate greater than 5 cfm) sampling system is used. However, it may be considerably lengthened for very low particulate concentrations (concentration less than 0.1 grain per SDCF), or due to statistical considerations previously discussed.

3. Particle Size:

The size distribution of particles present in the duct influences the diligence with which isokinetic sampling must be followed, and the efficiency characteristics of the collection device. If essentially all the particles are less than three to five microns in size, it is not necessary to sample at isokinetic conditions, such as for oil-fired power boiler flue gases [16]. However, it is normally a good idea to maintain isokinetic conditions within plus-or-minus ten percent because the particle size distribution in the duct is usually not known, and may contain numerous large particles [7].

4. Collection Efficiency:

An additional consideration in terms of particle size involves the efficiency of the collection device. ASME Performance Test Code No. 27 [8] requires that the separation efficiency of a particle collection system be 99.0 percent by weight or greater for the particles present in the duct, considering those particles 1.0 microns or larger in size. This requirement will be more thoroughly covered in the section on collection devices.

5. Amount Collected:

An additional consideration involves the collection of a sufficient amount of material during sampling to facilitate accurate analysis by gravimetric or chemical techniques. Total particulate determinations are normally made by weighings, where the scale should be accurate to plus-or-minus 0.5 percent of the weight of dust caught in the collection device [8]. This requires use of an analytical balance accurate to plus-or-minus 0.1 milligram plus accurate and consistent techniques when weighing heavy objects (often 40 to 80 grams) such as alundum thimbles or evaporating dishes. This arises because of the inherent problems in taking the small difference between two large numbers. Moisture in the ambient air can cause serious variations in weighings by adsorption on surfaces. Therefore, it is necessary to use a consistent weighing procedure employing drying of the object to be weighed; followed by placing in a desiccator, both for a given period (often one hour each) followed by rapid weighing.

It is not necessary to use such a precise balance when weighing filter bags or cyclones used with high volume sampling systems, but the amount of material collected should be at least four times as great as the tare weight of the collection device [7]. Additional problems are presented in attempting to perform chemical analyses of materials comprising only a portion of the total particulate present, particularly if present only in trace quantities. Elaborate and complicated extraction and analytical procedures may become necessary, such as in determination of polynuclear hydrocarbon emissions from combustion sources. A Soxhlet extraction may be performed on filtration devices and media, while conventional chemical analyses may be performed on water solutions from wet impingers, as described in *Standard Methods* [17].

6. Sampling Rate:

The sampling rate used depends on the nozzle size, the flow characteristics of the sample train, and the capacity of the prime mover. The lower limitation in sampling rate is statistical phenomena, which consider that the larger the sample taken in a given time interval the more likely that it will be representative of the duct being sampled. Also, the greater the amount of material collected, the less the chance that any errors in weighing will become significant. The upper limitation is related to the physical characteristics of the system because it is necessary to maintain isokinetic conditions. The maximum flow rate for a given sampling train depends on the flow resistance and capacity of the prime mover for any given nozzle size.

All of the above must be considered in designing a sampling train for particulate collection from a given duct. It is also important to remember that the pitot tube should be placed adjacent to the sample probe for velocity measurement at the point of sampling. This allows making the necessary adjustments in sampling rate for maintaining isokinetic conditions during the test period.

V. COLLECTION DEVICES:

Several techniques may be used for particulate collection, which include separation from the gas stream by wet impingement, filtration, centrifugal force, electrostatic precipitation, and thermal precipitation. These devices may be used singly or in series for particle collection in sampling trains. Each of the devices has operating characteristics which make it suitable for collection of different size ranges of particles and for particular sources conditions.

A. Requirements:

Pertinent requirements for the separation efficiency of particle collection devices (or a series of devices) are quoted as follows in ASME Performance Test Code No. 27 [8]:

> "1. This code is designed to cover tests on any gas stream confined in a duct and carrying particles 1.0 micron and larger."

> "58. The filtering device should be located in front of the measuring and exhausting devices. By agreement of the parties to the test, equipment which is built otherwise may be used, provided it is fully described in the test report. For separating the dust from the gas, a filtering material is preferred. However, other types of separating devices may be used if their efficiency of separation can be demonstrated to be above 99.0 per cent under test conditions."

"59. The filter may be made of any material which has a filtering efficiency of 99.0 per cent by weight for dust of the approximate size analysis to be encountered during the test, and which will withstand the temperature, humidity, and stress to which it will be subjected during the test."

The particle collection system should be able to remove 99.0 percent or greater of the particles present in the duct. This condition implies the use of a multistage particle collection system, using different types of collection devices to remove different sized particles. The collection system should be arranged so that the largest particles are removed first and the finest particles last to minimize increases in resistance to flow.

Addition of a final collection device to remove particles greater than a specific size, such as 1.0 micron, provides for consistent interpretation of results obtained. Use of a Greenburg-Smith impinger [7] or a heated paper thimble of a specified pore size [10] as the final particle collection device has proved successful for consistent removal of particles larger than a given size, often one micron in diameter. It is normally a good practice to remove all but the smallest particles of interest in the collectors upstream of the final collection device to minimize plugging, and to provide a relatively constant pore size (filter) or orifice (impinger) opening during the entire test. This minimizes the restriction of net pore size and resultant increase in resistance to flow caused by buildup of particles on the filter, resulting in collection of progressively smaller particles as the sampling period progresses. An additional reason is that increase of pressure drop during sampling requires frequent adjustment of the sampling rate to maintain isokinetic conditions even though the stack is at relatively constant velocity.

In summary, the collection system used should be able to meet the following requirements:

1. It should be able to remove 99.0 percent by weight or greater of particles present in the flue gas.
2. The final collection device in the sampling train should be able to remove particles larger than 1.0 microns in size from the gas stream, or other particle size specified for the purposes of the test.

B. Wet Impingement:

Wet impingement is used for collection of particulate materials in a liquid phase, usually distilled water [18]. Stack gas flows through a tube extended to near the bottom of a bottle, and is then caused to flow upward through the liquid by a vacuum exerted on the exit tube placed above the liquid level. The gas stream impinges against a flat surface at the bottom in leaving the immersed inlet tube exit and causes the gas to break into small bubbles. Particles are then removed by change in the direction of gas flow and contact with the liquid phase.

It is often necessary to cool the gas stream prior to entering the impingers to minimize evaporation of the absorbing liquid. This is done by having sufficiently long sampling lines between the probe and the first collection stage, and placing the impingers in a cold water or ice bath. This assures a relatively cool and constant liquid temperature in the impingers during the test. Placing the impingers in a plastic pail with cold water has been found satisfactory for most applications. It is also important that the sample probe and connecting tubing slant slightly downward toward the first collection stage to assure drainage of condensed moisture from the flue gas into the collector to minimize losses.

1. Parameters:

Particle collection efficiency is enhanced by maximizing the degree of contact between gas and liquid phases, by increasing the surface area of the gas bubbles formed per unit volume of flue gas and the height of liquid in the impinger. The ratio of surface area exposed per unit volume of gas increases as the bubble size decreases so it is advantageous to break up the gas bubbles. There should be a small diameter tip at the end of the inlet tube and a minimal clearance between the impinger tip and the flat surface below the tip to maximize bubble shearing. Many of the larger particles are unable to negotiate the sudden 180 degree change in direction at the bottom of the impinger because there is insufficient force to propel them upward with the gas molecules. Other particles are trapped and subsequently absorbed by contact of gas surfaces with the liquid as the bubble rises towards the surface. This phenomenon is enhanced by having as high a liquid level as possible in the impinger. However, the practical limitations in maximizing these parameters are increases in pressure drop across the impinger and carryover of the liquid. Thus, there are practical limits to the minimum inlet tip bore, the clearance between impinger tip and flat surface, and the maximum liquid level.

2. Devices:

Two types of impingers are commonly used in particulate sampling. The Greenburg-Smith impinger has a small orifice of about 2.3 millimeter diameter, and approximately five millimeters clearance between the end of the impinger tip and the attached splash plate [19] on the upper piece. The upper section is inserted into the lower section by means of a ground glass fitting at the top. The capacity of the impinger is 500 ml, and it is normally filled with either 100 or 250 ml of distilled water or other liquid. It is a highly efficient collector for particulate materials because it provides for high impaction efficiency plus considerable contact-between gas and liquid phases. The Greenburg-Smith impinger is capable of collecting 90 to 99 percent *by count* for particles greater than or equal to 1.0 micron in size [20]. This efficiency has been found to be relatively independent of the nature of the particle being collected [21]. However, there is also a substantial pressure drop across the impinger of approximately three inches of mercury at a flow of one cfm. The normal range of sample flow rates is 0.5 to 0.8 cfm when using the Greenburg-Smith units, as meter vacuum becomes excessive if several are used in series at the higher flow rates [10]. The impinger tip may also be subject to plugging by large particles. Therefore, it finds most common use as the final particle collection stage for particles larger than one micron in size for sample flow rates less than one cfm.

A second device is commonly referred to as a gas washing bottle. It is similar to the Greenburg-Smith impinger except for the straight tip at the bottom of the inlet tube. Capacities are 125, 250, or 500 ml and they are normally used half-full. There is normally a clearance of approximately ¼ inch between the tip and the bottom of the impinger. The gas-liquid contact is not as extensive because there is no orifice and splash plate to produce the extremely small bubbles found with the Greenburg-Smith unit. However, the pressure drop is considerably less, being approximately one inch of mercury at one cfm and two inches at 1.8 cfm. This allows use of considerably higher sampling rates and arrangement of three or four units in series, where the last bubbler is normally dry to catch liquid carryover. Most of the material is normally trapped in the first bubbler but variable amounts of material pass through, depending on the sampling time and the size range of the particles in the flue gas.

A good check of collection analysis to determine efficiency is whether significant portions of material are present in the last gas-liquid contact phase in comparison to the total amount of material collected. If more than one percent of the total material collected is found in the last impinger, it may indicate the passage of particles through the sampling train without being collected. It could also indicate the presence of condensable vapors. This is particularly a problem when small particles of one to five microns in size or below comprise a significant portion of those present. This was recently shown by Miller and Abrams [22] in comparing emissions of burning waste wood alone versus waste wood plus fuel oil, where the size range of particles produced by burning fuel oil were probably smaller than flyash from combustion of waste wood. The train used in both samples consisted of an alundum thimble followed by four 250 ml gas washing bottles, three of which were filled with 150 ml each of water, and the last one was dry. Less than 0.5 percent by weight of the total was collected in the last scrubber when burning waste wood only. However, three percent by weight was present in the last scrubber when burning fuel oil plus waste wood, indicating that not all the material was being collected, as shown in Table 22. It is noted that the amount and nature of material collected changes in passage through the sampling train. The major portion of nonvolatile materials are collected in the thimble, and greater portions of organic materials are collected in the scrubbers. It appears that the particles of a volatile (organic) nature are of a smaller size range.

The reduced efficiency of gas washing bottles in collecting small particles can be partically overcome by substituting more stages. However, it is normally good practice to use a Greenburg-

Smith impinger, membrane filter, or paper thimble as the final contact stage to assure that particles larger than one micron in size are collected.

A third type of wet collector is commonly referred to as a knockout bottle. It consists of a gallon bottle with a straight tipped glass tube inserted into the liquid to a depth of about ½ inch below liquid level. It is useful for condensing the major portion of moisture from the stack gas without significantly raising the liquid level because of the large cross-sectional area of the bottle. This provides a minimum change in pressure drop across the collection system, making for less correction in the sampling rate to maintain isokinetic conditions. It also acts as a first stage collector to remove the larger particles and act as a type of classifier. The three types of impingers are illustrated in Figure 55.

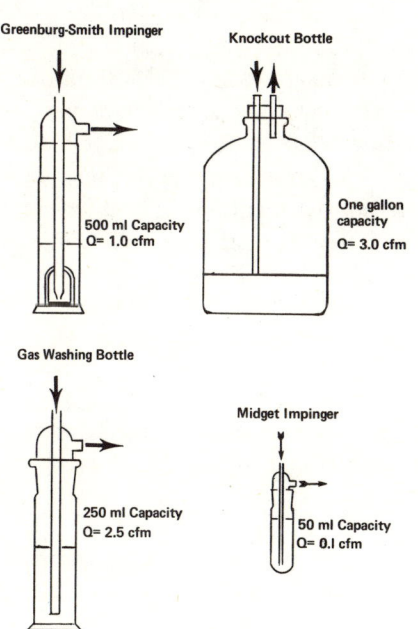

Fig. 55.

Other devices may also be used but with limited effectiveness. Midget impingers of 0.1 CFM capacity are normally too small to provide the flow rates necessary for collecting adequate amounts of material, and in maintaining isokinetic conditions. Fritted glass tipped impingers are subject to plugging by large particles, and are not normally used.

3. Analyses:

It is first necessary to remove all particulate materials deposited in the collection device, sampling lines, and the probe, prior to analysis. This includes a thorough washing of the sample probe, and all tubing connecting the impingers, with distilled water to remove any particles adhering to the inner walls, which often requires three or four consecutive washings. It is often necessary to make additional washings with acetone if a significant portion of the particulate material is organic in nature because of its relative insolubility in water. It is noted that the Los Angeles County [10], Public Health Service [7], and Bureau of Mines [5] make use of glass liners inside of stainless steel probes. The glass has smoother walls with fewer and shallower cavities than stainless steel, thus promoting easier and more efficient cleaning [12]. Experience by the authors indicates that deposited particles can be more easily removed from glass or Teflon than from stainless steel tubing, and the degree of cleaning can be more easily observed.

The washings of the probe and tubing should be collected and then analyzed separately from the material collected in the impingers. This is so as not to interfere with moisture content determination by the condensation method, and to attempt to classify the particles collected in different sections of the sampling train.

A known volume of liquid is added to each impinger prior to sampling. The total liquid volume in each collector is then measured after concluding the test to determine the amount of water vapor condensed from the flue gas for moisture content determination by the condensation method. Normally, liquid in the last gas-liquid collection stage is analyzed separately to provide an estimate of collection efficiency [22].

Analysis of the collected liquid samples then proceeds based on the constituent analyses desired. Total particulate and total nonvolatile particulate analyses are then made on aliquots from the well-mixed liquid solutions by gravimetric techniques. The liquid aliquots are evaporated to dryness in tared crucibles at either room temperature or 105°C, depending on the amount of volatile organic material present. In exploratory tests on sources it is best to run both to determine whether this problem exists. Samples may be evaporated at 105°C (220°F) in drying ovens if there are negligible quantities of highly volatile materials present. Vacuum evaporation at room temperature may also be used.

Analysis for the nonvolatile solids fraction is then made by placing the crucible containing the evaporated solids in a muffle furnace at 600°C (100°F) for one or two hours. The purpose of this heating step is to drive off the volatile organic material and leave a nonvolatile ash which is presumably inorganic in nature. The nonvolatile portion is the amount of material remaining, while the volatile portion is expressed as the difference between the two values. It is also possible to use solvent extraction to selectively remove organic materials from the distilled water prior to ashing, such as benzene [10]. A similar gravimetric analysis is then performed on the organic liquid by evaporating to dryness in a tared container.

It is necessary to use careful and consistent weighing techniques with an analytical balance to obtain accurate results. The crucible or weighing dish should be carefully cleaned by successive acid, soap and water, and distilled water washings before and after use. The crucible should be placed in a desiccator for one hour prior to weighing in each determination. Crucibles heated to 1000°F in muffle furnaces should be allowed to cool before removal to minimize the thermal stress and possible cracking. Also, the fusion and volatilization temperatures of inorganic materials should be considered in selecting the maximum temperatures during ashing to minimize the possibility of losses. Frequently, exploratory tests are required in these determinations.

Chemical analyses may be made on specific constituents by wet chemical analyses on aliquots of the liquid samples collected. Procedures are specified in Standard Methods [17], or by means of spectroscopy, spectrophotometry, atomic absorption spectrometry, coulometry, or other instrumental methods. This provides for a material balance check on the gravimetric analyses, as well as analyses of indicator constituents present. Possible analyses include sodium, potassium, iron, calcium, magnesium, nitrate, sulfide, sulfate, chloride, phosphate, fluoride, carbonate, hydroxide, and other ions, as well as solution pH. An advantage of wet impingement is that a wide variety of conventional direct analyses in water are readily available and useful. Corrections should be made for sulfur trioxide when sampling power boiler flue gases where sulfur-containing fuels are being burned. This is facilitated by parallel gaseous analyses for sulfur trioxide and sulfate determinations in the collection liquid.

4. Evaluation:

Particulate collection by wet impingement is a useful method, particularly for three different types of sources. Combustion source flue gases containing significant portions of organic materials are subject to volatilization and possible combustion if left inside a thimble or sample probe in the duct [10]. Therefore, rapid removal and collection of the particles in a cooled medium outside of the duct is advantageous, as is the case with pathological waste and other incinerator gases. This is consistent with the definition of particulates as being so at standard conditions. This method is useful for sources where the main portion of the particles are inorganic in nature and highly soluble in water. An example is collection of sodium sulfate particles from the flue gases from a Kraft pulp mill recovery furnace [23] [24]. A third type of source where wet impingement is very useful is for flue gases containing significant amounts of water vapor, and being at or near saturation. The previously mentioned Kraft recovery furnace, and flue gas streams following scrubbers are particularly suitable.

The method allows for collection of particles down to one micron in size when using a Greenburg-Smith impinger as the last collection stage. However, sampling rates are limited to one cfm at the meter when using Greenburg-Smith units, and two cfm with gas washing bottles because pressure drops become excessive at greater flow rates. A wide range of chemical and physical analyses may be performed directly for constituents present in the liquid phase by conventional analyses as defined in *Standard Methods* [17]. However, the amount of material collected in the sampling train is normally limited to two to five grams, depending on the type and number of collection stages. The accuracy of gravimetric analysis may also be limited by the necessity for taking the small difference between two large numbers when weighing heavy crucibles.

Wet impingement collection techniques are discussed in published material from the following organizations: National Council for Air and Stream Improvement [23], National Air Pollution Control Administration [7], Los Angeles County Air Pollution Control District [10], Dade County Pollution Control Authority [15].

C. Filtration:

1. Principles:

Filtration involves flow of a fluid stream through a porous medium which allows for passage of the gas stream but removal of the particles present. The primary mechanisms of particulate collection by filtration are inertial impaction, direct interception, and diffusion [25]. The efficiency of particle collection normally increases during the sampling period when using filtration techniques because of filling of void spaces in the porous medium and buildup of a particle layer on the inlet surface. However, this results in an increased resistance to flow and often necessitates frequent adjustments in flow rate during the test period in order to maintain isokinetic sampling conditions.

The major variables affecting the ability of filtration devices to collect particles are the approach velocity (which affects penetration), the cross-sectional area of filter surface available, the pore size of the filter, the depth of the filter, and the packing density of the medium. Different types of filtration devices may be used to collect particles over a wide size range at nearly any desired collection efficiency for a variety of gas flow rates. The size range of particles penetrating the filtration unit is a matter of primary consideration in sampling train design.

2. Materials:

Factors to consider in selecting filter media are their respective collection efficiencies and allowable flow rates, as well as their limitations in withstanding flue gas temperatures, moisture, and corrosive conditions [18]. Collection efficiency tends to increase during the sampling period, and may eventually reach the point where flow resistance is so great as to cause excessive meter vacuum and to impair adequate gas flow through the train. The allowable flow rate varies with the pore size, surface area, thickness, and type of filter medium being used. Filtration media such as alundum or paper thimbles may be used for low flow rate systems of one to two cfm while bag filters are used with high flow rate systems of 50 to 80 cfm [8] [10] [26].

Temperature, moisture and corrosive conditions in flue gases can interfere with the successful utilization of filtration techniques in particulate sampling. ASME Performance Test Code No. 27 [8] requirements state:

"59. The filter may be made of any material which . . . will withstand the temperature, humidity and stress to which it will be subjected during the test."

The maximum temperature of the filter material should not be exceeded, to prevent losses in weight or possible destruction of the medium. A list of maximum temperatures for filtration media is presented in Table 23 [8].

An additional consideration with temperature is the maximum temperature to be withstood by the particulate material collected, without weight losses. This would result from volatilization or combustion of the organic materials present, or fusion and subsequent formation of very small amounts of fumes of inorganic constituents at high temperatures. The material would then escape collection in the filtration medium and result in particle losses. The phenomenon would be most pronounced where particles were caused to remain inside the duct in either a collection device or sample probe for an extended period during a test at elevated temperatures. Stack temperature influences the location of the filtration unit, either internal to (inside) or external of (outside) the duct. The Los Angeles County manual [10] recommends placing the filter outside of the duct and maintaining it at a maximum temperature of 300 to 400°F to minimize the above losses, particularly if the flue gas temperature is 700°F or above. This is achieved by using an extended sample line to cool the gases to about 400°F, and connecting it to the fiber which is maintained in an insulated, electrically heated box. Normally, Vycor or Pyrex glass lines are used in the sample line to facilitate ease of cleaning. This procedure is useful because it reduces the retention time of particles in the duct to ten seconds or less, thus minimizing the possibility of losses due to elevated temperatures. Several other procedures also recommend external location of the filtration assembly to minimize these weight losses at elevated flue gas temperatures [3] [7] [11] [13] [22] [27]. However, the maximum temperature at which losses do not occur should be investigated and wet impingement techniques used in place of filtration, if the necessary temperature is below the dew point temperature of the flue gas. Careful sample line cleaning procedures are also necessary to minimize wall losses.

However, internal location of the filtration

unit is useful if the volatile organic portion of the particulate is negligible and the inorganic material has a high fusion temperature. It eliminates the necessity of maintaining heated sample lines external to the stack and minimizes thermal deposition of particles on the probe walls caused by cooling outside of the duct. This minimizes hazards to operating personnel in wet weather and the difficulties in probe wall cleaning. Internal location of the filter, particularly alundum thimbles, is used in several procedures [6] [14] [15] [27] [28] [29] [30] [31]. However, the possible losses at high temperatures versus particle deposition should be carefully considered in any consideration of internal versus external location of filtration device.

There is also a lower operating limit to temperature when using filtration due to condensation of moisture, which interferes with the collection and even results in the destruction of paper filter media by clogging. Filtration is normally a dry collection method which requires that the gas be above its dew point temperature at all times [18]. This often requires heating of the filter, when it is located outside of the duct, to 250 to 300°F to prevent condensation [7] [10]. The sampling lines between the stack and the filter also must be heated.

An additional problem results from the presence of acidic gases such as sulfur dioxide and sulfur trioxide in the flue gas streams, particularly coal and oil-fired power boilers, which can cause rapid deterioration and rupture of certain filtration materials [32]. Nylon, cotton, and cellulose ester membrane filter materials are particularly subject to this phenomenon and should not be used in these applications [7] [18]. The presence of sulfur trioxide in quantities greater than 10 to 20 ppm by volume increases the apparent dew point of the flue gas to approximately 240 to 280°F. It is then necessary to prevent condensation of sulfuric acid mist in the filter [33].

3. Collection Devices:

Numerous devices are used for collection of particulate materials from flue gas streams by means of filtration techniques [2] [13] [20] [34] [35]. These include alundum and paper thimbles, glass fiber, paper, and membrane filter, sheets, packed bed, and cloth bag filters. The type or types used depends on the analysis being performed, the sample flow rate, pressure drop involved, the operating characteristics and limitations of the unit, and the composition of the flue gas stream. Several types of filtration devices and holders are shown in Figure 56.

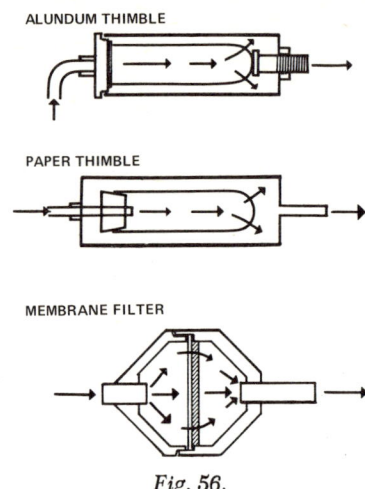

Fig. 56.

a. Paper Thimbles—Paper thimbles are useful for a limited number of particulate sampling applications up to a maximum temperature of 250 to 300°F [10] [15] [36]. They are inexpensive and efficient in collecting particles down to less than one micron in size [15] [20]. However, they lose their strength when wet, and must be kept above the dew point of the gas stream. They are also hygroscopic, and exhibit a strong affinity for organic materials. They may prove useful as the final particle collection stage for particles above one micron (or other size) following impingers or other collectors. However, they must be kept dry by heating at all times [10].

b. Alundum Thimbles—Alundum thimbles are made from a special aluminum oxide material and, when sealed in holders employing asbestos gaskets, are able to withstand temperatures up to 1,000°F. Their ability to withstand high temperatures has allowed them to be used as internal filtration stages in numerous sampling procedures [6] [7] [13] [15] [27] [28] [30] [31]. They are also relatively nonhygroscopic and may be weighed with sufficient accuracy on an analytical balance. Initial work establishing their suitability for particulate sampling was done by Anderson [37]. Several different grades of thimble porosities are available for filtration of different size particles. A summary of the retention capabilities of various grades of thimbles is shown in Table 24.

However, alundum thimbles present problems because they are heavy in relation to the amount of material collected (40 to 60 grams tare weight versus one to two grams of sample collected), thus requiring use of an analytical balance. The chance for large relative error is increased because it is necessary to take the small weight difference between two larger numbers. The thimbles are relatively expensive but may be reused if cleaned properly after use by consecutive washings with hot dilute hydrochloric acid, tap water, and distilled water [10]. They also create a substantial pressure drop caused by high resistance to flow. The increase in resistance to flow can be minimized by placing a layer of Pyrex glass wool on the inside of the thimble to assist in particle collection. Insertion of the glass wool is facilitated by wrapping the wool on a test tube and placing inside the thimble, then removing the test tube [6] [10].

The alundum thimble may be placed either internal to [6] [27] or external to the duct [10]. The choice depends on the source temperature, the type of thimble being used, and the amount of volatile organic material present.

c. Analytical Filter Papers—These papers are readily available in a wide selection of grades for laboratory use. They are inexpensive, of good wet strength, low ash content, and chemical purity, and normally are used for specific contaminants such as lead or beryllium [18]. However, they are not usually used for particulate sampling because of a tendency for ignition at low temperatures, their hygroscopic nature, and the high resistance to gas flow. However, they may find limited application as final particulate filters following other collection devices in a sampling train. Smith and Surprenant [39] made a detailed study of these materials and found that Whatman Filter numbers 42, 44, and 50 had the greatest collection efficiency among the analytical filter papers. A summary of their data for pressure drop in inches of water and penetration by dioctyl phthalate test aerosols of 0.3 microns size in percent by count is presented in Table 25.

d. Glass Fiber Filters—The glass-fiber filters normally used for ambient air sampling are also useful in some source test applications. They are relatively nonhygroscopic and able to withstand temperatures exceeding 600°F [20], and have a high collection efficiency exceeding 99.9 percent by count on the DOP test aerosol of 0.3 micron size previously mentioned, at a linear velocity of 28 feet per minute and a pressure drop of 4.4 inches of water [18]. This corresponds to 1.0 cfm through a 2.5 inch diameter filter. They can be used as second stage filters following alundum thimbles, where both are located inside the stack [6] [9]. However, it is necessary to consider the possible adverse effects of maintaining the collected particles inside the duct at elevated temperatures for extended periods. It is also possible to utilize them as a final stage particulate collector for particles above 0.3 microns in diameter.

However, the filters are subject to problems with tearing, and plugging if the gas falls below the dew point. Therefore, it is necessary to heat the unit if used following impingers. Also, they can not be used to determine materials present in the filter pad, such as sodium. Glass fiber filters can also be made in the form of thimbles and have been successfully used [9].

e. Membrane Filters—Membrane filters provide a very effective method for particle collection as they combine high collection efficiency with relatively low resistance to gas flow [40]. Membrane filters consist of a porous cellulose ester fibre with a large number of relatively uniform, very small pores. Commercially available grades range from pore sizes of 0.01 up to 5.0 microns, where the appropriate grade should be selected according to the minimum particle size to be collected [4] [42]. They are very useful as a final particle collection stage since the pore size can be carefully selected as desired. Membrane filters are also useful for particle size determination, which will be discussed in a subsequent section. A summary of the variation in pressure drop for several grades of membrane filters is shown in Figure 57.

The major problem with membrane-type filters is that the maximum operating temperature is about 180°F [20]. This limits their usefulness because the filter must also be maintained above the dew point of the gas stream. They also cannot be used in flue gas streams where substantial portions of sulfur dioxide and trioxide, and other acidic gases are present.

f. Packed Bed Filters—Packed filters can be used to remove particles from gas streams. They normally make use of glass wool material,

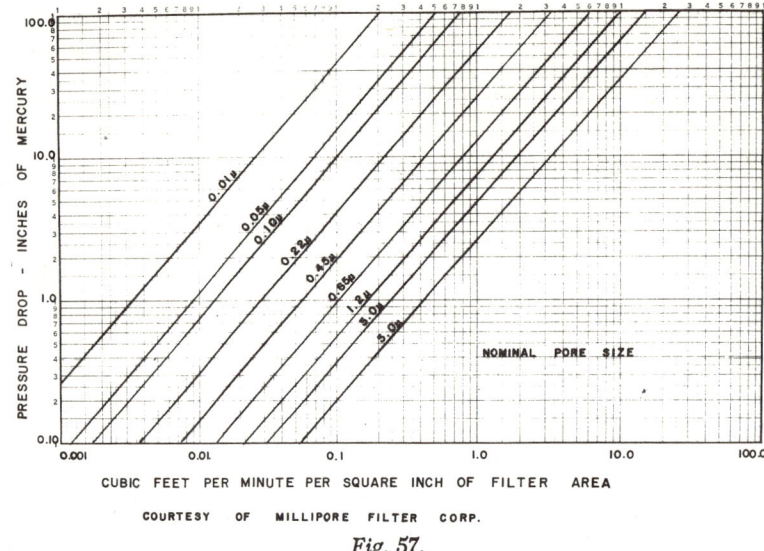

Fig. 57.

although other materials have also been used [2]. Warshavsky [43] makes use of an internally located filter holder assembly, packed with silica wool placed between two glass screens, as a first stage collector to remove particles from the exhaust gases of a coking oven. British Standard 3405 [31] recommends use of commercially available packed glass wool filter cartridges for particle collection from power boiler flue gases in an externally located filter assembly. The San Francisco Bay Area manual [6] describes the use of an internally located filter for particle collection in saturated ducts following scrubbers. This is one of the few applications of filtration where the gas stream is wet.

The method is potentially useful for collection of large particles, but may present difficulties for small particles below five microns in size. It may prove useful as a primary collection method if the material is to be washed from the glass wool, as with soluble ionic analyses.

g. Bag Filters—These are often used for collection of large quantities of sample material in high volume flow rate equipment. The first successful application was described by Hardie [44] in collection of cinder emissions from a coal-fired power boiler. However, limitations include maximum allowable temperature of the bag material, and the possible presence of moisture and acidic gases. Pressure drops vary up to more than one inch of mercury at flow rates of 50 to 60 cfm. It is also necessary to collect at least four times as much material as the tare weight of the bag, to minimize any weighing errors. The penetration of particles can also be a problem depending on the flow rate used. It may be necessary to place a glass fiber filter as a second stage collector. Normally, the bags can remain above the dew point by the high gas flow rates used, but care should be taken to assure that the filtration unit remains above the flue gas dew point at all times. It should also be assured that the bag not be damaged by excessive temperatures. The bag filter units are useful in collecting large amounts of sample at flow rates of 20 to 80 cfm [8] [10] [43]. It is also necessary to maintain the bags inside airtight containers to minimize air leakage under vacuum.

4. Analyses

Total particulate analyses with filtration units are normally made by means of gravimetric techniques, where the filters are at least dried in a desiccator for one hour prior to weighing on an analytical balance. Alundum thimbles should be placed in an oven at 105°C for one hour, then desiccated and weighed, as previously discussed. Soxhlet extractions may be performed to collect constituents in a liquid phase to facilitate chemical analyses [6] [10].

5. Evaluation:

Filtration provides a highly versatile and useful technique for particle collection. The use of membrane filters in final particulate collection trains provides an excellent method of removal of particles greater than a given size but limits the maximum flow rate because of pressure drop. However, it is necessary to maintain the unit above the gas dew point by heating, especially if used following wet impingers. The location of alundum thimbles internal or external to the duct depends on the nature of the particles being collected. The filter medium should normally never be used above its maximum operating temperature or below the dew point of the flue gas. The pressure drop across the filter also is an important consideration in sampling train design.

Filtration techniques are utilized by numerous organizations, including the following: 1) Dade County Pollution Control Authority [15]; 2) Los Angeles County Air Pollution Control District [10]; 3) Bay Area Air Pollution Control District [6]; 4) British Standards Institution [31]; 5) British Coal Utilization Research Association [29]; 6) American Petroleum Institute [30]; 7) American Society of Mechanical Engineers [8] [35]; 8) National Council for Air and Stream Improvement [23]; 9) Incinerator Institute of America [45]; 10) National Air Pollution Control Administration [46]; 11) Air Pollution Control Association Foundry Committee [47].

D. Centrifugal Collection:

1. Principles

Particulate collection is achieved in centrifugal-type devices by causing the incoming gas stream to flow in a spiral pattern in the form of a confined vortex. Centrifugal forces tend to drive the particles towards the outer walls because of greater inertia which makes them less able to negotiate changes in direction than the lighter gas molecules [48]. There are two types of particle collectors which employ centrifugal force, one of which has no moving parts. For the stationary type the gas stream is normally introduced tangentially at the top of the unit and caused to flow downward in a vortical pattern to the inlet of the concentric exit tube extending through the top of the collector. Particles are caused to strike the outer walls and are subsequently collected after falling through the conical section in the lower part of the unit. The device provides a useful method for collecting particles larger than approximately five microns in size in a dry state, with a low pressure drop of three to five inches of water [18] [20]. It is particularly useful for high volume sampling systems where the gas flow rate exceeds five cfm [7] [10] [31]. These systems are particularly useful for flue gases where it is necessary to collect a large amount of particulate material, such as high concentration (greater than 1.0 gr/SDCF) sources containing a high portion of large particles greater than 100 microns in size.

The rotating type makes use of a conically-shaped revolving head inside a stationary outer cover. The gas is introduced axially at the tip of the cone and particles are caused to deposit on the outer surface by the action of the rotating cone. The device is primarily a research instrument, and will be described in the section on particle size analysis [49].

2. Devices:

Cyclone collectors have been used for particle collection in high volume sampling systems, frequently as a first stage collector for particles larger than five microns. They allow collection of large amounts of material without any significant increase in pressure drop during the sampling period, the only limitation being in the size of the jar attached below the cyclone. They may be used either internal or external to the duct, depending on the size of the duct, the configuration of the probe assembly, and the nature of the material being collected. It is also possible to use the pressure drop across the cyclone as a measure of gas flow rate once the sampling train is calibrated [3] [10] [29]. A filter of relatively high efficiency is often placed following the cyclone to allow collection of many of the smaller particles not removed in the first stage. The use of glass fiber paper is particularly applicable [29] [31].

Several investigators have made use of the cyclone collector as a particulate removal device. Stern [50] used a large external cyclone for cinder collection from a coal-fired power boiler in a high volume sampling system (10 to 80 cfm) using a vacuum cleaner motor as a prime mover. Pressure drop across the cyclone was used as the basis of gas flow measurement, and a calibration procedure utilizing an orifice was described. Stairmand [2] designed five different-sized cyclones for sampling applications covering a flow range from two to 130 cfm. He found that they were 95 to 97 percent efficient (weight basis) in the collection of total particulate emissions from the flue gases from pulverized coal power boilers. A small glass cyclone has also been designed to collect particles greater than five microns in size at low sampling flow rates of 0.3 to 0.5 cfm [10].

It is common practice to place a cyclone as a primary collector in series with a secondary filtration stage, particularly in combustion and incinerator flue gases. Holton and Schultz [51] employed a high volume (3 to 20 cfm) sampling system on an externally-located cyclone-bag filter combination for particle collection from the flue gases of a pulverized coal boiler. Sample flow rate was determined by measurement of pressure drop across a calibrated variable throat venturi section, and isokinetic sampling conditions were adjusted by velocity measured with a standard pitot tube placed adjacent to the sample probe. However, it was necessary to heat the probe, cyclone, and bag filtration units prior to commencement of the test, requiring electric heaters of 3,000, 500, and 1,000 watts, respectively, for the three units for a gas flow of 20 cfm. A similar system has been devised for sampling boiler flue gases [64] for a flow range of 5 to 20 cfm. Cyclone-filter bag combinations have also been successfully used for sampling refuse incineration flue gases, where the sampling rate can be as high as 50 to 80 cfm. It is also necessary to heat the collection system above the dew point of the flue gases to prevent condensation [3] [10] [52]. Smith [7] describes a three-stage particulate collection system employing a cyclone, coarse fritted glass filter, and wet impingers in series. However, the sampling train is designed for a maximum of 1.0 cfm. Stern [50] utilized an electrostatic precipitator following a cyclone as a second stage collector when testing flue gases from a coal-fired boiler. The advantage of the precipitator is that it combines high collection efficiency with low head loss.

Devices have been constructed which employ a cyclone-filter combination as a single unit. Zimmerman [53] placed a glass wool filter sheet between two metal screens atop a cyclone for removal of small particles. Rosin and Rammler [54] modified this design by placing four filters in parallel to reduce the pressure drop caused by resistance to flow with only one filter in line. Both of the above devices are placed external to the duct because of their size and must be heated if moisture condensation is a potential problem. Hawksley, Badzioch, and Blackett [29] [55] designed a cyclone-filter combination which can be placed directly in the duct, as the cyclone is placed axially to the sample probe. It has been successfully utilized in a high volume sampling system of 5 to 20 cfm for coal-fired boiler flue gases, and may be modified for location external to the stack. It is illustrated in Figure 58.

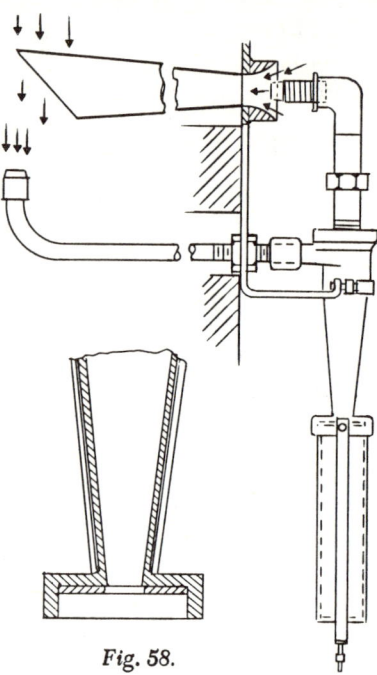

Fig. 58.

Courtesy of: Mr. A. C. Wilson, Airflow Developments, Ltd., Leatherhead, Surrey, England.

3. Analysis:

Material collected in the cyclone jar is normally analyzed by gravimetric means as previously described. Particle size analyses may also be made on the collected material in some applications.

4. Evaluation:

Cyclone collectors provide a useful method for removing particles greater than 5.0 microns in size, particularly for high concentration sources where large amounts of material must be collected. They can be used in high-volume sampling systems because of their low resistance to flow. Cyclone devices are normally located external to the duct being sampled, and should be heated if moisture condensation poses a potential problem. Their main use is as a primary collector for large particles.

E. Electrostatic Precipitation:

1. Principles:

Electrostatic precipitation involves imparting an electrical charge to a group of particles, which then migrate to a surface of opposite charge where they are collected and subsequently removed. Particles may become charged by friction, direct dissociation, or by action of an outside electrical field in which the particles are subjected to bombardment and diffusion of gas ions [56]. The third method is used for most electrostatic-type sampling devices where a corona discharge phenomenon is used. A negative charge is imparted to a wire placed axially inside a cylinder which is a positively charged, and a voltage difference of approximately 10,000 to 20,000 volts and about 0.5 to 1.0 milliamperes current (d.c.) is maintained. The gas stream is caused to flow through the cylinder, and the particles are charged by the ions from the corona discharge, and migrate to the inner surface of the cylinder [25] [56]. The particles adhere to the inner surface until the end of the test and may then be removed. Normally, a metallic foil liner is placed on the inner surface of the cylinder to facilitate ease of removal and subsequent analysis. Settling and inertial effects on the larger particles assist in the collection process. A thorough discussion of the principles of electrostatic precipitation is presented by White [57].

2. Variables:

The efficiency of particle collection by electrostatic precipitation is influenced by the properties of the particles and the flue gas stream, the design of the collector and the way it is operated, and the gas flow rate through the unit. Properties of the particles which influence collection are their bulk density, concentration in the gas stream, surface area exposed, size distribution, and electrical resistivity. Normally, the gas must be above its dew point in passing through the unit, as moisture droplets significantly reduce the collection efficiency by lowering the voltage between wire and cylinder. Efficiency is also temperature-dependent at higher temperatures because of "back ionization."

The diameter and length of the collection cylinder determine the superficial loading velocity through the unit and the retention time inside the cylinder. The superficial loading velocity is normally 30 to 100 cfm per square foot (feet per minute) in commercially available units to facilitate high collection.

The electrodes should be spaced as close as possible to facilitate collection efficiencies [56]. The length of the cylinder (usually about twelve inches) must be sufficient to collect the charged particles and should be increased as gas flow rate increases. Robinson [58] found that at a superficial velocity of about 80 feet per minute, essentially all the particles were deposited within ten inches of the upstream end of the cylinder. Most of this material was found within the first three inches.

It is also necessary to maintain a sufficient voltage difference between the wire and cylinder to maintain collection efficiency. The necessary voltage varies with the characteristics of the individual unit and current drawn but should be at least 14,000 volts [18]. The efficiency depends on inlet particulate loading and the nature of the particles being collected. High inlet concentrations tend to reduce efficiency by causing sparking and reduced voltage. This may be partially alleviated by use of a first stage collector such as a cyclone. Some aerosols may be resistant to placement of a charge on their surfaces, such as certain types of organic materials. Particle size distribution also affects collection efficiency because the degree of charge depends on particle surface area, where the surface area exposed per unit mass of particles increases as the particle size decreases [59].

All of the above factors result in a device with a high collection efficiency over a wide size distribution [60]. The device is a high-volume, low-pressure-drop unit with a wide potential usage. However, voltage reductions caused by high-moisture-content flue gases, particulate inlet loads, and high electrical resistivities of certain dusts limit the usefulness of the method. It is also necessary to maintain the temperature of the probe and collector above the dew point of the flue gas at all times, and maintain the high voltage necessary for operation of the collector. This involves movement of heavy and cumbersome equipment, as well as creation of potential safety hazards in rainy weather.

3. Applications:

Electrostatic precipitation has been utilized in several source monitoring applications, although not with great frequency. The Los Angeles County District Manual [10] and the APCA TA-5 Committee Report [6] indicate the only organizations which prescribe its use in certain situations.

The first reported use of electrostatic precipitation for particle collection was in 1919 [62]. Stern [50] describes one of the earliest uses of an electrostatic sampler as a second stage collector following a cyclone to remove the small particles below five microns in size. It consisted of a cylinder with a wire electrode placed inside, and was designed for gas flow rates between 10 and 80 cfm. The combination unit was able to obtain essentially 100 percent efficiency in particulate collection for the flue gases from a coal-fired power boiler.

Barnes and Penney [63] developed a cylindrical electrostatic collector with a maximum flow rate of three cfm which would be primarily used as an ambient air monitoring device. However, it could be used for source test applications by use of a cyclone or other primary collector upstream of the precipitator. Subsequent tests on this unit by Schadt and Cadle [21] with test aerosols of stearic acid, sodium chloride, and glycerol indicated collection efficiencies exceeding 99.0 percent by weight during all tests. Flow rates were varied from 0.2 to 2.5 cfm, and voltage was varied from 8,000 to 16,000 volts without significant effect on collection efficiency.

A cylindrical electrofilter [64] has been developed for use on certain sources which operate at a maximum flow rate of 1.0 cfm at 20,000 volts. It is useful as a single stage collector, and provides for a maximum collection of about two grams of material. Most of the material is collected in the initial upstream section of the cylinder.

Rounds and Matoi [65] developed a truncated cone electrostatic collector to provide for more uniform dust deposition along the length of its inner wall. It allowed for collection of a maximum of approximately ten grams of material at a flow rate of 1.5 cfm at 15,000 volts potential difference. It provided for essentially 100 percent collection of particulate materials from the flue gases of open hearth furnaces in steel mills. A Whatman No. 42 filter paper behind the collector did not indicate the presence of red iron oxide dust by either visual or gravimetric analysis.

Leonard [66] described the successful use of an electrostatic precipitation sampler for sampling the flue gases from a Kraft pulp mill recovery furnace. The unit operated at a maximum flow of one cfm with a potential difference of 18,000 volts. It was not possible to measure detectable quantities of sodium sulfate in wet impingers placed after the electrostatic collector during a two hour sampling test on a recovery furnace.

Stairmand [2] described three different precipitation units used to collect particles from the flue gases of coal and oil-fired boilers, and mists from acid manufacturing process offgases. The devices all employed a one-inch (ID) tube 16 inches long, with a copper sheath to collect particles inside the glass and a tungsten electrode wire. The units were all heated to prevent condensation and used a potential difference of 15,000 volts. Performance tests showed the units to have greater than 99.0 percent collection efficiency (weight) for all particles larger than 0.2 microns in size at a pressure drop of less than 0.1 inch of water.

Lippman [67] developed a high-volume, two-chamber, electrostatic sampler (0-27 cfm) which was designed for ambient air sampling. It could be adapted for source sampling by placing a cyclone as a primary collection device upstream of the collector. However, the high superficial velocity of over 1000 feet per minute results in a slightly decreased collection efficiency of 96.0 percent by weight for particles less than one micron in size.

Binek [68] investigated four different types of electrostatic samplers for particulate collection, and found two of these to have collection efficiencies greater than 99.0 percent by weight in ambient air studies.

Electrostatic precipitators can produce collection efficiencies exceeding 99.0 percent by weight for particles larger than 0.2 to 1.0 microns in size at pressure drops below 0.1 inch of water gauge. This makes the job of maintaining isokinetic sampling rates much easier. However, it is necessary to have a superficial velocity less than 80 to 100 feet per minute and a stable power supply to provide a stable source of potential difference of 15,000 volts. Agglomeration of particles during electrostatic precipitation limits its potential usefulness for particle size determinations.

Problems associated with moisture condensation require heating the unit to above the dew point of the flue gas to prevent sparking; heating to 300°F is normally sufficient. It is often necessary to provide a primary collector upstream of the precipitator to prevent excessive loading and subsequent sparking. The precipitation units normally require a heavy power supply and cumbersome equipment, which may also present problems in both transport and in rainy weather.

There is also the possibility of chemical changes or oxidation if organic particles are present. The operating characteristics of several units have been listed in Table 26.

4. Analysis:

Following the sampling, the collection tube may be carefully weighed to determine the amount of material collected. Handling must be gentle to avoid knocking off material. Subsequent analysis may then be made by carefully removing the collected particles, or by other methods. Techniques for gravimetric and chemical analyses are as previously described for alundum thimbles.

F. Thermal Precipitation:

1. Principles:

Thermal precipitation acts in a similar fashion to electrostatic precipitation except that thermal forces act to remove particles from flowing gas streams instead of electrical forces. Thermal forces cause small particles to flow from zones of hotter to colder temperatures. The flow of the gas phase from the warmer to colder side of the particle has the net effect of pushing it towards the colder surface because of greater kinetic energy exerted by gas molecules on the warmer side. These thermal forces are of small magnitude, and are over and above any convection effects in the gas stream, which limits the flow rate through the sampler to a maximum of 0.5 liters per minute [56].

It had been observed by early investigators such as Aitken [69], Einstein [70], and Epstein [71], that a clear area developed around a hot object placed in a dusty air space. Rosenblatt and LaMer [72] found that the velocity for particle movement under the influence of a thermal field was directly proportional to the thermal gradient. Further work by Saxton and Ranz [73] indicated that the magnitude of thermal force acting on a particle was directly proportional to the particle diameter and the temperature gradient, and inversely proportional to its thermal conductivity. The range of particle collection for thermal precipitators is approximately 0.001 to 100.0 microns in size, with nearly 100 percent efficiency for particles between 0.1 and 10.0 microns. Particles less than 0.1 micron approach the mean free path of gas molecules and may be driven randomly by molecular motion with little net displacement [56]. However, for particles larger than five to ten microns in size the kinetic energy of molecular bombardment may become nearly the same in all directions, again with little net displacement.

Primary variables are the temperature gradient per unit length across the collector, the flow rate through the unit, the diameter and length of the collection section, and thermal conductivity of the particles. The gradient per unit length is normally not greater than 3000°C per centimeter while the flow rate must be kept between ten and 500 milliliters per minute, depending on the device. It is desirable to maintain as large a diameter as possible to minimize the superficial velocity but maintain a sufficient thermal gradient. The length is normally six to twelve inches to allow deposition of the particles. It is not feasible to collect particles with high thermal conductivities because of the reduction in the thermal forces. The devices are useful for microscopic examination of particles because they provide for gentle deposition without shearing of the particles. However, the low flow rates involved limit their use to research applications and certain particle size determinations.

2. Devices:

The simplest type of thermal precipitator involves passage of a gas stream past a heated electric wire. This repels the particles and causes them to be collected on a glass slide or other cooled surface placed near the wire [74]. However, flow rates are restricted to 10 to 20 ml per minute.

Bredl and Grieve [75] designed a cylindrical unit with a central heating element with a concentrically located aluminum collection surface for particle deposition. The system had a maximum flow rate of about 3,000 ml per minute, and provided a mean thermal gradient of about 2,200°C per centimeter between the core and outer wall. This required maintaining the central core 100°C warmer than the temperature of the flue gas with the added requirement that the cooled surface be above the dew point of the flue gas. Sufficient thermal repulsion to allow complete particle collection on the foil was provided without moisture condensation on the cooled foil collection surface, which would destroy the sample. The inner core of the unit was tapered so as to provide a space between the hot and cold surface of 1.43 millimeters at the inlet and

0.16 millimeters at the exit. This provided for deposition of the larger particles first with a more intense thermal force at the rear to deposit the very small particles. This provided for a certain degree of particle size classification on the foil for subsequent microscopic examination.

Another type of precipitator incorporates an electrically heated plate placed above a water-cooled plate [76]. Gas is introduced at the top of the cold plate and caused to flow between the hot and cold surfaces to deposit the particles. The unit provides for flow rates from 100 to 500 milliliters per minute and is useful for bacteria and liquid aerosol sampling because the temperature gradient is limited to 3200°C per centimeter [77].

An additional thermal precipitation phenomenon, called thermophoresis, occurs in sampling probes after the flue gases are drawn outside of a hot duct. The air cools the metal surface and causes particle deposition on the inner wall of the tube. This may require heating the probe or locating the collection system near the duct to minimize weight losses. Probe walls should be thoroughly cleaned by brushing or washing after completing sample collection.

The operating characteristics of several thermal precipitation devices are listed in Table 27.

G. Simplified Methods:

Additional techniques are available for total particulate determination in flue gas streams. Certain of these methods, such as impaction, are used primarily for particle size determination and will be discussed in the appropriate section. The following discussion is related to short-cut methods which provide approximate results with a minimum of time and expense.

1. Filtration:

The methodology of particulate sampling by filtration techniques has been previously described. Katz and Verrochi [78] devised a simplified procedure to evaluate the collection efficiencies of particle collection devices for flue gases from coal-fired power boilers. It involves withdrawal of flue gas samples from the respective inlet and outlet gas streams of the collector through membrane filters in the sampling train at the same flow rates for 10 to 15 minutes. The weights of material collected on the filters are then compared to provide an index of collection efficiency.

2. Dry Collection:

The principle involved is to allow particles from the flue gas stream to flow directly into the collection device without the aid of a prime mover. It is possible to locate the collection device either internally (inside the duct) or externally, and determine weight of material collected per unit time.

Lees and Morley [79] used tubular samplers placed inside the duct with openings facing the direction of flow to collect particles by impingement inside the cylinder. The method was used to estimate the collection efficiency of flyash collectors on flue gases from coal-fired power boilers and other sources by simultaneous location at the respective inlets and outlets for the same time interval, followed by weighing upon removal. Evaluation of the method was made by locating a sampling probe connected to a sampling train adjacent to the tubular sampler, and withdrawing flue gas at isokinetic velocity for a given time interval. Results showed that the tubular sampler collected from 18 to 33 percent of the amount of flyash collected by the sampling train at isokinetic velocity. The method would provide only a very rough approximation of the particulate emission levels from a source because resistance to impact pressure by the nonflow device tends to prevent collection of the smaller particles.

A somewhat more refined device for estimating particulate losses from stack gas collection devices has been developed by the Central Electricity Generating Board in Great Britain [80] [81]. It involves placing a sampling probe in the flow stream and connecting it to an externally located cyclone for collection of the particulate materials present in a glass jar. The exit gases are then caused to flow back into the duct by the aspiration effect of passing the flue gas of conical ejector tube to create sufficient vacuum for flow. The device has been claimed to operate within 12 percent of isokinetic sampling rate for a velocity range of 1000 to 6000 feet per minute if a suction of greater than one inch of water is maintained. Sampling nozzle size may be changed for different velocity and load conditions. The device provides a rapid means for estimating particulate losses and combustion conditions by either visual observation and has been extensively used in England on coal-fired power boiler flue gases. A diagram of the device is shown in Figure 58.

3. Adhesive-Coated Cylinders:

Gruber and Schumann [82] placed cylinders coated with adhesive paper in refuse incinerator exhaust gases for about 30 seconds each, to estimate particulate losses. The numbers of particles deposited per square inch per minute are counted with a microscope and are used as the index of particulate emission rate for a municipal emission regulation. Subsequent work indicated that 7.0 centimeter diameter cylinders were the most convenient for field use [83], though smaller diameter cylinders displayed higher collection efficiencies. The method provides a rapid and potentially useful technique for estimating emissions of particles greater than 20 microns in size.

H. Summary:

Performance characteristics of the various types of collection devices are listed in Table 28. The values should be used as approximate since there are variations among different types of collection devices in the particular classes. The minimum particle size collected refers to the smallest particle removed with an efficiency of greater than 99 percent by weight. Values listed may be subject to variation depending on the type of collector and the sampling rate used. Pressure drops for impingers in terms of sampling rate are shown in Figure 59.

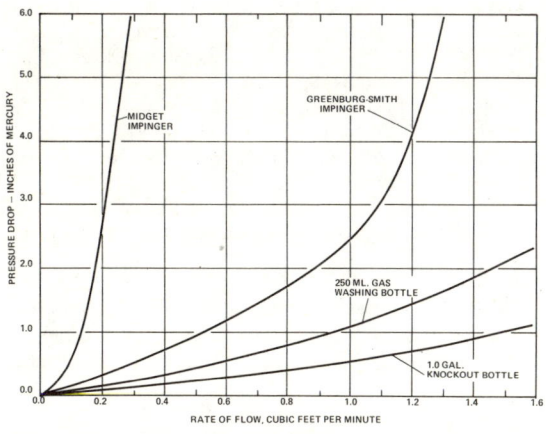

Fig. 59.

I. REFERENCES

1. Wilcox, J. D., "Isokinetic Flow and Sampling," *Journal of the APCA*, Vol. 5, No. 4, pp. 226, 245, February 1956.
2. Stairmand, C. J., "Sampling of Dust Laden Gases," *Transactions of the Institution of Chemical Engineers*, Vol. 29, No. 1, pp. 15-44, January 1951.
3. Rehm, F. R., "Test Methods for Determining Emission Characteristics of Incinerators," *Journal of the APCA*, Vol. 15, No. 3, pp. 131-135, March 1965.
4. Evans, D. G., "A Water-Cooled Probe for Sampling Gases from Shaft Furnaces," *Journal of the Institute of Fuel*, Vol. 37, No. 278, pp. 108-112, March 1964.
5. Smith, J. F., Hultz, J. A., and Orning, A. A., "Sampling and Analysis of Flue Gas for Oxides of Sulfur and Nitrogen," U. S. Bureau of Mines Report of Investigations No. 7108, 1968.
6. Wolfe, E. A., ed., "Source Test Methods," Bay Area Air Pollution Control District, San Francisco, California, 1961.
7. Smith, W. S., Martin, R. M., Durst, D., and Hyland, R., "Stack Gas Sampling Improved and Simplified with New Equipment," Paper No. 67-119. Presented at the 60th Annual Meeting of the APCA, Cleveland, Ohio, June 13, 1967.
8. "Determining Dust Concentration in a Gas Stream," Performance Test Code 27-1957, American Society of Mechanical Engineers, New York, New York, 1957.
9. "Simplified Methods for Measurement of Grit and Dust Emissions from Chimneys," British Standard 3405, British Standards Institution, London, England, 1961.
10. Devorkin, H., Chass, R. L., Fudurich, A., and Kanter, C. V., Holmes, R. G. ed. "Source Testing Manual," County of Los Angeles, Air Pollution Control District, Los Angeles, California, 1965.
11. "Flue Dust Sampling Instructions—Cyclone Method," Babcock and Wilcox Co., Alliance, Ohio, 1968.
12. Personal communication with Mr. Walter S. Smith, U. S. Public Health Service, National Air Pollution Control Administration, Durham, North Carolina, 1968.
13. Hardie, P. H., "Resumé of Methods for Measuring Flue Dust," *Transactions of the American Society of Mechanical Engineers*, Vol. 59, No. 10, pp. 355-358, October 1937.
14. "Test Methods for Gas Volume and Dust Sampling Determinations," Research-Cottrell, Inc., Bound Brook, New Jersey, 1968.
15. Rivera, M. F., "Incinerator Source Testing Manual," Dade County Pollution Control Authority, Miami, Florida, September 1968.
16. Parker, G. J., "Some Factors Governing the Design of Probes for Sampling in Particle and Drop-Laden Streams, *Atmospheric Environment*, Vol. 2, No. 5, pp. 477-490, September 1968.
17. "Standard Methods for the Examination of Water and Wastewater," 12th edition, American Public Health Association, New York, New York, 1965.
18. Clayton, G. D., ed. in "Stack Sampling," Ch. 9 in *Air Pollution Manual*, Part 1–Evaluation, American Industrial Hygiene Association, Detroit, Michigan, 1961.
19. Greenburg, L., and Bloomfield, J., "The Impinger Dust Sampling Apparatus as used by the United States Public Health Service, *Public Health Reports*, Vol. 47, No. 12, pp. 654-675, March 18, 1932.
20. Bloomfield, B. D., "Source Testing," Ch. 28 in Stern, A. C., *Air Pollution*, Vol. 2, 2nd ed., pp. 487-536, Academic Press, Inc., New York, New York, 1968.
21. Schadt, C., and Cadle, R. D., "Critical Comparison of Collection Efficiencies of Commonly Used Aerosol

Sampling Devices," *Analytical Chemistry,* Vol. 29, No. 6, pp. 864-868, June 1957.
22. Miller, A. M., Brown, J., and Abrams, R., "Applied Techniques of Analyses of Stack Emissions," Presented at the West Coast Regional Meeting of the National Council for Air and Stream Improvement, Portland, Oregon, October 2, 1968.
23. "Manual for the Sampling and Analysis of Kraft Mill Recovery Stack Gases," National Council for Air and Stream Improvement, Atmospheric Pollution Technical Bulletin No. 14, New York, New York, October 20, 1960.
24. Banciu, I., "Determination of Sodium Salt Losses in the Recovery Boiler of a Sulfate Pulp Mill, *Celuloza Hirtie* (Bucharest, Rumania), Vol. 15, No. 7, pp. 253-257, July 1966.
25. Caplan, K. J., ed., "Performance Testing," Ch. 12 in *Air Pollution Manual, Part 2–Control Equipment,* American Industrial Hygiene Association, Detroit, Michigan, 1968.
26. Jacobs, M. B., *The Chemical Analysis of Air Pollutants,* Interscience Publishers, Inc., New York, New York, 1958.
27. "Method of Testing Dust from Extraction Plant and the Emission from Chimneys of Electric Power Stations," British Standard 893, British Standards Institution, London, England, 1940.
28. Haaland, H. H., ed., "Methods for Determination of Velocity, Volume, Dust and Mist Content of Gases," Bulletin WP-50, Western Precipitation Corp., Los Angeles, California, 1968.
29. Hawksley, P. G. W., Badzioch, S., and Blackett, J. H., *Measurement of Solids in Flue Gases,* British Coal Utilization Research Association, Leatherhead, Surrey, England.
30. "Sampling and Analysis of Waste Gases and Particulate Matter," *Manual on the Disposal of Refinery Wastes,* Vol. 5, American Petroleum Institute, Inc., New York, New York, 1954.
31. "Simplified Methods for Measurement of Grit and Dust Emissions from Chimneys," British Standard 3405, British Standards Institution, London, England, 1961.
32. Baum, F., and Reichardt, I., "Measuring Errors During Determination of the Dust Content in Waste Gases by Adsorption of Sulfur Trioxide on Measuring Filters," *Staub,* Vol. 27, No. 9, pp. 18-22, September 1967.
33. Rendle, L. K., "Measurement of the Flue Gas Solids Burden from Oil-Fired Installations with a Silica Wool Filter," *Journal of the Institute of Fuel,* Vol. 37, No. 276, pp. 11-14, January 1964.
34. Rossano, A. T., and Schell, N. E., "Procedures for Making an Inventory of Air Pollutant Emissions," *Journal of the APCA,* Vol. 8, No. 2, pp. 147-152, August 1958.
35. "Dust Separating Apparatus," Performance Test Code 21-1941, American Society of Mechanical Engineers, New York, New York, 1941.
36. Hemeon, W. C. L., and Haines, G. F., "The Magnitude of Errors in Stack Dust Sampling," *Air Repair,* Vol. 4, No. 3, pp. 159-164, November 1954.
37. Anderson, E., "On the Quantitative Determination of Industrial Gas Dispersoids," *Transactions of the American Institute of Chemical Engineers,* Vol. 34, No. 2, pp. 589-601, February 1938.
38. Personal communication with Mr. Harold H. Haaland, Western Precipitation Corp., Los Angeles, California, 1969.
39. Smith, W. J., and Surprenant, N. F., "Properties of Various Filtering Media for Atmospheric Dust Sampling," *Proceedings of the American Society for Testing and Materials,* Vol. 73, pp. 1122-1135, Philadelphia, Penn., 1953.
40. Goetz, A., "Application of Molecular Filter Membranes to the Analyses of Aerosols," *American Journal of Public Health,* Vol. 43, No. 2, pp. 150-159, February 1953.
41. Fitzgerald, J. J., and Detwiler, C. G., "Optimum Particle Size for Penetration Through the Millipore Filter," *AMA Archives of Industrial Health,* Vol. 15, No. 1, pp. 3-8, January 1957.
42. First, M. V., and Silverman, L., "Air Sampling with Membrane Filters," *AMA Archives of Industrial Hygiene and Occupational Medicine,* Vol. 7, No. 1, pp. 1-11, January 1953.
43. Warshavsky, T. P., Kogan, L., Levin, E. D., and Shevchenko N. S., "An Installation for the Determination of the Dust Concentration in Coke Oven Gases," *Coke and Chemistry,* (Moscow, USSR) No. 8, pp. 18-20, 1958, Trans. in USSR Literature on Air Pollution and Related Occupational Diseases," by Levine, B. S., ed., Vol. 5, U. S. Dept. of Commerce, Washington, D. C., January 1961.
44. Hardie, P. H., "Cinder and Flyash Measurements," *Combustion,* Vol. 6, No. 9, pp. 10-16, March 1935.
45. "Incinerator Testing," Bulletin T-6, Incinerator Institute of America, Inc., New York, New York, 1968.
46. "Specifications for Incinerator Testing at Federal Facilities," U.S. Public Health Service, National Air Pollution Control Administration, Durham, North Carolina, October 1967.
47. Weber, H. J., ed., "Air Pollution Problems of the Foundry Industry: Information Report No. 4– Instruments and Techniques for Measuring Air Pollution Emissions," Technical Manual No. 1, Air Pollution Control Association, pp. 49-55, 1963.
48. Caplan, K. J., "Source Control by Centrifugal Force and Gravity," Ch. 28 in Stern, A. C., ed., Air Pollution, Vol. 2, 1st ed., Academic Press, Inc., New York, New York, 1962.
49. Goetz, A., Stevenson, O., and Preining, O., "The Design and Performance of the Aerosol Spectrometer,"*Journal of the APCA,* Vol. 10, No. 10, pp. 378-383, October 1960.
50. Stern, A. C., "The Measurement and Properties of Cinders and Flyash," *Combustion,* Vol. 4, No. 12, pp. 35-47, June 1933.
51. Holton, W. C., and Schultz, E. J., "Some Notes on Dust Sampling Equipment and Techniques," *Transactions of the American Society of Mechanical Engineers,* Vol. 75, No. 10, pp. 1327-1331, October 1953.
52. Drinker, P., and Hatch, T., *Industrial Dust,* 2nd ed., McGraw-Hill Book Co., New York, New York, 1954.
53. Zimmerman, E., "Measurement of Flue Dust in Stack Gases," *Zeitschrift des Vereines deutsches Ingenieure,* Vol. 75, No. 16, pp. 481-486, April 18, 1931.
54. Rosin, P., and Rammler, E., "The Practices of Flue Dust Measurement," *Braunkohle,* Vol. 34, pp. 505-546, 1935.
55. Hawksley, P. G. W., Badzioch, S., and Blackett, J. H., "Improved Sampling Equipment for Solids in

Flue Gases," *Journal of the Institute of Fuel,* Vol. 31, No. 4, pp. 147-156, April 1958.
56. "Collection of Atmospheric Particulates," Ch. 2 in McCrone, W., Draftz, R., and Delly, J., *The Particle Atlas,* Ann Arbor Science Publishers, Inc., Ann Arbor, Michigan, 1967.
57. White, H. J., *Industrial Electrostatic Precipitation,* Addison-Wesley Publishing Co., Reading, Massachusetts, 1963.
58. Robinson, M., "Miniature Electrostatic Precipitator for Sampling Aerosols," *Analytical Chemistry,* Vol. 33, No. 1, pp. 109-113, January 1961.
59. Grindell, D. H., "An Electrostatic Dust Monitor," *Proceedings of the Institution of Electrical Engineers,* Vol. 107, Part A, No. 34, pp. 353-365, August 1960.
60. Giuntini, J., and Godard, L., "Simple Procedure for Capture of Dusts by Thermal or Electrostatic Precipitation," *Geofisica Pura Applicato,* Vol. 50, No. 4, pp. 42-45, Sept.-Dec. 1961.
61. Mitchell, R., and Engdahl, R., "A Survey of Improved Methods for the Measurement of Particulate Concentrations in Flowing Gas Streams," *Journal of the APCA,* Vol. 13, No. 11, pp. 558-562, November 1963.
62. Lamb, A., Wendt, G., and Wilson, R., "Portable Electrical Sampler for Smokes and Bacteria," *Transactions of the American Electrochemical Society,* Vol. 35, pp. 357-370, 1919.
63. Barnes, E. C., and Penney, G. W., "An Electrostatic Dust Weight Sampler," *Journal of Industrial Hygiene and Toxicology,* Vol. 20, No. 3, pp. 259-265, March 1938.
64. "Methods for Determination of Velocity, Volume, Dust and Mist Content of Gases," Bulletin WP-50, 6th ed., Western Precipitation Corp., Los Angeles, California, 1958.
65. Rounds, G. L., and Matoi, H. J., "Electrostatic Sampler for Dust-Laden Gases," *Analytical Chemistry,* Vol. 27, No. 5, pp. 829-830, May 1965.
66. Leonard, J. S. "Stack Emission Sampling," *Tappi,* Vol. 49, No. 10, pp. 84A-88A, October 1966.
67. Lippman, M., DiGiovanni, H., Cravitt, S., and Lilienfeld, P., "Lightweight High-Volume, Electrostatic Survey Sampler," *American Industrial Hygiene Association Journal,* Vol. 26, No. 5, pp. 485-489, September-October 1965.
68. Binek, B., Spurny, K., and Pixova, J., "Electrostatic Impingers for Absorption of Samples of Aerodispersed Harmful Matter," *Pracovni Lekar* (Prague), Vol. 15, No. 10, pp. 415-419, November 1963.
69. Aitken, J., "On the Formation of Small Clear Spaces in Dusty Air," *Transactions of the Royal Society of Education,* Vol. 32, p. 239, 1884.
70. Einstein, A., "On the Theory of Radiometer Forces," *Zeitschrift fur Phyzik,* Vol. 27, No. 1, pp. 1-4, January 1924.
71. Epstein, P., "On Resistance Experienced by Spheres in their Motion through Gases," *Physical Review,* Vol. 23, No. 6, pp. 710-733, June 1924.
72. Rosenblatt, P., and LaMer, V. K., "Motion of a Particle in a Thermal Gradient; Thermal Repulsion as a Radiometer Phenomenon," *Physics Review,* Vol. 70, No. 5, pp. 385-395, September 1946.
73. Saxton, R. L., and Ranz, W. E., "Thermal Force on an Aerosol Particle in a Temperature Gradient," *Journal of Applied Physics,* Vol. 23, No. 8, pp. 917-923, August 1952.
74. Watson, H. H., "The Dust-Free Space Surrounding Hot Bodies," *Transactions of the Faraday Society,* Vol. 32, No. 8, pp. 1073-1084, August 1936.
75. Bredl, J., and Grieve, T., "A Thermal Precipitator for the Gravimetric Estimation of Solid Particles in Flue Gases," *Journal of Scientific Instruments,* Vol. 28, No. 1, pp. 21-23, January 1951.
76. Kethley, T., Gordon, M., and Orr, C., "A Thermal Precipitator for Aerobacteriology," *Science,* Vol. 116, No. 3014, pp. 368-369, October 3, 1952.
77. Wright, B. M., "Gravimetric Thermal Precipitator," *Science,* Vol. 118, No. 3054, p. 195, August 14, 1953.
78. Katz, J., and Verrochi, W. A., "A Filter Test Method for Rapid Weight Distribution of Dust in Gas Streams to Evaluate Dust Collectors," *Journal of the APCA,* Vol. 18, No. 6, pp. 401-402, June 1968.
79. Lees, B., and Morley, M., "A Routine Sampler for Detecting Variations in the Emission of Dust and Grit," *Journal of the Institute of Fuel,* Vol. 33, No. 229, pp. 90-94, February 1960.
80. "The Cegrit Automatic Dust Sampler," Report No. PL 15A, Air Flow Developments, Ltd., High Wycombe, Bucks, England, September 1966.
81. "Improvements in or Relating to Apparatus for Extracting Samples of Gas-Borne Dust," British Patent No. 872,904 London, England, July 12, 1961.
82. Gruber, C. W., and Schumann, C. E., "The Use of Adhesive-Coated Paper for Estimating Incinerator Particulate Emissions," *Journal of the APCA,* Vol. 12, No. 8, pp. 376-378, August 1962.
83. Pritchard, W., Schumann, C., and Gruber, C., "Selection of a Suitable Adhesive-Coated Cylinder for the Collection of Airborne Particulates," *American Industrial Hygiene Association Journal,* Vol. 28, No. 6, pp. 517-522, November-December 1967.

Table 22. Distribution of Particle Collection in a Sampling Train.

Collection Device	Total Particulate Weight mg	Total Particulate Percent of Total	Nonvolatile Particulate Weight mg	Nonvolatile Particulate % Ash in Sample
1. Waste Wood Only				
Alundum Thimble	209.1	80.0	170.0	81.0
Line Washings	26.5	10.0	15.2	60.0
Scrubber No. 1	16.6	6.5	1.5	9.0
Scrubber No. 2	3.7	1.5	0.8	22.0
Scrubber No. 3	3.5	1.5	0.4	11.0
Scrubber No. 4	1.8	0.5	0.0	0.0
Total Scrubber	25.6	10.0	2.7	11.0
TOTAL	261.2	100.0	187.9	72.0
2. Waste Wood and Fuel Oil				
Alundum Thimble	193.9	51.0	145.4	75.0
Line Washings	29.3	7.0	16.2	55.0
Scrubber No. 1	105.3	28.0	22.3	21.0
Scrubber No. 2	19.3	5.0	1.4	7.0
Scrubber No. 3	21.3	6.0	1.0	5.0
Scrubber No. 4	13.1	3.0	0.5	4.0
Total Scrubber	159.0	42.0	25.2	16.0
TOTAL	382.2	100.0	186.8	43.0

Courtesy of Mr. A. M. Miller, Scott Paper Company, Everett, Washington, 1968 (22).

Table 23. Maximum Operating Temperatures for Filter Media.

Material Used	Maximum Temperature °F
Filter Paper	150
Cellulose Ester	180
Cotton	190
Wool	235
Felt	250
Nylon	275
Paper Thimble	300
Orlon	350
Glass Wool	650
Glass Fiber	800
Alundum Thimble	1000

Data taken in part from ASME Performance Test Code No. 27, p. 11 [8].

Table 24. Particle Retention Characteristics of Clean Alundum Thimbles.

Thimble	Grade	Particle Retention ($\geq$ microns)
RA 84	Fine	0.1
RA 320	Fine	0.1
RA 360	Medium	5.0
RA 321	Medium	5.0
RA 225	Medium	5.0
RA 98	Coarse	20.0
RA 322	Coarse	20.0
RA 766	Coarse	30.0

Data supplied by: Mr. H. H. Haaland, Western Precipitation Corp., Los Angeles, California, 1969 [38].

Table 25. Effect of Flow Rate on Pressure Drop[a] and DOP Smoke Penetration[b] for Various Air Sampling Media.

Flow Rate, linear ft per min		AEC No. 1	CWS No. 6	HV 70 9 mil	HV 70 18 mil	Hurlbut Glass Paper	Whatman Chemical Filter Papers							S & S No. 604	Membrane Filters		AEC Mineral Filters			MSA Type "S"	MSA 1106
							No. 1	No. 4	No. 32	No. 40	No. 41	No. 41H	No. 42	No. 44	No. 50		"HA"	"AA"	Glass-Asbestos	All-Glass		
5	Penetration	0.022	0.015	5.3	0.47	0.001	73.0	84.0	3.7	32.0	89.0	93.0	5.0	6.5	30.0	93.00	0.002	0.002	0.029	0.008	45.0	—
	Pressure drop	0.7	0.67	1.1	1.2	1.05	1.9	0.5	7.2	2.5	0.35	0.45	8.7	7.7	9.5	0.35	5.4	2.2	0.7	0.75	0.3	—
10	Penetration	0.036	0.023	5.0	0.53	0.001	68.0	81.0	3.0	26.0	84.0	89.0	2.0	4.5	13.0	90.0	0.002	4.3	0.052	0.020	50.0	—
	Pressure drop	1.45	1.45	2.2	2.45	2.2	3.75	0.95	14.6	5.1	0.68	0.9	17.0	15.3	17.7	0.7	10.9	0.002	0.071	1.45	0.6	—
20	Penetration	0.045	0.04	3.5	0.65	0.003	43.0	77.0	0.45	16.0	77.0	81.0	0.75	1.4	3.0	85.0	0.01	0.002	0.071	0.038	52.0	—
	Pressure drop	2.9	2.9	4.6	4.9	4.4	7.7	1.95	27.3	10.7	1.35	1.9	33.0	28.6	35.2	1.45	21.6	8.5	3.0	3.05	1.05	—
28	Penetration	0.055	0.057	2.0	0.69	0.005	27.0	73.0	0.35	8.0	75.0	76.0	0.22	0.5	0.9	79.0	0.015	0.01	0.073	0.05	52.0	—
	Pressure drop	4.2	4.05	6.3	6.9	6.1	10.6	2.8	38.0	15.0	2.0	2.7	45.5	40.0	48.5	2.1	31.0	11.8	4.25	4.25	1.6	0.03
50	Penetration	0.045	0.045	1.7	0.45	0.005	11.0	62.0	0.3	15.0	67.0	65.0	0.35	0.2	0.15	67.0	0.015	0.015	0.08	0.051	51.0	4.4
	Pressure drop	6.7	7.5	9.4	13.0	10.8	19.6	5.5	69.4	25.3	2.8	5.5	81.0	71.0	86.0	3.9	59.6	24.5	7.8	7.7	3.0	—
100	Penetration	0.031	0.037	0.2	0.1	0.005	1.2	25.0	—	0.23	3.8	34.0	—	—	—	39.0	—	0.02	0.04	0.025	45.0	—
	Pressure drop	13.3	17.0	21.8	27.0	19.8	40.5	12.0	—	54.0	44.0	11.5	—	—	—	8.5	—	39.0	16.0	15.2	6.5	—
150	Penetration	0.021	0.018	0.1	0.025	0.003	0.3	18.1	—	—	29.0	21.0	—	—	—	18.0	—	—	0.018	0.013	34.0	—
	Pressure drop	22.5	25.5	34.5	38.2	32.5	60.0	—	—	—	12.5	17.0	—	—	—	15.0	—	—	26.7	25.0	10.8	—
200	Penetration	0.011	0.01	—	—	—	—	—	—	—	15.0	13.0	—	—	—	7.0	—	—	—	—	28.0	—
	Pressure drop	29.5	35.0	—	—	—	—	—	—	—	17.8	24.0	—	—	—	22.7	—	—	—	—	16.3	—

[a] Pressure drop in inches of water.
[b] DOP smoke penetration in percent (di-octyl phthalate particles 0.3 μ diameter, 50 micrograms per liter of air by count.
Data from Smith and Surprenant, "Properties of Various Filtering Media for Atmospheric Dust Sampling," ASTM, 1953 [39]. Courtesy of American Society for Testing and Materials.

Table 26. Operating Characteristics of Electrostatic Samplers.

System	Diameter inches	Length inches	Voltage volts	Maximum Flow Rate ft³/min	Superficial Velocity ft/min	Maximum Wt. Collected grams	Collection Efficiency % (1)
Western Precipitation	1.5	12.0	20,000	1.0	82	2.0	99.0
Mine Safety Appliance	—	—	15,000	3.0	—	1.0	99.0
Weyerhaeuser (Leonard)	1.7	6.0	18,000	1.0	60	2.0	99.5
Drummond	1.0	12.0	5,000	0.2	30	1.0	99.0
Stairmand	1.0	16.0	15,000	0.6	110	50.0	99.0
Del Electronics (Lippman)	1.7	12.0	—	27.0	1670	—	96.0
Kaiser (Rounds)	3.0(2)	12.0	15,000	1.5	30(3)	10.0	99.5

Notes: (1) The collection efficiency is in percent by weight for particles larger than approximately 0.5 microns in size.
(2) This is the diameter at the inlet to the truncated cone.
(3) The superficial velocity at the inlet (largest diameter) of the truncated cone.

Table 27. Operating Characteristics of Thermal Precipitators.

Device	Maximum Flow Rate ml/min	Temperature Gradient °C/cm	Gap cm	Collection Surface
Hot Wire	20	3000	0.025	Slide
Circular Plate	100	2500	0.038	Slide
Thermopositor	400	—	—	Slide
Tapered Axial	3000	2200	0.079	Foil
Konisampler	50	3000	—	Slide

Table 28. Properties of Collection Devices Used in Particulate Sampling.

Device	Flow Rate cfm	Pressure Drop in. Hg	Minimum Particle Size Collected microns	Maximum Temperature °F	Material Collected grams	Applications
Filtration						
Alundum Thimble	0.5-2.0	3-8	1.0-10.0	1000	1-2	Primary Internal Collector
Paper Thimble	0.5-2.0	1-4	0.5-5.0	300	1.0	Final Collector
Glass Fiber Pad	1-80	0.1-0.2	0.3	800	0.2	Final Collector
Membrane Filter	0.1-1.5	1-5	0.1-1.0	180	0.1	Final Collector and Particle Size
Bay Filter	10-80	0.1-0.3	1-5	250	100	Secondary Collector
Impingers						
Greenburg-Smith	0.5-1.0	2-8	1.0	150	1-2	Final Collector
Gas Washing Bottle	0.5-2.5	1-3	5.0	150	1-2	Intermediate Collector
Knockout Bottle	1.0-5.0	1-2	10.0	220	1-5	Primary Collector
Cyclone						
Small	0.3-1.0	0.3	3.0	1000	5	Primary Collector
Medium	3.0-20.0	0.3	5.0	1000	20	Primary Collector
Large	10.0-80.0	0.3	5.0	1000	100	Primary Collector
Electrostatic-Precipitator	1.0-5.0	0.05	0.20	300	1-2	Secondary Collector
Thermal Precipitator	0-0.1	0.05	0.01	500	0.05	Particle Identification & Sizing

CHAPTER 8

PARTICULATE SAMPLING TRAINS

I. SPECIFIC ORGANIZATIONS

A number of different sampling train arrangements have been employed for a variety of sources because no one system has yet been proved feasible to cover all situations. A variety of sampling train configurations have been utilized by several organizations and regulatory agencies for numerous total particulate sampling applications of source gas emissions. These employ most of the different types of collection devices previously mentioned, either singly or in combination. It is important to emphasize that most of these employ multistage collection systems to facilitate more efficient collection and to attempt to make a classification of particles by size. Most systems employ a pitot tube in parallel and adjacent to the sample probe to assist in maintaining isokinetic sampling conditions. A summary of the particulate collection systems employed by several organizations is listed in Table 29. Included are the types and arrangement of collection devices, the use of heated probes and sampling lines, the ranges of flow rates, pressure drops and nozzle sizes, and the means used to assure isokinetic sampling conditions. The types of sources where these sampling trains find application are also included. Values have been estimated for parameters not given by the organizations listed.

```
        Abbreviations for Tables 29A and 29B
AT      — Alundum Thimble
Cond.   — Moisture Condenser
Cycl.   — Glass Cyclone
DGF     — Disc Glass Fritted Filter (USPHS)
ESP     — Electrostatic Precipitator
GCT     — Glass Cloth Thimble
GFF     — Glass Fiber Filter
GSI     — Greenburg-Smith Impingers
GWB     — Gas Washing Bottle
GWF     — Glass Wool Packed Filter
KNB     — Glass Knockout Bottle (1.0 gal)
MF      — Membrane Filter
PF      — Paper Filter
PT      — Paper Thimble
WTM     — Wet Test Meter
```

Classification of sampling trains has been in terms of two flow rates for the sake of simplification. Low flow rate systems are useful over a wide range of conditions and may be used for hot-dry ducts, wet-saturated ducts, as well as high concentration sources. They are not limited by pressure drop characteristics which would necessitate requiring excessively large or cumbersome equipment. They are also useful for taking integrated samples for extended periods. However, high flow rate systems facilitate collection of large amounts of sample in short time intervals. This is very useful in taking samples during boiler sootblowing and other operations which last for only a short time, and also allows taking a number of samples covering an extended period. They are very useful for particulate sampling in boilers and incinerators but must be kept heated if moisture condensation poses a potential problem to prevent rupture or losses in the filtration device of the train. Details regarding individual sampling trains should be consulted in the appropriate references.

Diagrams of several commonly used particulate sampling trains are illustrated in Figures 60 through 65. The sampling train described by Smith [15] makes use of a connection ("umbilical") cord which allows the collection system to be located remotely from the flow measuring section and prime mover. The collection system and probe may then be located at the sample port on a stack while the heavy pump and meter are at ground level.

II. SPECIFIC CONTAMINANTS

A. Polynuclear Hydrocarbons:

Certain polynuclear aromatic hydrocarbon compounds have been found to act as carcinogenic agents, and are therefore of interest in air pollution studies. Hangebrauck [25] devised a sampling train for collecting particulate materials from combustion sources by withdrawal of flue gases at isokinetic velocities through a three-stage collection system. The gases first passed through a series of three bubblers immersed in an ice water bath to remove large particles and condensate; then consecutively through two U-tubes in a dry ice-alcohol bath, and a glass fiber filter. The total amount of ma-

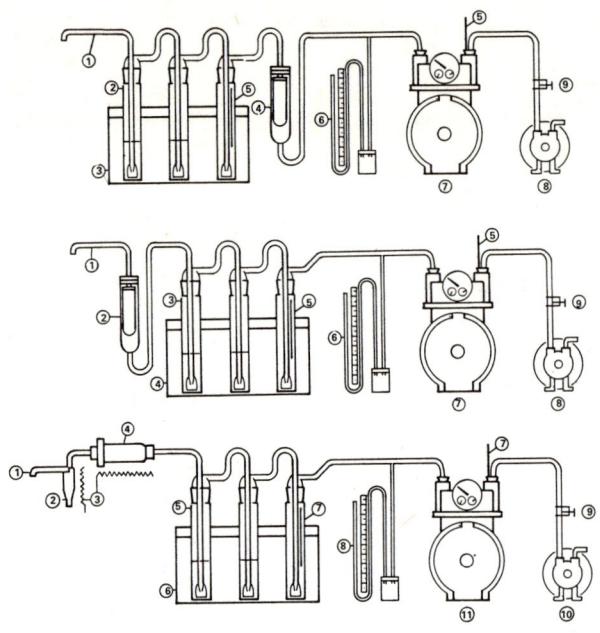

Fig. 60.

Courtesy of: Mr. Howard Devorkin, Los Angeles County Air Pollution Control District, 1969 (11).

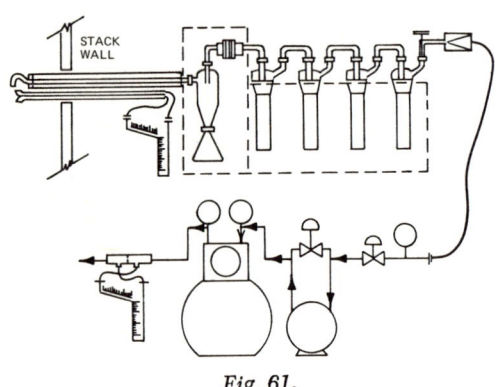

Fig. 61.

Courtesy of: Mr. Walter S. Smith, National Air Pollution Control Administration, 1969 (15).

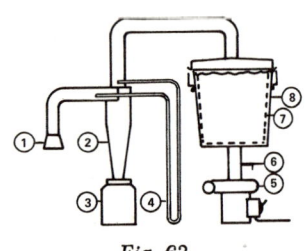

Fig. 63.

Courtesy of: Los Angeles County Air Pollution Control District (11).

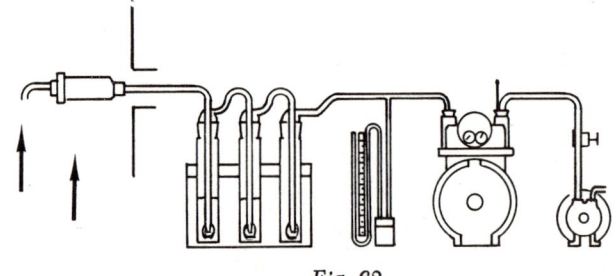

Fig. 62.

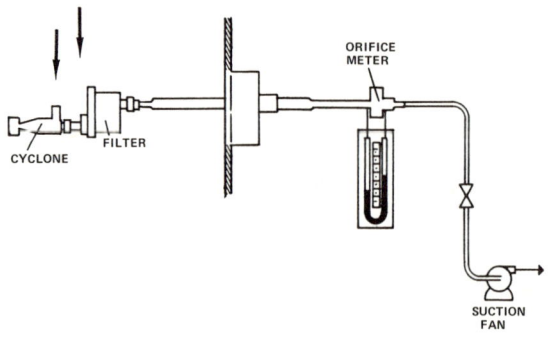

Fig. 64.

Courtesy of: British Standards Institution, B.S. 3405 (7).

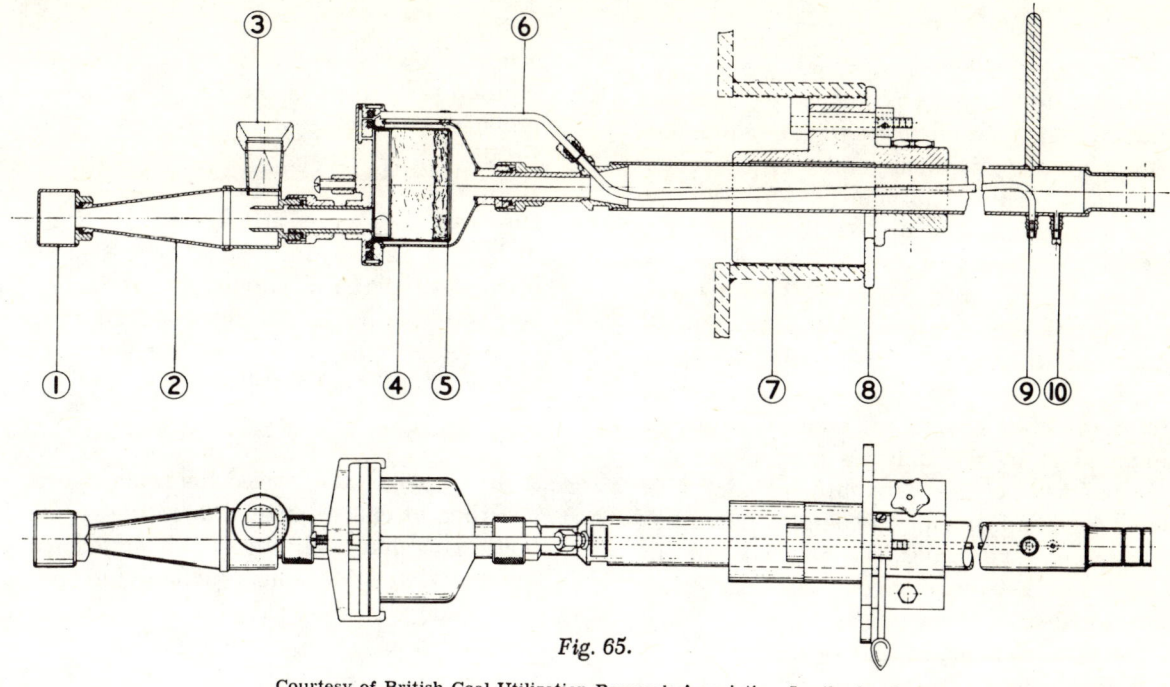

Fig. 65.

Courtesy of British Coal Utilization Research Association, Leatherhead, Surrey, England (24).

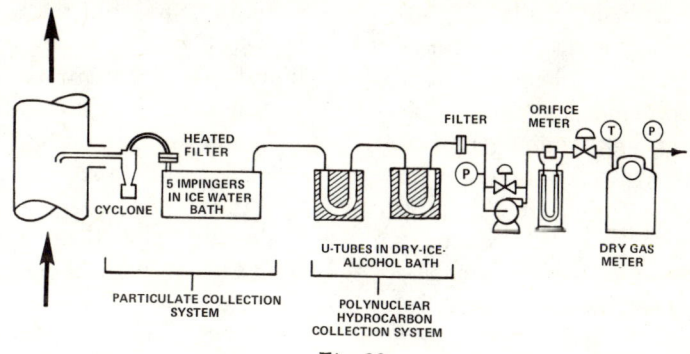

Fig. 66.

Courtesy of: *Journal of the Air Pollution Control Association*, Article by Hangebrauck, Von Lehmden, and Meeker, July 1964 (25).

terial collected was first determined and the contents of each stage then subjected to a benzene extraction. Subsequent identification and analysis of the individual compounds present was made by means of paper chromatography and ultraviolet plus visible range spectrophotometry [26]. A diagram of the sampling train used is shown in Figure 66.

Similar techniques have been used for sampling of organic materials from sources such as incinerators. Filtration followed by extraction in organic solvents and subsequent analysis has been extensively used by Tebbens [27] in analyzing flue gases from combustion processes. Both paper thimbles and membrane filters were used as the primary collection devices, and the samples were analyzed by extractions with diethyl ether, and dilute sodium hydroxide or hydrochloric acid solutions. Freeze-out techniques were also employed for organic liquids in a procedure similar to that previously described by Hangebrauck [28]. Analyses for individual organic constituents could then be made by ultraviolet or infrared spectrophotometry, or spectrofluorometry.

An additional method involves collection of organic materials in impingers containing distilled water or organic solvents. The liquid phase may then be analyzed for specific materials by a series of selective extractions employing ether, dilute caustic and acid [28]. It has been developed primarily for analysis of organic constituents from incinerator flue gases. Gas chromatography also provides a potentially useful method for analysis of these materials.

B. Fluoride Compounds:

Particulate and gaseous compounds are often emitted from phosphate fertilizer, aluminum, and steelmaking operations, and pose a potential threat to vegetation and animals. Most work to date has been devoted to detection of fluorine-containing materials in the ambient air and on the leaf surfaces of vegetation. However, two recent articles have been devoted to sampling for fluoride-containing materials for industrial sources. Ott and Hatchard [29] determined total fluoride emissions from sources in an aluminum plant by withdrawing flue gas samples at isokinetic velocities through a series of three Greenburg-Smith impingers containing five percent sodium hydroxide. The method did not differentiate between gaseous and particulate fluorine materials, however.

Dorsey and Kemnitz [30] devised a method for differentiating between gaseous and particulate fluorine-containing materials in tests run on sources in phosphate fertilizer plants. It employed withdrawal of the flue gas at the isokinetic rate through a heated cyclone and filter assembly to remove particulate materials followed by two Greenburg-Smith impingers containing 0.1 N sodium hydroxide solution to absorb the gaseous fluorides. The sample probe employs a stainless steel nozzle followed by a glass tubing insert upstream of the cyclone. The purpose of this glass insert is to convert the highly reactive hydrogen fluoride to the less reactive silicon tetrafluoride gas which could pass through the cyclone and filter without reaction, for subsequent absorption in the caustic solution. The reaction of HF with the sample probe is as follows:

$$4HF + SiO_2 \rightarrow SiF_4 + 2H_2O \quad (96)$$

However, it is necessary for the first impinger to have a straight tip instead of a tapered tip because of the susceptibility to plugging. This occurs because silica is reformed and precipitated by reaction of silicon tetrafluoride with the caustic solution as follows:

$$SiF_4 + 4NaOH \rightarrow SiO_2 + 4NaF + 2H_2O \quad (97)$$

A dry impinger was placed following the second absorber to collect any liquid carryover, and a bottle containing silica gel then added to protect the meter. Maximum gas flow for optimum conversion was about 0.8 cfm. A diagram of the sampling train is shown in Figure 67.

Following collection, the total amount of particulate materials collected is determined by weighing, and the fluoride materials extracted by washing with water or caustic solution. The liquid samples may then be distilled in the presence of perchloric acid to minimize possible interference caused by carbonate, sulfate, chloride, sodium, or other ions. Unfortunately, the distillation step is fairly time-consuming. Following distillation, samples may be analyzed by titration with thorium nitrate [31], or by the alizarin-zirconium [32] or zirconium-SPADNS [33] spectrophotometric methods. Specific references should be consulted regarding the suitability of these methods [34].

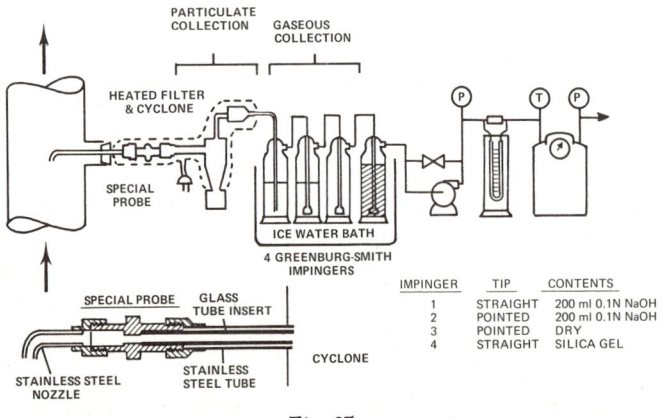

Fig. 67.

Courtesy of: *Journal of the Air Pollution Control Association*, Article by Dorsey and Kemnitz, January 1968 (30).

C. Mists, Tars and Liquid Droplets:

Liquid droplets pose a special problem in sampling because it is often difficult to separate them from other particles or gaseous constituents present in the flue gas. Several methods have been developed which employ filtration techniques for mist collection. The filters must normally be heated to prevent condensation of water as this would cause material to be washed out of the filter. Wet impingent techniques may also prove useful for collecting liquid materials if the presence of gases does not interfere with the analyses.

Nonaqueous mists such as oil or tar droplets may be collected by drawing flue gas through a filter paper. The materials present may be determined by weighing or by solvent extraction of the filter [18]. Use of a so-called "tar camera" to determine the collection efficiencies of removal devices for these materials employs the above principle. It provides a qualitative estimate of efficiency of the collector by visual observation of the filter. Normally, one cubic foot of gas is drawn through the filter on collector inlet samples, and 10 to 20 cubic feet on the exit [14]. Collection of these samples by wet impingent into distilled water is also possible.

Sampling for aqueous acid mists such as nitric and sulfuric acid droplets is similar to other types of particulate sampling. However, precautions must be taken to assure that the oxides of nitrogen or sulfur do not interfere with the respective analyses. This normally precludes the use of wet impingement techniques for collection; therefore filtration and centrifugal collection devices are commonly used. It is also necessary to heat the filtration and cyclone units to prevent condensation of moisture or formation of additional acid mist from sulfur trioxide [35]. The Public Health Service has made use of the same sampling train for sampling of both nitric and sulfuric acid mist in flue gases from these sources [36] [37]. Gases are withdrawn at isokinetic velocities through a glass probe into a glass cyclone to remove particles larger than three to five microns in size. The gases are then drawn through a two-layer glass fiber filter pad to remove the smaller particles. The entire section is heated to prevent moisture condensation. It is useful to place impingers containing water or another absorbing liquid following the filter to facilitate analyses of gaseous constituents and also to protect the flow measuring units from corrosive gases. A typical train for sampling acid mists is shown in Figure 68.

A second method employs withdrawal of flue gas samples at isokinetic velocities through a series of four Gooch funnels packed with inert asbestos wool to remove the acid mist. The entire arrangement is kept heated to prevent condensation of moisture [14]. Several additional methods for sampling sulfuric acid mist are described in the recent Public Health Service Bulletin [37].

The characteristics of the flue gas stream and the nature of the particles to be collected determine the sampling train to be used to collect particulate materials from a given source. The temperature and moisture content of the flue gas influence the location and types of collection devices used and the necessary precautions to be taken. These include the use of water-cooled probes to prevent particle combustion, and heated sampling lines and collection units to minimize condensation. The presence of sulfur dioxide plus other corrosive or otherwise hazardous substances in large concentrations in the flue requires use of a collection system such as scrubbers to prevent damage to the meter, and the use of safety equipment such as gas masks by operating personnel. The nature, size

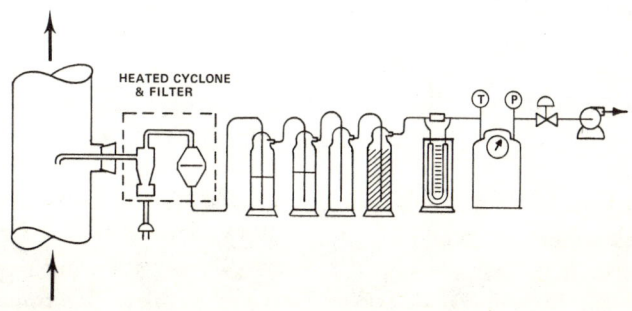

Fig. 68.

range, and concentration of particles in the gas stream influence the type, arrangement, and number of collection devices in use.

III. COMBUSTION SOURCES

A. Coal-Fired Boilers:

Coal-fired boilers emit both incombustible flyash and organic material (unburned and partially burned fuel) over a wide size range. Flue gas temperatures normally are about 450°F, with a moisture content of about ten percent by volume. Particulate concentrations may be as low as 0.10 grains per SDCF following high efficiency collectors, or as high as five grains per SDCF in the boiler exit stream, depending on the ash content. Early work by Hardie [38] and Caldwell [39] utilized several sampling trains in parallel at a given stack location to determine particulate loadings. The sampling trains employed high volume withdrawal of particles into glass cloth filter bags, as described in ASME Performance Test Code No. 27 [4]. Similar high volume systems have been used by Wasser [40], Allner [41] and Noss [42]. All these systems employ a cyclone followed by a cloth filter, and make use of pressure null balance for estimating isokinetic sampling conditions. A similar system employing an alundum thimble and a pressure null-balance is described in British Standard No. 893 [6]. In a recent study on coal-fired boilers, the conventional U.S. Public Health Service sampling train was used, employing a cyclone, filter and wet impingers [43]. Isokinetic sampling was maintained by placing an S-type pilot tube adjacent to the sample probe.

B. Oil-Fired Boilers:

The particulate concentrations emitted from oil firing are considerably lower than those for coal-burning, being on the order of 0.02 to 0.10 grains per SDCF. The materials are primarily organic in nature with both organic and acid mists present. The ash content of fuel oil is much lower than coal, and the resulting flyash emissions are thus far lower. The particles are often less than five microns in size and thus require use of efficient filtration or impingement collection devices. Collection devices should be located externally because of high temperatures inside the duct. Sampling methods for oil-fired boilers have been described by Stein [44], Schule [45], and the Los Angeles County Test Manual [11].

C. Wood-Fired Boilers:

Waste wood-fired boilers emit particulate matter similar to coal, except that the ash content of wood is normally less than coal, and there is more organic material present in the material collected. Temperatures of flue gases are about 400°F, the moisture content ranges from five to fifteen percent by volume. Particulate loadings range from 0.05 grains per SDCF following some collectors up to three grains per SDCF in some boiler exit gases, depending upon the ash content of the wood and the boiler firing conditions. Previous work is extremely limited from these sources, but Miller and Abrams [46] successfully used an internally-located alundum thimble followed by a series of gas washing bottles for collection. An alternative sampling train would involve use of a glass lined probe, an externally located alundum thimble to minimize combustion and a series of gas washing bottles.

D. Refuse Incinerators:

Refuse incinerators are of either single or multiple chamber design, and burn materials which are highly variable in composition, ash, and moisture contents. Particles emitted may be fairly large, thus requiring a minimum probe size of ½ to ¾ inch and use of a high flow rate sampling system (Q greater than 3.0 cfm). A significant portion of the particulate material collected may consist of volatile organic material; therefore, it is normally advisable to locate the collection devices external to the duct because of the high temperatures normally found in the flue gases (700-1800°F). It may also become necessary to make use of water-cooled probes to prevent volatilization or combustion of particulate materials deposited on the inner boundary of the sample probe and to avoid thermal damage to the probe. The former problem is particularly pronounced for pathological waste incinerators, where both the fuel burned and material collected are organic in nature. Concentrations are highly variable depending on the type of fuel burned and the firing conditions, ranging from 0.1 to 3.0 grains per SDCF.

Several sampling methods have been devised for particulate determinations in flue gases from refuse incinerators. Rehm [21] [47], the Incinerator Institute of America [22] [48], and Walker [49] have utilized a high-rate sampling system employing a single-stage cyclone followed by a cloth bag filter as a collection system, and a pressure null balance for estimating isokinetic sampling conditions. The system also provides for use of a water-cooled probe for flue gases where the stack temperature is greater than 900°F. It provides for external location of the collection system to minimize volatilization, and a high sampling rate, but the pressure null balance does not always represent isokinetic conditions, especially at velocities less than 600 feet per minute [50] [51]. The system has been illustrated in Figure 69.

The recent Public Health Service manual on incinerator testing [16] specifies use of the conventional USPHS sampling train [15] [52]. The collection system consists of a cyclone followed by a filter and a series of impingers containing distilled water. It is a low-flow-rate system with the collection system located externally, and isokinetic sampling maintained by a pitot tube located adjacent and in parallel to the sampling train. However, it may have problems in collecting large particles, as the low flow rate of 1.0 cfm does not allow use of large nozzles of greater than ½ inch diameter in most circumstances. The use of impingers as the final collection step is specified because of the possibility that organic substances present in the vapor state at stack temperatures may condense to form particulate materials when the plume is cooled to atmospheric conditions. From the section on definitions, particulate matter is defined as a substance being present as a liquid or solid aerosol (particle) at standard conditions of dry gas at 60°F and 29.92 inches of mercury.

A similar source test procedure is specified by Dade County, Florida, with the exception that an alundum thimble is placed internally within the stack and followed by a series of impingers on the outside [8]. The system has a maximum flow rate of one cfm and isokinetic sampling is assured by a pitot tube in parallel to the sample probe. The maximum probe diameter is only ½ inch, and location of the thimble internally may cause volatilization and combustion of the particulate material collected in the hot flue gases.

The procedure used by the Los Angeles County Air Pollution Control District makes use of the advantages of both of the previously mentioned systems, based on previous work by Kanter, Lunche, and Fudurich [53]. Gas is withdrawn at 60 to 100 cfm into a cyclone and then a cloth filter bag for particulate collection where isokinetic sampling is maintained by placing a pitot tube adjacent to the probe [11].

Yocom, Hein, and Nelson [54] devised a freeze-out method for sampling particulate and volatile organic materials being evolved from combustion in a single-chamber backyard incinerator. The sampling train consisted of five traps in series, two encased in an ice water bath, one at a controlled intermediate temperature, and the last two in a dry ice bath. The liquid condensed in each trap was first extracted and analyzed by the scheme previously discussed under organic materials, and also analyzed by infrared absorption spectrophotometry.

E. Wigwam Burners:

Wigwam burners are large single-chamber incinerators used for burning waste wood and bark from lumber mill operations. They are truncated cones with a hemispherical screen at the top, where wood is burned from the pile at the bottom and the flue gases pass outward through the screen. The fuel burned is nonuniform in quality and moisture content, and combustion conditions are rarely optimum. The par-

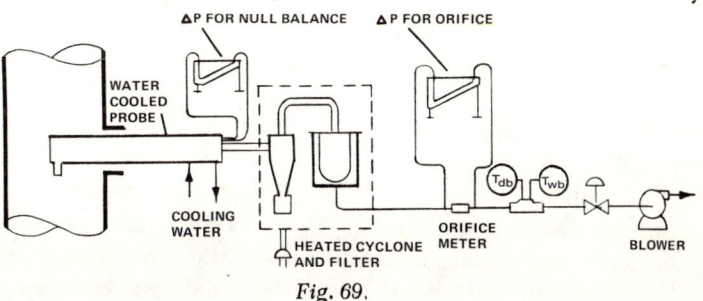

Fig. 69.

Courtesy of: The American Society of Mechanical Engineers (4) and the Incinerator Institute of America (22).

ticles emitted cover a wide size range; some are large flakes while others are fine partially-burned sawdust. Since the sampling location at the screen on top of the burner presents difficulties because of excessive temperatures, long probes must be extended from the ground. The velocity at the dome exit fluctuates rapidly because of the highly variable draft in the burner, and the wind also may create nonuniformities if the probe is not placed properly within the screen.

Three recent studies at Oregon State University have been the primary efforts towards developing sampling procedures for these sources. Boubel and Northcraft [23] used a modified high-rate sampling system (10 to 40 cfm) for sampling wigwam burner flue gases. Gas was withdrawn from the dome through large diameter nozzles of one to four inch diameter through a long external sample line, extending to the ground. It then passed through a cyclone followed by two metal screens placed in series. Isokinetic sampling was estimated by means of a pressure null balance system with a standard pitot tube placed adjacent to the probe. However, it was discovered that the velocity was not only highly variable but also below the 600 feet per minute level where substantial deviations from isokinetic conditions have been found to occur [50] [51].

Boubel and Wise [55] [56] performed a survey of a number of existing wigwam burners by sampling at a presumed velocity of 600 feet per minute, which had been found to be the approximate average dome exit velocity from the previous study [23]. Therefore, no attempt was made to obtain exact isokinetic sampling conditions during the tests. Sufficient 15 foot lengths of four inch aluminum conduit were used to place the one inch nozzle at the edge of the dome screen. Gas was withdrawn through the pipe and into a wire screen placed in an ice bath to cool the gas stream, then passed through two impingers, the first containing distilled water and the second dry. A membrane filter was placed following the impingers as a final collection stage.

Hyde [57] has developed a modification of the above system for sampling wigwam burner offgases by a thermal null balance system for estimating isokinetic sampling conditions. Here, two tubes of 7/8 inch diameter, each facing into the gas stream, have hot wire anemometers in their respective centers. One of these tubes is used to withdraw flue gas to the sampling train, while in the other gas only passes through a short length. If velocities at the two adjacent points (separated by about two inches between axial centers of the tubes) are presumed equal, the velocities will be equal if the signals from the hot wire anemometers are the same. The sample flow rate can be adjusted during the test to keep the signals in balance. Limited tests have shown the method to be effective for establishing isokinetic sampling conditions at velocities down to 200 feet per minute. The flue gas is drawn into the tube down to ground level, and into a cyclone for removal of the large particles. It is then passed through an alundum thimble into a series of three impingers containing distilled water in an ice bath. The cyclone and filter could be kept heated if necessary, and the sampling rate was between 0.5 and 1.5 cfm. A diagram of the system is shown in Figure 70.

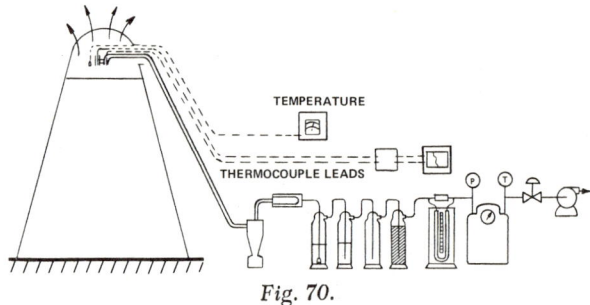

Fig. 70.

Courtesy of: Mr. Paul Hyde, Oregon State University, October 1968 (57).

F. Agricultural Field Burning:

Emissions of particulate and gaseous materials during agricultural field burning have been the subject of increasing interest in the past several years [58]. While information and sampling methods have been limited to date, three recent studies have provided estimates of particulate and gaseous materials. Meland and Boubel [59] used a high volume sampler located on an elevated tripod above a plot to be burned for estimating particulate emissions by drawing air up through the glass fiber filter pad.

Darley [60] made use of a flat table for burning gases in the laboratory, and placed a conical hood over the table to conduct the gases to a central duct located at the tip of the cone. Velocities were measured inside the circular duct, and gas samples were drawn through a probe into a filter located outside the duct, at about 50 cfm. Portions of the gas stream were conducted to continuous hydrocarbon, carbon monoxide, and nitrogen dioxide analyzers after passing through the filter. All quantities were measured during the entire length of a particular burn.

Boubel, Darley, and Schuck [61] used both methods in making further studies on emissions from agricultural field burning operations. Particulate emissions were monitored in the field, again by a high volume sampler on a tripod employing a glass fiber filter, and facing downward towards the rising gases from a plot being burned. An additional train drawing gas through a membrane filter was also used to make particle size range determinations. Plots of grass were harvested, and the material taken to the laboratory for tests on the burning table previously described, as gaseous analysis with continuous monitoring instruments in the field would be extremely difficult.

IV. INDUSTRIAL SOURCES

The numerous and widely varied industrial sources require use of variety of different sampling techniques for the different types of particulate materials present. Industrial sources to be discussed include pulp and paper, ferrous and nonferrous metallurgical operations. Individual references should be consulted for other sources, as these are intended as examples only. Relationships should be drawn for other sources with characteristics similar to those mentioned where applicable.

A. Pulp and Paper:

Particulate sampling from sources in Kraft and Sulfite process pulp and paper mills presents special problems because of the high moisture contents of 25 to 30 percent by volume normally present in the flue gases. Wet impingement sampling is particularly suitable for these sources because of the high solubility in water of the particulate materials present.

1. Lime Kilns:

Lime kilns in Kraft pulp mills are the least complicated source to sample because the particulate materials are nearly all inorganic sodium and calcium salts. Flue gases leaving the lime kilns upstream of the scrubber have high particulate loadings of ten to thirty grains per SDCF, and are at gas conditions of about 25 percent moisture and 500°F. Flue gas characteristics are similar to those following dry process cement kilns except that the moisture content is higher. The sampling train used normally employs a large probe of 3/8 or 1/2 inch and a coarse grade alundum thimble located internally to remove most of the particulate material present. It is followed by four gas washing bottles, three containing distilled water, and a paper thimble as a final collection device. Normally, it is not possible to collect samples for more than ten or fifteen minutes because of the high loadings present. Significant quantities of lime dust containing calcium are found in these samples.

In passing through the lime kiln scrubber, the flue gas is cooled to essentially saturated conditions of approximately 30 percent moisture by volume at a temperature of 160 to 165°F. The particulate concentration is reduced to between 0.05 and 0.4 grains per SDCF, depending on the inlet loading and the efficiency of the scrubber. The material present is mainly the finer sodium particles. A sampling train includes a knockout bottle, four gas washing bottles con-

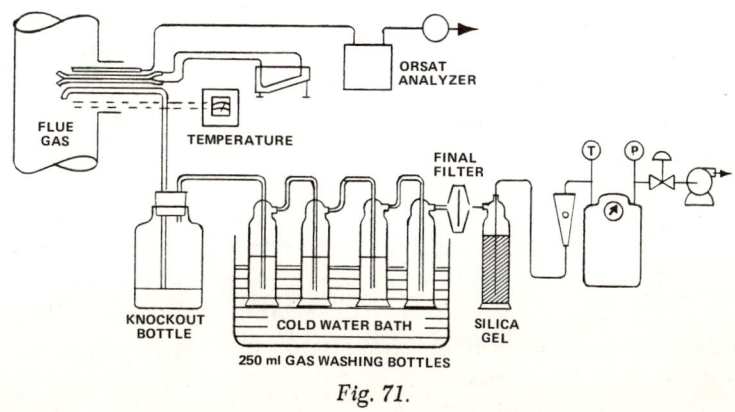

Fig. 71.

taining distilled water, a fifth dry one, and a paper thimble as a final collection stage. An alternative method employs use of a heated fine grade alundum thimble located internally following the scrubber [12]. A diagram of the sampling train is shown in Figure 71. A rough check for collector efficiency is to measure the gas temperature following the scrubber as it is directly proportional to the degree of gas-liquid contact and hence efficiency.

2. Kraft Recovery Furnaces:

Particulate material from Kraft mill recovery furnace flue gases is primarily sodium sulfate, with some sodium carbonate and organic material also present. The flue gas conditions presented depend on the location of sampling and the type of direct contact evaporation system employed for black liquor. Wet impingement methods have been found useful by several authors for sampling particulate emissions because of the high moisture contents normally present in flue gases from recovery furnaces [15] [62] [63].

Flue gases being emitted from recovery furnaces have high particulate loadings of five to ten grains per standard cubic foot at conditions of approximately 600°F and 22 percent moisture by volume. There are often large flake-like particles present, which may require use of a ½ inch or larger probe to obtain a representative sample, and possibly a high-volume sampling system if velocities in the duct exceed 2,500 fpm. This would employ a cyclone, a heated felt filter bag, and a heated glass fiber filter in series at flow rates of five to twenty cfm at meter conditions. Low-rate sampling trains which might be used include: (a) an alundum thimble located either internally or externally followed by a train of impingers, or (b) an externally located cyclone followed by a series of impingers. The purpose of using the thimble or cyclone is to selectively remove the larger particles upstream of the impingers to avoid wall losses or over-loading the impingers. The characteristics of the flue gases following either a cyclone or cascade—type direct-contact evaporator with a precipitator are similar to those of the furnace exit. The difference is that the particulate concentrations are reduced to three to five grains per standard dry cubic foot, the temperature reduced to 300°F, and the moisture content increased to 30.0 percent or more. The methods used for sampling these sources are similar to those for the furnace exit.

Particulate sampling following the precipitator is similar to that before the precipitator, except that the concentration is considerably lower, ranging from 0.05 to 0.40 grains per SDCF. It is normally necessary to take a sample of one hour or longer in duration to collect a sufficient amount of material. Normally, a low-rate sampling system is sufficient as the precipitator will remove most of the particles greater than five to ten microns in size. Several sampling systems have been used for these sources. First, an alundum thimble located internally followed by a series of impingers has been extensively used [13] [18]. It is possible to eliminate the thimble and use a glass-lined probe connected to a knockout bottle and a series of impingers. However, it should be cautioned, when using impingement into distilled water, of the possible interference of soluble acidic gases when sufficient quantities of these substances are present. These materials dissolve into solution to show up in ionic analyses and also in total solids determinations. Sulfur dioxide and sulfur trioxide from the flue gas form sulfite and sulfate ions, respectively, in solution, while nitrogen dioxide may form nitrate ion. Sodium and pH analyses plus ionic balances should be run on the solution collected. It is also possible to run simultaneous gaseous analyses and make corrections for these constituents in terms of chemical equilibrium considerations. Leonard [64] successfully used an electrostatic precipitator for sampling Kraft recovery furnace flue gases following a precipitator recovery unit. Evaluation of the unit did not indicate a detectable amount of sodium ion when a scrubber was placed following the unit. However, the maximum flow rate was limited to approximately 0.5 cfm at meter conditions.

Following secondary scrubbers, the gas stream is cooled to essentially saturated conditions at a temperature between 160°F and 170°F. Following fresh water-showered scrubbers, sampling trains employing wet impingement for particle collection are particularly suitable because they provide for collection of the liquid and particles without the necessity of heating the collection system. It is often desirable to place a one gallon knockout bottle ahead of the impingers to collect the major portion of the liquid and large particles with a minimum of change in pressure drop. This is because the cross-sectional area of the one gallon container is much greater than the impingers. Therefore, the same amount of liquid can be condensed with less pressure drop during the sampling period. A diagram of the sampling train has been illustrated in Figure 71.

Particulate sampling following venturi-type recovery units is complicated by the entrainment of black liquor droplets which contain sodium salts in the flue gases. Therefore, it is a good idea to locate sampling ports at least five to ten diameters downstream from the cyclone separator of the unit to minimize the possibility of droplet collection in the sampling train. This is particularly important in horizontal ducts. The flue gas condition is normally 185°F, saturated at approximately 40.0 percent moisture by volume, at concentrations of 0.4 to 1.0 grains per SDCF. The normal procedure is to use a scrubbing train similar to that previously described, or use a heated, internally-located alundum thimble as a first stage collector. Here the gas temperature is normally low enough to prevent volatilization or combustion of the particles present. However, liquor droplets may prove to be a problem.

3. MgO Recovery Furnaces:

Determination of particulate emissions from magnesium base sulfite recovery furnaces is perhaps the most difficult of any source in the pulp and paper industry. The material consists of fine magnesium oxide and hydroxide particles as well as sulfite and sulfate salts of magnesium, sodium, and calcium following the last liquid scrubbing stage. The situation is complicated by the presence of sulfur dioxide in concentrations of from 100 to more than 1000 ppm, and significant quantities of sulfur trioxide in a saturated gas stream at about 125°F. The presence of the sulfur oxide gases renders the otherwise desirable wet impingement techniques useless because of the absorption of the gases into the solution to form sulfite and sulfate ions. These would then be indicated as total solids in solution though not present as such in the flue gas, to yield selectively high results. These sources bear a marked similarity to flue gases from sulfuric acid manufacturing processes.

It is therefore necessary to make use of dry collection techniques, such as filtration, to collect particulate materials from the flue gases. In any case, the collection system must be kept heated to prevent condensation and subsequent losses by rupture or leaks in the filtration units. One possible sampling train employs a heated sample probe, with the gases then consecutively passing through an externally-located heated fine grade alundum thimble followed by a heated glass fiber filter. The external location is favored to allow time for entrained droplets to evaporate upstream of the filtration device. Certain types of filters cannot be used because they are subject to deterioration by acidic gases. The filters should be maintained at a temperature above the acid dew point of the gas stream to prevent formation of sulfuric acid mist from the sulfur trioxide present. Impingers containing sodium hydroxide and water should then be placed following the last filtration unit remove the acidic gases and protect the flowmeter from corrosion. A knockout bottle can be placed before the impingers to condense most of the moisture. Separate tests should be made for gaseous constituents because the flow-rate used is normally from 0.5 to 2.0 cfm in particulate sampling, which is much higher than normally used in gaseous collection. A diagram of the sampling train is shown in Figure 72.

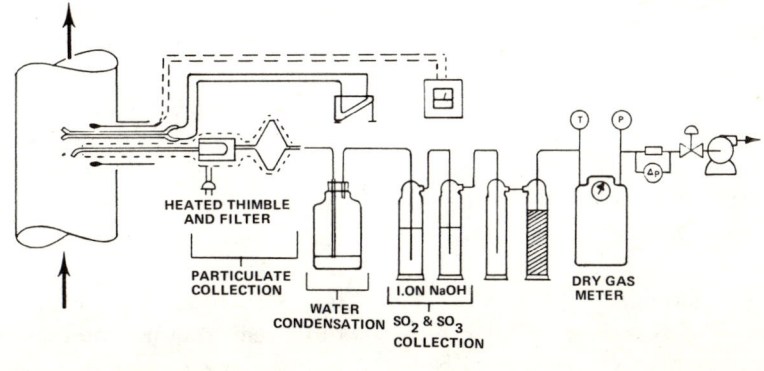

Fig. 72.

STAGE	COLLECTOR
1	Alundum Thimble
2	Glass Fiber Filter

The use of cyclones as primary collection stages is problematical [65]. They are able to remove mist droplets but these may consist of entrained water droplets containing dissolved sulfur dioxide and trioxide. Subsequent analyses may indicate these as total particulates, as previously discussed. Therefore, filtration is probably a more suitable technique but it should be emphasized that sampling techniques for these sources are in need of further development.

B. Metallurgical Operations:

Flue gases from steel mills, iron foundries, and other ferrous metallurgical operations contain iron oxide and other particulate materials which are particularly suitable for collection by filtration and electrostatic precipitation techniques. Filtration is useful because the flue gases are relatively low in moisture content, normally not exceeding five to ten percent by volume, except following scrubbing devices [66]. The particles also are suitable for efficient collection by electrostatic precipitation because of their low electrical resistivity. The major problems are that flue gases from certain sources are at high temperatures exceeding 800°F, and contain acidic sulfur dioxide and sulfur trioxide [9]. The high temperatures may necessitate use of water-cooled probes and external location of collection devices in sampling trains. The acidic gases prevent use of filtration devices which are subject to attack, such as membrane filters, or wet impingement techniques where these gases would be analyzed as particulate though not present as such in the flue gases. Particulate concentrations range from 0.02 to 0.05 grains per standard dry cubic foot from sources following high efficiency collection devices, to as high as seven to ten grains per standard dry cubic foot in the exit gases from basic oxygen furnaces [67].

1. Ferrous Metallurgy:

In one of the earliest articles on particulate sampling, Brady and Touzalin [68] made use of an externally located paper thimble for particle collection from blast furnace flue gases. They were perhaps the first to recognize the importance of isokinetic sampling rate in particulate measurement, and used a graphical method for its determination. Smith, Rounds, and Matoi [9] used a low-rate sampling system with an externally located alundum thimble packed with glass wool, followed by two dry impingers packed with glass beads to collect particulate materials. Schneider [69] also employed an alundum thimble but used a pressure null balance system for particulate measurement on open hearth furnaces. Weber [2] described the use of externally located alundum and paper thimbles for particle collection in flue gases from iron foundry cupolas as being superior to wet impingement for these sources because a dry sample is recovered for direct weighing. He also described methods for temperature measurement in hot flue gases. Rounds and Matoi [10] developed an electrostatic sampler using a truncated cone collecting tube, as previously described in the section on electrostatic precipitation collection devices. It is often useful to place a series of impingers following the particle collection devices to facilitate moisture content determination by condensation, and also to protect the flow meter from corrosion by removal of the acidic gases often present.

2. Nonferrous Metallurgy:

Information regarding specific sampling techniques for nonferrous metallurgical operations other than aluminum processing is extremely limited. Determination of particulate and gaseous fluoride emissions from aluminum and phosphate fertilizer operations has been previously discussed [29] [30]. Particulate materials are evolved from processing of ores of copper, lead, and zinc in the form of oxides of arsenic, antimony, lead, zinc, and copper [70]. Gas streams from these sources normally have relatively low moisture contents but have high sulfur dioxide levels of 10,000 to 60,000 parts per million by volume and are sometimes at high temperatures exceeding 1000°F. Sampling for particulate materials is sometimes complicated by the large size of smelter stacks, whose internal diameters may range from 20 up to 60 feet. This makes flow measurement and particulate sample traverses across ducts extremely difficult by necessitating use of several sample ports around the flue, combined with very long extended pitot tubes and sample probes, which are awkward and unwieldy on platforms at elevated locations. The chances for particulate losses may be increased by particle deposition on walls of the long sampling tubes, and by the fact that often the elevated flue gas temperatures may require external location of the collection devices used. Also low velocity profiles of below 600 feet per minute may prevail in portions of these large ducts, thus requiring use of thermal null balance devices to achieve isokinetic sampling, and large nozzle sizes of greater than one-half inch diameter, as well as possibly a high-rate sampling system. Possible solutions include the use of several sampling trains for simultaneous sampling at different ports in the stack, or making particulate determinations for the flue gas in the breeching prior to entering the stack. The problems associated with sampling in horizontal ducts and possible bends must be considered in these instances, however.

In sampling from these sources, it is first necessary to selectively remove the particles from the sampled gas stream in devices which do not react with the gaseous components. Normally, the collection system should be externally located at a great enough distance from the duct to allow cooling of the gas stream to about 300°F. One low-rate sampling train would employ a cyclone followed by an alundum thimble and glass fiber or glass disc filter, or an electrostatic precipitator. A series of impingers containing sodium hydroxide and distilled water should be placed following the collection devices to remove the sulfur dioxide and sulfur trioxide and protect the flow measuring device. A high-rate sampling system might employ a cyclone followed by a bag filter constructed of a material capable of withstanding the high sulfur dioxide concentrations, and a glass fiber filter as a final collection stage. It is noted that tests would have to be run on individual bag materials to determine their suitability in these applications. The excessive pressure drops associated with wet impingers preclude their use for SO_2 scrubbing in high-rate systems. Use of a calibrated orifice flow meter constructed of corrosion-resistant stainless steel as a flow measuring instrument, and a water aspirator as a prime mover, may be suitable for high-rate sampling systems where high sulfur dioxide concentrations may be encountered. Diagrams of these sampling trains are shown in Figure 73. It should also be noted that persons attempting to work on such sources should exercise extreme caution with regard to the presence of arsenic and high concentrations of sulfur dioxide in the flue gas, and keep all ports closed tight except when in use!

One early study on a copper smelter described the use of filtration techniques for particle collection. Harkins and Swain [71] used a series of fritted glass filter discs to collect the particulate materials, followed by a series of impingers containing sodium carbonate solution to absorb the sulfur dioxide present and protect the meter. A standard pitot tube was placed in parallel with the sample probe to provide for flow measurement during the sampling period.

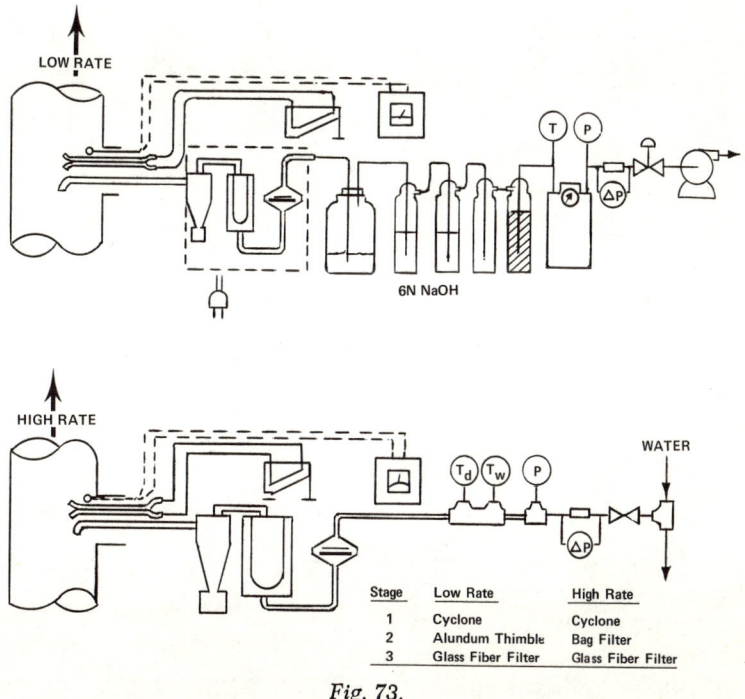

Fig. 73.

3. Burning Insulated Copper Wire:

Particulate materials are also emitted from furnaces used to burn insulation off of copper wire. Kaiser [72] used a heated, externally-located paper thimble, followed by a series of impingers containing distilled water to sample the exit gases from a scrubber following one of these furnaces. He found that approximately sixty percent by weight of the material collected in the train was found in the thimble, and about twenty percent in the probe washings. A summary of the results for three individual tests is listed in Table 30. It is noted that use of gravimetric weighing of a single thimble does not necessarily provide an accurate measurement of the particulate material present in a given duct. However, no attempt was made to classify the material by particle size.

V. PERFORMANCE TESTING

It is often necessary to determine the removal efficiency for air pollution control devices on the basis of total weight and also by particle size. This involves determination of both the inlet and outlet loadings of the collector in the flue gas. The collection efficiency of a control device is defined as follows:

$$E = \left[\frac{ER_{In} - ER_{Out}}{ER_{In}} \right] \times 100 \quad (98)$$

ER_{In} = Amount of particulate to collector in lb per hour
ER_{Out} = Amount of particulate from collector in lb per hour
E = Removal efficiency in percent

The two types of testing are laboratory and field studies [73]. Laboratory studies involve construction of prototype models of control equipment for tests in the laboratory. This gives a laboratory estimate of field conditions. The other type is to make an actual test on the unit. Further discussion will be confined to the second alternative.

The Industrial Gas Cleaning Institute [12] recommends a procedure for evaluating the collection efficiency of scrubbers, electrostatic precipitators and other duct collectors used as collection devices following process units which involves simultaneous sampling of inlet and outlet gas streams. The method makes use of filtration through an alundum thimble or other device or devices which meet the previously specified types in ASME Performance Test Code No. 27 [4]. Flue gases at the collector inlet are sampled by drawing them through an internally located alundum thimble because they are hot enough so that condensation does not occur. At the collector outlet, the gases are also drawn through an alundum thimble, (which must be heated) following a scrubber to prevent condensation in the thimble and potential losses. The sampling trains used are shown in Figure 74.

However, the method has three serious shortcomings. Internal location of the thimble at the collector inlet may create losses of material collected over an extended period during the test, through volatilization or combustion of some of the particles, depending on the gas temperature, particularly for those of an organic nature. This phenomenon may also occur in flue gases following a dry collector such as a cyclone, precipitator, or bag filter if the temperature is high enough. Second, the pores of the thimble progressively become filled during the run, causing it to become a more efficient collector. Third, the thimble may not be of the necessary collection efficiency, and there is no provision for a second stage collector as a safety factor. This may become a problem following the collection devices.

It should be emphasized that final collection stage should be added to insure an overall collection efficiency of 99.0 percent or greater of particles greater than a given size. This device should normally be placed following other collection devices so as to minimize the change

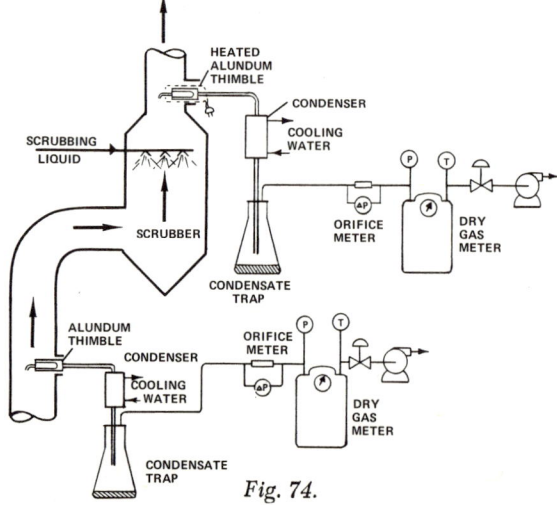

Fig. 74.

Courtesy of: Industrial Gas Cleaning Institute, Inc., Rye, New York, 1969 (12).

in net pore size during the test caused by progressive particle buildup. Membrane filters [1] are useful for this purpose as well as Greenburg-Smith impingers [4] [5], glass fiber filters [16] or paper thimbles [11]. The average porosities of these collectors are listed in Table 31. The average pore size should be specified for the purposes of the test. There is a definite need to make this a more rigorous standard in particulate sampling to provide for more uniform interpretation of results.

Membrane filters offer perhaps the most uniform method for interpreting results when used as final collection stages following primary and secondary collection devices. The filters should be of sufficient diameter to minimize pressure drops, and may need to be heated to prevent condensation. It may be necessary to select filter media capable of resisting the action acidic gases, particularly if impingers are not used. They may also need to be protected against flue gas temperature greater than 180°F. The use of cartridge thimble-type membrane filters of 1.0 micron pore size should be considered as final collection stages. The use of glass fiber filter paper is also feasible where it is desired to collect particles larger than 0.3 microns in size. The above provide possible methods for making interpretation of particulate sampling results much more uniform.

A recent publication by the Verein Deutschen Ingenieure (VDI or Union of German Engineering) describes the methodology for flow measurement, particulate sampling, and determination of the collection efficiencies of particulate cleaning devices [74]. Four different sampling trains are employed, with maximum sampling rates of 2.5, 6.0, 30, and 55 cubic feet per minute at meter conditions, respectively. These provide for determination of particulate concentrations in flue gases under a wide range of conditions, as previously discussed. A pitot tube is located in parallel to the sample probe to provide a continuous check for maintaining isokinetic conditions during the sampling period.

The individual sampling trains are constructed as follows. An orifice flow meter is used for measuring the rate of flow in the three systems with the highest flow rates (6, 30, and 55 cfm), with air or steam aspirators being used as the prime movers. The collection devices for the two highest flow rate systems (30 and 55 cfm) employ a cyclone followed by a filter. The system rated at six cfm employs a filter bag assembly for particle collection. The low flow rate system (2.5 cfm) employs a thimble followed by two impingers for particle collection, a dry gas meter for flow measurement, and a vacuum pump as a prime mover. Each of these trains is similar to others previously described.

The Gas Purification Institute of the Soviet Union [75] employs a sampling train similar in concept to the one utilized by the Industrial Gas Cleaning Institute in the United States for evaluating the collection efficiencies of particulate cleaning devices. The sampling train employs a heated, externally-located, conical paper filter packed with glass wool as the particle collection device. The filter holder is constructed of glass to facilitate visual observation of the filter during the sampling period. The probe and filter holder are electrically heated to a temperature of above 212°F to prevent condensation of moisture, but below 300°F to prevent thermal damage to the filter. An orifice meter is used to measure the rate of gas withdrawal from the flue gas. The system is illustrated in Figure 75.

The sampling procedure is as follows. Flue

1. SAMPLING TRAIN

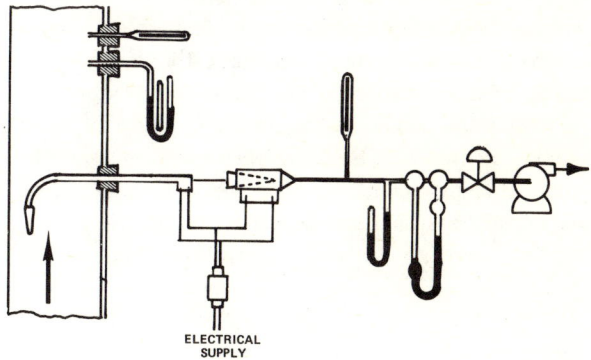

2. FILTER ASSEMBLY

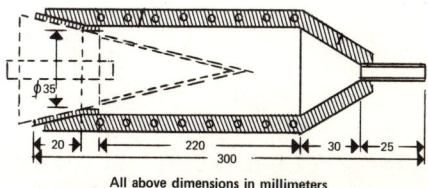

All above dimensions in millimeters

Fig. 75.

Excerpted from: Uzhov, V. N., NIIOGAZ (State Scientific Research Institute of Gas Purification for Industry and Sanitation), Moscow, USSR (75).

gas is withdrawn at an isokinetic rate of between one and two cubic feet per minute during a sampling period of from ten to twenty minutes. Normally, the sample is taken at the point of average velocity in the flue. The filter is weighed before and after the sampling period to ascertain the amount of particulate material collected from the gas stream withdrawn. Subsequent knowledge of the total gas volume sampled plus the amount of material collected allows calculation of the particulate concentration in the flue gas. Samples taken simultaneously from the respective inlet and outlet gas streams of a flue gas cleaning device provide for determination of its particulate collection efficiency.

VI. PARTICLE SIZE ANALYSIS

A. Problems Involved:

One of the most difficult tasks in particulate sampling is accurate determination of particle size distribution in flue gases. There are four reasons why particle size determination in flue gas streams is much more complicated than for ambient air measurements. First, the high temperatures of many flue gas streams necessitate external location of the collection device. This may induce error because of the tendency for the larger particles to deposit on the walls following bends because they are less able to undergo changes in direction than the gas molecules. Smaller particles may also tend to deposit on the inner walls of the probe outside of the stack because of thermal precipitation effects once the probe becomes cooled. Second, the high moisture contents of flue gases, particularly when following scrubbers, produce condensation of water droplets, which tends to wash out the particles and cause rupture or leakage through or past filtration devices. This necessitates heating of the probe and collection device to overcome any tendency for condensation. Third, the particulate concentrations in flue gases are on the order of 1,000 to 50,000 times higher than for ambient air. This requires extremely short periods so as to collect a small enough amount of material on a filter or other device to allow identification and sizing of individual particles. The fourth reason involves the tendency of certain particles to either agglomerate or break apart under certain conditions, thus making their characterization difficult if not impossible.

B. Particle Characteristics:

The size of particles is usually the property on which their classification is based, because it has such a significant effect upon their motion. However, additional properties to be considered are their respective specific gravity, shape, physical structure, surface properties, pore structure, and chemical composition. Recent texts by Cadle [76], and Orr and Dalla Valle [77] provide extensive discussions of these properties.

Classification of particles by size is the most common method in use, and several techniques are available for characterization on this basis. Classification may be made for particles of any given size, or any given size range, on the basis of number by count, by mass, by surface area, or by volume [78]. For particles of uniform shape and specific gravity, it is possible to convert from one base to another. However, this situation does not normally occur in reality, so extensive calibrations must normally be made for most applications. The statistical problem in particle classification is one of considerable magnitude, so it is common to interpret particle size distributions into finite size range increments. Size distributions are normally plotted on either a frequency or cumulative distribution as shown in Figure 76. The configuration of these curves is largely dependent on the size increment chosen,

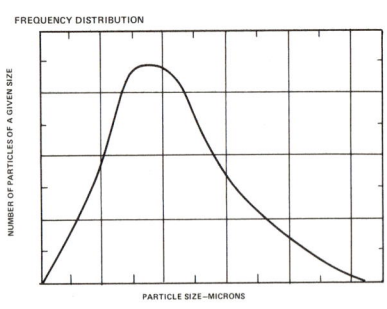

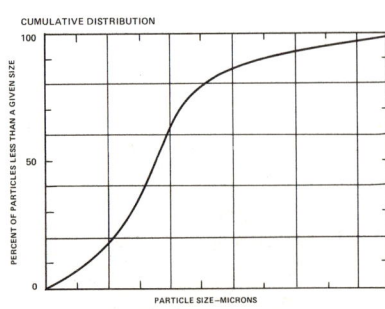

Fig. 76.

so great care must be exercised in the selection of individual increments. The most common representation of particle size distribution is on the basis of logarithmic probability, as shown in Figure 77. This type of plot is the basis for several relationships for representing particle size distribution, including the Rosin-Rammler [79], Roller, and Nukiyama equations [80]. The previously mentioned texts provide extensive discussions of the statistical ramifications of particle size distribution [76] [77] [78].

C. Measuring Techniques:

There are two approaches to particle size analysis. The first involves counting and measuring the size and other properties of each particle individually, which is direct, but can be very time-consuming. The second involves measurement of one or more properties of an aggregate of particles without attempting to determine each individually. It is necessary to consider differences in particle shape, structure, and configuration in this case, however [78]. Measurement techniques to be discussed include single and multiple stage impaction, centrifugal collection, and filtration, plus a brief discussion of continuous monitoring systems. A summary of techniques for particle size analysis is listed in Table 32.

Many of these techniques are primarily used for laboratory research, and are not designed for use in source testing applications. A recent article has made an extensive review of the methods available for particle size measurements in gas streams and other applications [80].

Microscopy is often used for identifying, sizing, and counting particles collected by impaction, filtration, or other techniques. The range of usefulness for the different types of microscopes is listed in Table 33.

The effect of particle diameter and specific gravity on terminal settling velocity based on Stokes' law, is shown in Figure 78 [98].

1. Greenburg-Smith Impinger:

Greenburg-Smith impingers employ direction of a high velocity jet against a plate surface to provide for separation of particles from a gas stream. Their greater momentum prevents the particles from negotiating changes in direction as easily as the gas molecules, and facilitates their collection. Greenburg and Bloomfield [81] [82] placed a dry glass slide on the impaction plate below the orifice of a dry impinger to collect particles for subsequent microscopic examination. In another case, they placed water in the impingers to provide for wet collection of particles which were insoluble in water. A subsequent evaluation of the collection characteristics of impingers was made by Hatch, Warren, and Drinker [83].

The method provides for collection of a large number and size range of particles in a single stage, but is subject to the following disadvan-

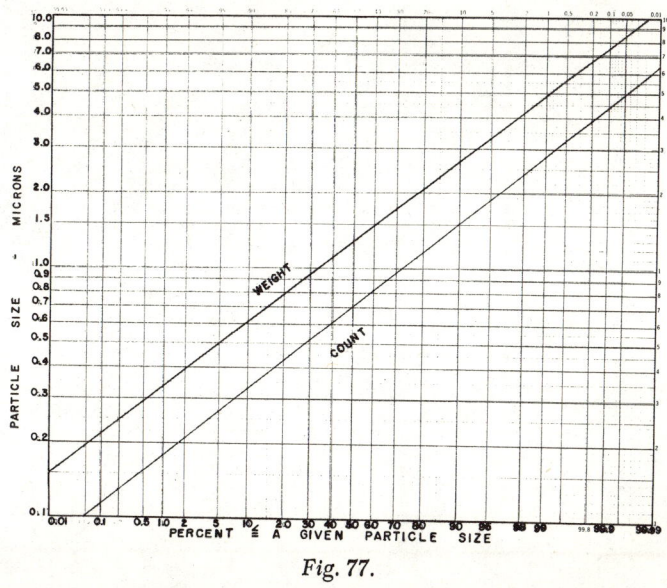

Fig. 77.

Note: Conventional Logarithmic-Probability Paper is used in making these plots.

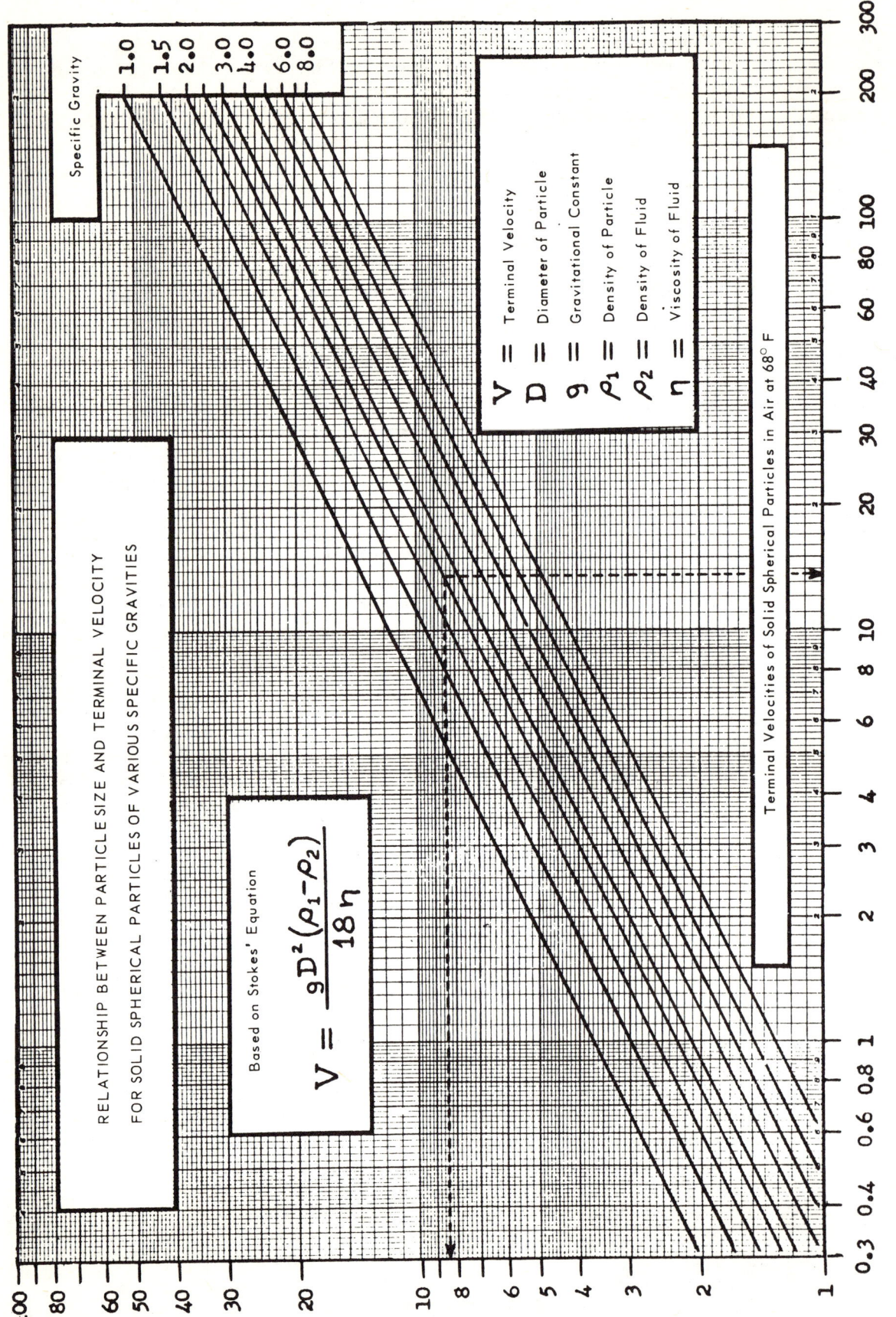

Fig. 78.

tages. When used for dry collection, the larger particles striking the collection plate may tend to break up, particularly if they are of a fragile structure, and smaller particles may become re-entrained in the gas stream and carried out in the exit. When used as a wet collector, the washing tends to collect the smaller particles but soluble materials tend to become dissolved and lose their identity. Therefore, the method is subject to serious limitations for particle size analysis.

2. Cascade Impactor:

Cascade impactors operate on the principle of impaction of particles on a dry surface, previously described for collection with the Greenburg-Smith impingers. The difference is that the device consists of two or more collection stages in which gas stream is forced to pass through a series of jets where particles are directed against collection slides placed perpendicular to the jets [84]. The jets are of progressively smaller areas, so that selectively smaller particles are collected in each successive stage. Ranz and Wong [85] made a theoretical and experimental study of particle collection by impaction, and found that experimental results for collection on individual slides were in close agreement with a mathematical relation used to describe collection efficiency. May [86] found that the average size particle retained on each of the four collection stages were twelve, four, two, and one microns, respectively. Lippman [87] found that the average size particles retained in four successive stages of a commercially available impactor were twelve, three, one, and one-half microns, respectively. Wilcox [88] designed a five-stage cascade impactor which was able to efficiently collect particles over a size range from 0.3 to 100 microns.

A recent article by Gussman and Gordon [89] describes the use of a cascade impactor for making particle size measurements in the flue gases from a horizontal aerosol tunnel, in attempting to simulate stack sampling conditions. The system illustrated in Figure 79 employs a stainless steel probe where the incoming gases are caused to flow toward the slide of the first collection stage without a change in direction. This causes the large particles which would otherwise be deposited on the probe walls, in rounding the initial bend, to be impacted onto the first slide. The gas stream is then directed out of the duct perpendicular to its initial direction of flow into a series of three successive cascade stages followed by a membrane filter as a final collection step. A critical orifice, and a vacuum pump then follow.

The device operates at a flow rate of 17.5 liters per minute [0.62 cfm], and has interchangeable probe sizes of ¼ inch or greater for sampling at velocities up to about 3,000 feet per minute. The slides are made of transparent Teflon, and coated with a liquid oil where subsequent microscopic analysis is used to determine the sizes of the individual particles. A supply of dry, filtered, compressed air is added continuously to the sample probe head adjacent to the first slide, to prevent entry of particles from the flue gas by acting as a blowback when the impactor is not actually in the sampling mode. Flow is maintained through the impactor system at all times, and the compressed air flow is stopped with a timer for a preset period, when it is desired to take a sample from the flue gas. Sampling times as short as 0.6 seconds can be used to allow collection of representative samples. The system is illustrated in Figure 79.

The approach used in the technique employing the air blowback to keep the system clean between samples, the location of the first stage inside the duct, and the multistage cascade impactor, is basically sound. However, selection of suitable sampling times, and use of the device in either extremely hot ducts or in saturated gas streams are problems which have not yet been solved. The possibility of an extended probe employing a water cooled head may be suitable for ducts at high flue gas temperatures. However, the device would prove bulky, and it would be necessary to coat the slides with a material of extremely low volatility when exposed to elevated temperatures, to enable collection of the particles. For flue gases saturated with water vapor, the sample probe head must be heated to prevent condensation. The impactor must not be heated to a temperature that will volatilize the oil placed on the plates or slides. This is particularly important if gravimetric analyses for particles by size are being made because volatilization of the oil will result in selectively low results.

Pilat, Ensor, and Bosch [90] [91] have recently developed an internally-located six stage cascade impactor for measuring the size distribution of particles between 0.5 and 20 microns

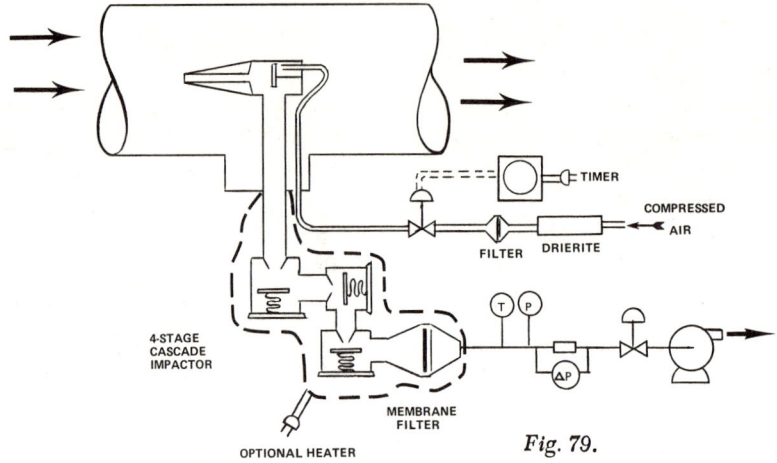

Fig. 79.

Courtesy of: *American Industrial Hygiene Association Journal*, Article by Gussman and Gordon, May-June 1966 (89).

in diameter, on both mass and count bases. The device has been successfully used to measure the size distribution of particles in the flue gas streams from both a coal-fired power boiler and a Kraft pulp mill recovery furnace. The sampling train employing the internally-located impactor is illustrated in Figure 80. A series of impingers placed in a cold water bath are used to condense moisture from the flue gas stream withdrawn, to protect the meter. The nozzle sizes used are either 3/16th or 1/4th inch inside diameter for sampling rates between 0.5 and 1.0 cfm at meter conditions, depending on the velocity in the flue gas at the point of sampling.

The time necessary for collection of an amount of material sufficient for accurate analyses varied inversely with particulate concentration. Low particulate concentrations of 0.1 grains per SDCF (60°F, 29.92 in. Hg, dry) or less required long sampling intervals of up to 30 minutes in order to collect amounts of material sufficient for accurate gravimetric analyses. Extended sampling periods are normally required in flue gas streams following high efficiency particulate collection devices and other sources where relatively low particulate concentrations are to be expected.

Particulate loadings of two to three grains per SDCF or greater necessitated sampling periods as short as 60 to 90 seconds. The large number of particles present per unit volume of gas resulted in rapid deposition of particles on the collection plates to form small mounds below the impingement jets. Extended sampling periods result in consecutive bombardment by particles, with subsequent breakup of the previously deposited particles, entrainment in the gas stream, and redeposition in following collection stages. The result of this mound formation and particle breakup is the attainment of an inaccurate and nonrepresentative sample. The problem is made especially acute in flue gas streams *upstream* of collection devices because the particulate concentration is relatively high [often greater than 1.0 gr/SDCF], and the stream also contains a significant propor-

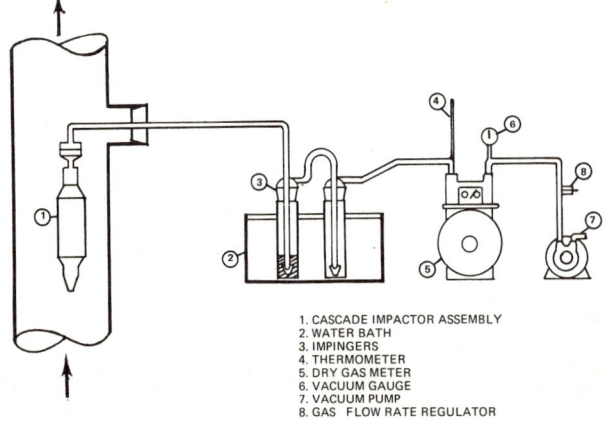

1. CASCADE IMPACTOR ASSEMBLY
2. WATER BATH
3. IMPINGERS
4. THERMOMETER
5. DRY GAS METER
6. VACUUM GAUGE
7. VACUUM PUMP
8. GAS FLOW RATE REGULATOR

Fig. 80.

Courtesy of: Dr. Michael J. Pilat and Mr. John C. Bosch, University of Washington, Seattle, Washington, 1970 (90) (91).

tion of relatively large particles which are subject to breakup and flaking off. As a result, it is highly advantageous to employ short sampling periods with the cascade impactor for flue gas streams containing relatively high particulate concentrations, such as on the inlet stream of control devices.

An approximate guide for the sampling period to be employed as a function of concentration is listed in Table 34.

The above tabulation is based on sampling rates of from 0.5 to 1.0 cfm at meter conditions with the above equipment, and should be used for estimating purposes only. To facilitate obtaining accurate results for particle size distribution, a batch particulate test should first be made on the source of interest to determine its approximate concentration.

The cascade impactor consists of six consecutive collection stages, each consisting of a perforated disc followed by a collection plate coated with an adhesive semi-liquid film. Flue gas is drawn through the probe into the impactor, passing through the holes in the discs, where the particles are caused to impinge on the plates. The particles impinge on and are collected by the adhesive material because their greater inertia renders them less able to negotiate changes in direction than the gas molecules. The gas stream then passes around the edge of the plate to the next disc. The holes in the discs are made progressively smaller in diameter so that selectively smaller particles are collected in each successive stage. The final collection stage consists of a sheet of glass fiber filter paper, which removes residual particles larger than 0.3 microns in size. It provides a means for gravimetric identification of the particles only, as the fibrous filter mat is not suitable for optical determinations. Membrane filters cannot normally be used for temperatures above 250°F because the acetate membrane is unable to withstand these conditions and head losses are more severe.

The impactor housing is three inches in diameter, approximately 15 inches long, and is constructed of lightweight aluminum to facilitate ease of handling in the field. The housing is constructed for ease of assembly and disassembly in the field. It is built with respect to the probe assembly so as to face axially into the gas stream, which eliminates the 90 degree bend upstream of the collection system that often causes deposition and subsequent losses of large particles. The coating medium for the plates is high temperature silicone stopcock grease, rated for temperature up to 400°F. Evaporation tests made by heating plates coated with measured amounts of this grease to varying temperatures in an oven showed negligible weight losses on the plates for temperatures up to 450°F [91].

The temperature range of usefulness for the cascade impactor has both upper and lower limitations. The lower limit is very important for flue gases containing high moisture contents near their dew points, such as following scrubbers. It is then necessary to enclose the impactor in an electrical heating jacket with an extended, heated inlet nozzle to evaporate any inlet water droplets, to prevent washouts. The upper temperature limit must be interpreted in terms of either the volatilization of organic materials in the collected particles, evaporation of the coating medium, or upper thermal stress limits of the metal used. For high temperature sources, a water-jacketed stainless steel housing is perhaps the most suitable unit. The optimum temperature range for most sources is to maintain the collector at between 250 and 350°F. However, saturated, low-temperature sources containing significant amounts of volatile organic materials present special problems. Removal of the device from the duct immediately after sample collection is a help in such circumstances.

Cascade impactors are potentially useful in source testing applications because they provide a means for classifying particles for both weight and count analysis on glass slides. However, there are several inherent disadvantages in using cascade impactors for particle size analyses in flue gases. Particles tend to break up at high velocities, making it necessary to use adhesives on the glass slides, and to restrict the sampling rate to about 0.5 cfm. The sampling time must also be extremely short, normally not more than 30 seconds in source measurements. The reason for the short sampling time is that there is a high probability of particles blowing off each collection stage. The chance of blowoff is increased as the sampling time increases, because mounds of particles tend to build up underneath the respective jets; individual particles then flake off from these mounds. Particle size analysis by count is extremely tedious because it involves light field microscopic counting of individual particles. Limited particle size frequency data results from use of cascade impactors because only four or five separation stages are available.

Additional problems when using cascade impactors for measuring particle size distribution occur for flue gases at either elevated temperatures or at saturation conditions. For hot flue gases, one is almost forced to locate the collector outside the duct to prevent burning up either the collector or the particles, or to prevent volatilization of the plate-coating material. There is then possible error in the larger particles depositing on the inner walls of the sample probe. In the case of saturated flue gases, water droplets may wash particles off the slides. This makes it necessary to heat the probe and impactor device to above the dew point to prevent condensation. It is desirable to locate the cascade impactor in the duct to minimize the potential chance for deposition of large particles on the walls of long sample probe tubes. However, internal location may require heavy and cumbersome heating or cooling systems for the impactor, with the possibility of having to cut sample ports of four inch or greater diameters.

3. Andersen Sampler:

The multistage impaction sampler developed by Andersen [92] is similar to the cascade impactor except that it has a series of six impaction plates arranged in series. The device is similar to a sieve tray column as there are a number of jets of varying sizes above each plate instead of just one. The jets become progressively smaller as the gas stream passes through the unit, so that the larger particles are collected first in the initial stages and the smaller particles in the later stages. A membrane filter is added following the device as a final collection stage.

A diagram of the device is shown in Figure 81. Analyses of particle sizes can be made either gravimetrically by weighing the trays before and after sampling, or by microscopic counting of individual particles collected on the plates. The size range of particles collected on each stage, as well as the diameter and the velocity through jets in each stage, are listed in Table 35. Velocities were computed based on a sampling rate of 1.0 cfm, which is the usual sampling rate.

Results show that, as in the case of the cascade impactor, there is a certain degree of overlap between stages with regard to the size of particles collected. These findings were verified in tests run by Flesch [93], who de-

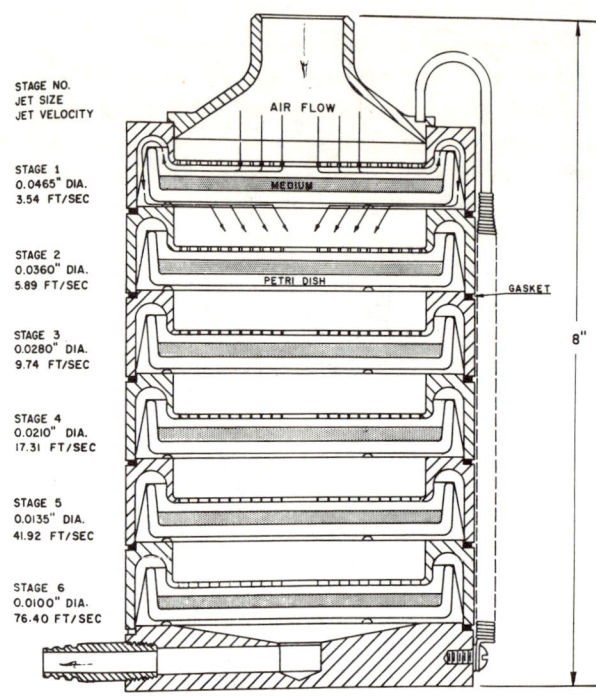

Fig. 81.

veloped a calibration procedure for the Andersen sampler using monodisperse aerosols. The device is potentially useful for particle size measurements from flue gases but so far its use has been mainly confined to ambient air, laboratory testing, and bacteriology work. It would have to be located external to the duct, perhaps following a sample probe as previously described, with cascade impactors. Heating of the sampler would be necessary if saturated gas streams were being sampled to prevent condensation. A relatively nonvolatile liquid would have to be used, particularly for hot flue gases when gravimetric determinations were being made.

4. Centrifugal Devices:

Cyclone collectors provide a means for collecting particles by causing the gas stream to rotate. The particles are less able to negotiate changes in direction than the gas molecules, and are subsequently collected by striking the walls. The efficiency of these devices can be greatly increased by having a rotating disc in the center, and the device can also act as a particle size classifier by forming it in the shape of the cone. Particles tend to deposit on the inner wall of the outer (stationary) surface, with larger particles near the apex and smaller particles further down, on the wider portion. A relatively device employing this principle was developed

Courtesy of: *Journal of Bacteriology*, Article by A. A. Andersen, November 1958 (92).

by Sawyer and Walton [94], who used a conical shaped unit with a sampling rate of 25 ml per minute with a head speed of 3,000 revolutions per minute. The device, known as the conifuge, provided for collection of particles from 30 down to 0.5 microns in size on two glass slides placed on opposite sides of the outer wall. A diagram of the unit is shown in Figure 82.

1. CONIFUGE (94):   2. SPECTROMETER (95):

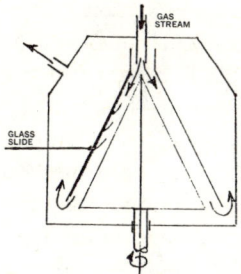 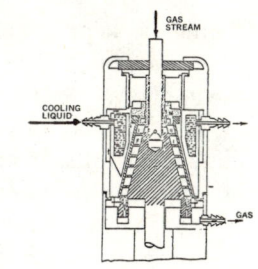

Fig. 82.

A more sophisticated device of similar design was constructed by Goetz and Kallai [95], who used a high speed centrifuge and a four channel helical groove assembly, also shown in Figure 82, to collect particles greater than 0.03 microns in size. Sampling rates of up to eight liters per minute have been used at rotation speeds of 24,000 revolutions per minute, which provided for forces acting to deposit the particles of up to 30,000 times that of gravity. The particles are precipitated on a foil enclosing the outer wall of the stationary [outer] section, and the device is essentially 100 percent efficient for all but the very small particles. However, there is no practicable use of this device in source test applications at present. A similar and more simplified system was designed by Kast [96].

A third unit was used by Okita and Yamashita [97], with limited success, for particle size measurement in Kraft pulp mill recovery furnace flue gases and oil mist particles. The device used was similar to the "Conifuge" previously described, except that both the inner and outer cones revolved in opposite directions. The device employed a sampling rate of 15 liters per minute at a rotation speed of 8,000 revolutions per minute for a sampling time of 15 minutes. Particles were deposited on two glass slides on the outer cone, and analyzed microscopically. The device had to be kept heated during the sampling period to prevent condensation.

The "Bahco" microparticle classifier employs a combination of elutriation and centrifugal force to measure particles according to their size distribution [98] [99]. A gas stream containing the particles is drawn into the unit and subjected to centrifugal force by rotation of the housing. The force of air movement (elutriation) is balanced by the rotary force (centrifugation), and particles larger than a certain size are caused to settle out. This fraction of larger particles is drawn to the outer part of the housing; collected, and weighed. The fine particles are carried on through the fan vanes. The particles which settle out depend on their size, density, and shape, as well as the air velocity through the unit. It is possible to change the terminal settling velocity (function of particle size) by changes in the air speed into nine different classifications for particles between one and 60 microns in size. However, it is necessary to have about ten grams of sample to obtain accurate weighings, and the unit must be calibrated with a test dust similar to the one being measured [100]. These requirements limit the usefulness of the technique.

5. Filtration:

Filtration involves deposition of particles on cylinders of a fibrous material, and relies on a combination of inertia, impaction, Brownian motion, and electrostatic attraction to remove particles from a gas stream. Fibrous filtration media cause the gas stream to flow through a fibrous mat, and particles are collected throughout the thickness of the filter. The devices provide for efficient collection of particles, but the fact that they are collected throughout the thickness of the filter makes particle size analysis by optical methods very difficult.

The membrane-type filter is most commonly used in particle size distribution measurements because it is basically a plastic type material with a large number of small and relatively uniform pores. Particles are thus collected mainly at the surface of the filter to facilitate direct observation with a microscope [101]. These materials are made from cellulose esters, nitrates, or other materials to provide for weight stability, chemical inertness, and good filtration characteristics. Microscopic examination then provides for identification and sizing of individual particles. However, membrane filters have a definite upper limit of temperature.

Limited use has been made of filtration techniques for particle size measurements. Katz and Smith [102] described a filtration method for particulate measurements by collection of dust on filter paper in mine air, which could be adapted for source sampling. Roesler [103] used a prefilter made of polyurethane foam, to remove large particles upstream of an Andersen sampler. The technique provides for removal of the larger particles so that particles in the so-called "respirable size range" can be collected and analyzed microscopically. The method is potentially useful for sources where it is desired to examine the small particles without many large particles being present.

Particle size measurements in flue gases following Kraft pulp mill recovery furnaces are extremely difficult because the gases are both hot and of high moisture content. Therefore, any cooling of the gas stream by withdrawal outside of the duct causes condensation of sufficient moisture to wash collected materials off of slides or filters. A method employing membrane filtration for particle size determination in Kraft recovery furnace flue gases has been recently developed by the National Council for Air and Stream Improvement [104] [105]. It employs preheating of the probe and filtration device to 200°F before beginning flue gas flow through the system. This temperature is above the dew point temperature, to prevent condensation, but below that which would cause thermal degradation of the filter paper.

The system used is shown in Figure 83, and consists of two parallel lines, one of which contains the membrane filter and the other a bypass valve. Flow resistance is set so as to be the same in both lines for the isokinetic sampling rate, prior to placing the probe into the duct. The device is set so that flow passes through the bypass line until it is desired to take a sample, then switched so that the gas stream passes through the membrane filter for collection of the particles. A flow rate of about one cfm is used, with a nozzle size of ¼ inch or greater, for a sampling time through the filter of 30 to 60 seconds. The bend in the sample probe should be as gentle as possible and the "Y" connection joining the bypass and filter lines should be of small enough angle to minimize deposition of large particles. The bypass and filter lines should be kept heated to prevent condensation by preheating the unit to 200°F before starting the gas flow.

The method proved successful in a limited number of tests. It provided for a rapid method of collection of samples for subsequent microscopic analysis. Problems were associated with the necessity for keeping the lines heated and deposition of the larger particles on the inner walls of the probe and at the "Y" connector. For flue gases at elevated temperatures it may also be limited by the maximum allowable temperature for the filter. This may be partially offset by short sampling periods and possible addition of cooling air as a diluent. The technology to overcome all these problems has not as yet been worked out to date.

6. Conclusions:

Particle size measurement in flue gases is perhaps the most difficult problem in source testing. Suitable methods for all sources are not as yet available, and considerable research will be needed to develop suitable techniques.

VII. REFERENCES

1. Clayton, G. D., ed., "Stack Sampling," Ch. 9 in *Air Pollution Manual, Part 1—Evaluation*, American Industrial Hygiene Association, Detroit, Michigan, 1961.
2. Weber, H. J., ed., "Air Pollution Problems of the

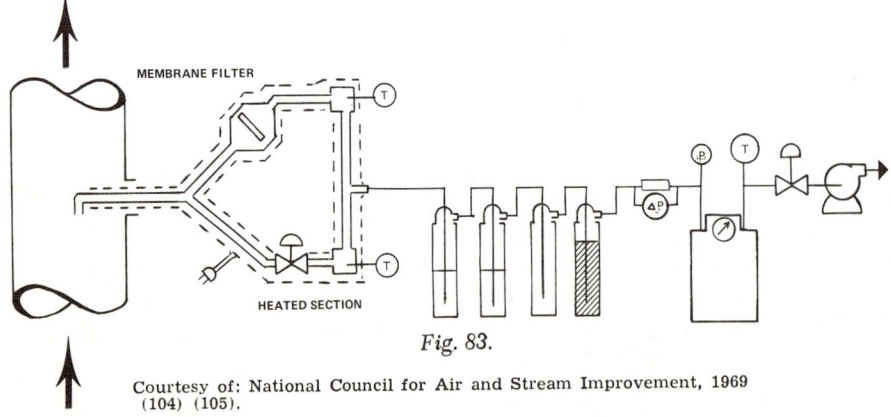

Fig. 83.

Courtesy of: National Council for Air and Stream Improvement, 1969 (104) (105).

Foundry Industry: Information Report No. 4—Instruments and Techniques for Measuring Air Pollution Emissions," Technical Manual No. 1, pp. 49-55, Air Pollution Control Association, 1963.
3. "Sampling and Analysis of Waste Gases and Particulate Matter," *Manual on the Disposal of Refinery Wastes,* Vol. 5, American Petroleum Institute, Inc., New York, New York, 1954.
4. "Determining Dust Concentration in a Gas Stream," Performance Test Code 27-1957, American Society of Mechanical Engineers, New York, New York, 1957.
5. Wolfe, E. A., ed., "Source Test Methods," Bay Area Air Pollution Control District, San Francisco, California, 1961.
6. "Method of Testing Dust Extraction from Plants and the Emission from Chimneys of Electric Power Stations," British Standard 893, British Standards Institution, London, England, 1940.
7. "Simplified Methods for Measurement of Grit and Dust Emissions from Chimneys," British Standard 3405, British Standards Institution, London, England, 1961.
8. Rivera, M. F., "Incinerator Source Testing Manual," Dade County Pollution Control Authority, Miami, Florida, September 1968.
9. Smith, S. H., Rounds, G. L., and Matoi, H. J., "Some Problems Encountered in Sampling Open Hearth Stacks," *Air Repair,* Vol. 3, No. 1, pp. 35-40, August 1953.
10. Rounds, G. L., and Matoi, H. J., "Electrostatic Sampler for Dust-Laden Gases," *Analytical Chemistry,* Vol. 27, No. 5, pp. 829-830, May 1955.
11. Devorkin, H., Chass, R. L., Fudurich, A., and Canter, C. V., Holmes, R. G., ed., "Source Testing Manual," County of Los Angeles, Air Pollution Control District, Los Angeles, California, 1965.
12. "Test Procedure for Gas Scrubbers," Publication No. 1, Wet Collectors Division, Industrial Gas Cleaning Institute, Inc., Rye, New York, November 1964.
13. "Manual for the Sampling and Analysis of Kraft Mill Recovery Stack Gases," National Council for Air and Stream Improvement, Atmospheric Pollution Technical Bulletin No. 14, New York, New York, October 20, 1960.
14. "Test Methods for Gas Volume and Dust Sampling Determinations," Research-Cottrell, Inc., Bound Brook, New Jersey, 1968.
15. Smith, W. S., Martin, R. M., Durst, D., and Hyland, R., "Stack Gas Sampling Improved and Simplified with New Equipment," Paper No. 67-119, Presented at the 60th Annual Meeting of the APCA, Cleveland, Ohio, June 13, 1967.
16. "Specifications for Incinerator Testing at Federal Facilities," U.S. Public Health Service, National Air Pollution Control Administration, Durham, North Carolina, October 1967.
17. Warshavsky, T. P., Kogan, L., Levin, E. D., and Shevchenko, N. S., "An Installation for the Determination of the Dust Concentration in Coke Oven Gases," *Coke and Chemistry,* (Moscow, USSR), No. 8, pp. 18-20, 1958, trans. by Levine, B.S. ed., in "USSR Literature on Air Pollution and Related Occupational Diseases," Vol. 5, U.S. Dept. of Commerce, Washington, D. C., 1961.
18. Haaland, H. H., ed., "Methods for Determination of Velocity, Volume, Dust and Mist Content of Gases," Bulletin WP-50, Western Precipitation Corp., Los Angeles, California, 1968.
19. "Flue Dust Sampling Instructions—Cyclone Method," Babcock and Wilcox Co., Alliance, Ohio, 1968.
20. "Sampling Studies on Emissions from Municipal Incinerators," Armour Research Foundation Project 832, Chicago, Illinois, 1961.
21. Rehm, F. R., "Test Methods for Determining Emission Characteristics of Incinerators," *Journal of the APCA,* Vol. 15, No. 3, pp. 131-135, March 1965.
22. "Incinerator Testing," Bulletin T-6, Incinerator Institute of America, Inc., New York, New York, 1968.
23. Boubel, R. W., Northcraft, M., Van Vliet, A., and Popovich, M., "Wood Waste Disposal and Utilization," Bulletin No. 39, Engineering Experiment Station, Oregon State University, August 1958.
24. Hawksley, P. G. W., Badzioch, S., and Blackett, J. H., *Measurement of Solids in Flue Gases,* British Coal Utilization Research Association, Leatherhead, Surrey, England.
25. Hangebrauck, R. P., VonLehmden, D. J., and Meeker, J. E., "Emission of Polynuclear Hydrocarbons and Other Pollutants from Heat Generation and Incineration Processes," *Journal of the APCA,* Vol. 14, No. 7, pp. 267-278, July 1964.
26. Tebbens, B. D., Thomas, J. F., and Mukai, M., "Hydrocarbon Synthesis in Combustion," *AMA Archives of Industrial Health,* Vol. 13, No. 6, pp. 567-573, June 1956.
27. Tebbens, B. D., Thomas, J. F., and Mukai, M., "Particulate Air Pollutants Resulting from Combination," Sanitary Engineering Research Laboratory, University of California, Berkeley, California, August 21, 1962.
28. Jacobs, M. B., *The Chemical Analysis of Air Pollutants,* Interscience Publishers, Inc., New York, New York, 1960.
29. Ott, R. R., and Hatchard, R. E., "Control of Fluoride Emissions at the Harvey Aluminum, Inc.—Soderberg Process Aluminum Reduction Mill," *Journal of the APCA,* Vol. 13, No. 9, pp. 437-443, September 1963.
30. Dorsey, J. A., and Kemnitz, D. A., "A Source Sampling Technique for Gaseous and Particulate Fluorides," *Journal of the APCA,* Vol. 18, No. 1, pp. 12-14, January 1968.
31. "Standard Sampling Techniques and Methods of Analysis for the Determination of Fluoride in Samples," Florida State Board of Health, Jacksonville, Florida, January 1966.
32. "Standard Methods for the Examination of Water and Wastewater," 12th edition, American Public Health Association, New York, New York, 1965.
33. Bellack, E., and Schouboe, P. J., "Rapid Photometric Determination of Fluoride in Water," *Analytical Chemistry,* Vol. 30, No. 12, pp. 2032-2034, December 1958.
34. Belcher, R., and West, T. S., "A Comparative Study of Some Lanthanum Chelates of Alizarin Complexone as Reagents for Fluoride," *Talanta,* Vol. 8, No. 12, pp. 853-870, December 1961.
35. Dooley, A., and Goodeve, C. F., "On Sulfuric Acid Mist," *Transactions of the Faraday Society,* Vol. 32, No. 8, pp. 1209-1218, August 1936.
36. "Atmospheric Emissions from Nitric Acid Manufacturing Processes," U.S. Public Health Service Publication No. 999-AP-27, Washington, D. C., 1966.
37. "Atmospheric Emissions from Sulfuric Acid Manufacturing Processes," U.S. Public Health Service

Publication No. 999-AP-13, Washington, D. C., 1965.
38. Hardie, P. H., "Cinder and Flyash Measurements," *Combustion*, Vol. 6, No. 9, pp. 10-16, March 1935.
39. Caldwell, W. E., "Characteristics of Large Hell Gate Direct-Fired Boiler Units," Transactions of the American Society of Mechanical Engineers, Vol. 56, No. 1, pp. 65-88, January 1934.
40. Wasser, R. W., "Sampling of Effluent Gases for Particulate Matter," *American Industrial Hygiene Association Journal*, Vol. 19, No. 6, pp. 469-476, November-December 1958.
41. Allner, W., "A New Method for the Determination of Dusts and Other Constituents in Industrial Gases," *Braunkohle*, Vol. 24, pp. 378-383, 399-402, 1925.
42. Noss, P., "Measuring Processes and Devices for Determination of Dust in Flowing Gas Streams," *Brennstoff-Warme-Kraft*, Vol. 4, No. 7, pp. 227-232, July 1952.
43. Gerstle, R., Cuffe, S., Orning, A., and Schwartz, C., "Air Pollutants from Coal-Fired Power Plants, Report No. 2," *Journal of the APCA*, Vol. 15, No. 2, pp. 59-64, February 1965.
44. Stein, J., Taylor, B., and Wade, G. R., "Techniques Employed in the Study of the Properties of Flue Gases from Large Oil-Fired Boilers," *Journal of the Institute of Fuel*, Vol. 34, No. 240, pp. 8-16, January 1961.
45. Schule, W., and Baum, F., "Investigation of Stack Emissions from Oil-Fired Boilers with Vaporizing Burners," *Staub*, Vol. 22, No. 5, pp. 200-203, May 1962.
46. Miller, A. M., Brown, J., and Abrams, R., "Applied Techniques for Analyses of Stack Emissions," Presented at the West Coast Regional Meeting of the National Council for Air and Stream Improvement, Portland, Oregon, October 2, 1968.
47. Rehm, F. R., "Incinerator Testing and Test Results," *Journal of the APCA*, Vol. 6, No. 4, pp. 199-204, February 1957.
48. "I.I.A., Incinerator Standards," Incinerator Institute of America, Inc., New York, New York, November 1968.
49. Walker, A. B., and Schmitz, F. W., "Characteristics of Emissions from Large, Mechanically-Stoked Municipal Incinerators," *Proceedings of the 1966 National Incinerator Conference*, pp. 64-73, American Society of Mechanical Engineers, New York, New York, 1966.
50. Hemeon, W. C. L., and Haines, G. F., "The Magnitude of Errors in Stack Dust Sampling," *Air Repair*, Vol. 4, No. 3, pp. 159-164, November 1954.
51. Dennis, R., Samples, W., Anderson, D., and Silverman, L., "Isokinetic Sampling Probes," *Industrial and Engineering Chemistry*, Vol. 49, No. 2, pp. 294-302, February 1957.
52. Rose, A. H., Stenburg, R. L., Corn, M., Horsely, R., Allen, D., and Kolp, P., "Air Pollution Effects of Incinerator Firing Practices and Combustion Air Distribution," *Journal of the APCA*, Vol. 8, No. 4, pp. 297-309, February 1959.
53. Kanter, C. V., Lunche, R. G., and Fudurich, A. P., "Techniques of Testing for Air Contaminants from Combustion Sources," *Journal of the APCA*, Vol. 6, No. 4, pp. 191-199, April 1957.
54. Yocom, J. E., Hein, G. M., and Nelson, H. W., "A Study of the Effluents from Backyard Incinerators," *Journal of the APCA*, Vol. 6, No. 2, pp. 84-89, August 1956.
55. Boubel, R. W., and Wise, K. R., "An Emission Sampling Probe Installed, Operated and Retrieved from Ground Level," *Journal of the APCA*, Vol. 18, No. 2, pp. 84-85, February 1968.
56. Boubel, R. W., "Particulate Emissions from Sawmill Waste Burners," Bulletin No. 42, Engineering Experiment Station, Oregon State University, Corvallis, Oregon, August 1968.
57. Hyde, P. E., "Particulate Sampling of Wigwam Waste Burners," Forest Research Laboratory, Oregon State University, Corvallis, Oregon, October 1968.
58. Bell, G. G., and Waggoner, N., "A Field Study of Air Pollution in the Lower Sacramento Valley During the 1960 Fall Season," California State Department of Public Health, Bureau of Air Sanitation, Berkeley, California, 66 pp., 1961.
59. Meland, B. R., and Boubel, R. W., "A Study of Field Burning Under Varying Environmental Conditions," *Journal of the APCA*, Vol. 16, No. 9, pp. 481-484, September 1966.
60. Darley, E., Burleson, F., Mater, E., Midelleton, J., and Osterli, V., "Contribution of Burning Agricultural Wastes to Photochemical Air Pollution," *Journal of the APCA*, Vol. 16, No. 2, pp. 685-690, December 1966.
61. Boubel, R. W., Darley, E. F., and Schuck, E. A., "Emissions from Burning Grass Stubble and Straw," Paper 66-28, Presented at the 61st Annual Meeting of the Air Pollution Control Association, St. Paul, Minnesota, June 25, 1968.
62. Banciu, I., "Determination of Sodium Salt Losses in the Recovery Boiler of a Sulfate Pulp Mill," *Celuloza Hirtie* (Bucharest), Vol. 15, No. 7, pp. 253-257, July 1966.
63. Collins, T. T., "Sampling and Analyzing of Sulfate Recovery Furnace Stack Gases," *Paper Industry and Paper World*, Vol. 29, No. 1, pp. 1437-1439, January 1948.
64. Leonard, J. S., "Stack Emission Sampling," *Tappi*, Vol. 49, No. 10, pp. 84A-88A, October 1966.
65. Goodeve, C. F., "The Removal of Mist by Centrifugal Methods," *Transactions of the Faraday Society*, Vol. 32, No. 8, pp. 1218-1221, August 1936.
66. Arbogast, A. H., "The Quantitative Determination of Dust in Gas," *Iron and Steel Engineer*, Vol. 35, No. 10, pp. 82-90, October 1948.
67. Schueneman, J. J., High, M. D., and Bye, W. E., "Air Pollution Aspects of the Iron and Steel Industry," U.S. Public Health Service Publication No. 999-AP-1, Cincinnati, Ohio, June 1963.
68. Brady, W., and Touzalin, L. A., "The Determination of Dust in Blast Furnace Gas," *The Journal of Industrial and Engineering Chemistry*, Vol. 3, No. 9, pp. 662-670, September 1911.
69. Schneider, R. L., "Engineering, Operation and Maintenance of Electrostatic Precipitators on Open Hearth Furnaces," *Journal of the APCA* Vol. 13, No. 8, pp. 348-353, August 1963.
70. Sheehy, J. P., and Lindstrom, C. A., "Mineral and Metallurgical Industry Emissions," Ch. 21 in Stern, A. C., ed., Air Pollution, Vol. 2, 1st ed., Academic Press, Inc., New York, New York, 1963.
71. Harkens, W. D., and Swain, R. E., "The Determination of Arsenic and Other Solid Constituents of Smelter Smoke," *Journal of the American Chemical Society*, Vol. 10, pp. 970-998, 1907.
72. Kaiser, E. R., and Toleiss, J., "Control of Air Pollution from the Burning of Insulated Copper

Wire," *Journal of the APCA,* Vol. 13, No. 1, pp. 5-11, January 1963.
73. Caplan, K. J., ed., "Performance Testing," Ch. 12 in *Air Pollution Manual Part 2—Control Equipment,* American Industrial Hygiene Association, Detroit, Michigan, 1968.
74. "Performance Measurements at Dust Collectors," VDI Standard 2066, translated and available as CFSTI 68-50469/3, U.S. Department of Commerce, Springfield, Virginia, 1966.
75. Uzhov, V. N., "Sanitary Protection of Atmospheric Air: Purification of Industrial Discharge Gases from Suspended Substances," Institute for Making Gas Purification Equipment, Moscow, U.S.S.R., translated by Levine, B. S., Publication No. OTS 59-21092, U.S. Department of Commerce, Office of Technical Services, Washington, D. C., 1959.
76. Cadle, R. D., *Particle Size Determination,* Interscience Publishers, Inc., New York, New York, 1955.
77. Orr, C., and DallaValle, J. M., *Fine Particle Measurement,* MacMillan and Company, New York, New York, 1959.
78. Lapple, C. E., "Particulate Size Analyses and Analyzers," *Chemical Engineering,* Vol. 75, No. 11, pp. 149-156, May 20, 1968.
79. Rosin, P., and Rammler, E., "The Practices of Flue Dust Measurement," *Braunkohle,* Vol. 34, pp. 505-546, 1935.
80. "Classification of Methods for Determining Particle Size, A Review," Analytical Methods Committee, *The Analyst,* Vol. 88, No. 3, pp. 156-187, March 1963.
81. Greenburg, L. G., and Smith, G. W., "A New Instrument for Sampling Aerosol Dust," U.S. Bureau of Mines Report of Investigations No. 2642, Washington, D. C., 1922.
82. Greenburg, L. G., and Bloomfield, J., "The Impinger Dust Sampling Apparatus as Used by the United States Public Health Service, *Public Health Reports,* Vol. 47, No. 12, pp. 654-675, March 18, 1932.
83. Hatch, T., Warren, H., and Drinker, P., "Modified Form of the Greenburg—Smith Impinger for Field Use, With a Study of Its Operating Characteristics," *Journal of Industrial Hygiene and Toxicology,* Vol. 14, No. 5, pp. 301-311, May 1932.
84. Sonkin, L. S., "A Modified Cascade Impactor. A Device for Sampling and Sizing Aerosols Below One Micron in Diameter," *Journal of Industrial Hygiene and Toxicology,* Vol. 28, No. 6, pp. 269-272, June 1946.
85. Ranz, W. E., and Wong, J. B., "Jet Impactors for Determining the Particle Size Distribution of Aerosols," *Archives of Industrial Health and Occupational Medicine,* Vol. 5, No. 5, pp. 464-477, May 1952.
86. May, K. R., "The Cascade Impactor: An Instrument for Sampling Coarse Aerosols," *Journal of Scientific Instruments,* Vol. 22, No. 10, pp. 187-195, October 1945.
87. Lippman, M., "Review of Cascade Impactors for Particle Size Analysis and a New Calibration for the Casella Cascade Impactor," *American Industrial Hygiene Association Journal,* Vol. 20, No. 5, pp. 406-416, October 1959.
88. Wilcox, J. D., "Design of a New Five-Stage Cascade Impactor," *Archives of Industrial Hygiene and Occupational Medicine,* Vol. 7, No. 5, pp. 376-382, May 1953.
89. Gussman, R. A. and Gordon, D., "Notes on the Modification and Use of a Cascade Impactor for Sampling in Ducts," *Journal of the American Industrial Hygiene Association,* Vol. 27, No. 3, pp. 252-255, May—June 1966.
90. Bosch, J. C., "Size Distribution of Aerosols Emitted from a Kraft Mill Recovery Furnace," Masters Thesis Submitted to the Department of Chemical Engineering of Washington, Seattle, Washington, 1969.
91. Pilat, M. J., Ensor, D. S., and Bosch, J. C., "Source Test Cascade Impactor," Submitted for Publication in *Atmospheric Environment,* December 3, 1969.
92. Andersen, A. A., "New Sampler for the Collection, Sizing, and Enumeration of Viable Airborne Particles," *Journal of Bacteriology,* Vol. 76, No. 10, pp. 471-490, November 1958.
93. Flesch, J., Norris, C., and Nugent, A., "Calibrating Particulate Air Samplers with Monodisperse Aerosols: Application to the Andersen Cascade Impactor," *American Industrial Hygiene Association Journal,* Vol. 28, No. 6, pp. 507-516, November—December 1967.
94. Sawyer, K. F., and Walton, W. H., "The Conifuge—A Size Separating Sampling Device for Airborne Particles," *Journal of Scientific Instruments,* Vol. 27, No. 10, pp. 272-276, October 1950.
95. Goetz, A., and Kallai, T., "Instrumentation for Determining Size and Mass Distribution of Submicron Particles," *Journal of the APCA,* Vol. 12, No. 10, pp. 479-486, October 1962.
96. Kast, W., "New Dust Measuring Apparatus for Rapid Determination of Dust Concentration and Particle Size Distribution," *Staub,* Vol. 21, No. 4, pp. 215-223, April 1961.
97. Okita, T., and Yamashita, S., "Measurement of Size Distribution of Sodium Sulfate Particles in the Flue Gas of a Kraft Pulp Mill," *Bulletin of the Institute of Public Health* (Japan), Vol. 16, No. 1, pp. 41-44, January 1967.
98. "Determining the Properties of Fine Particulate Matter," Performance Test Code 28-1965, American Society of Mechanical Engineers, New York, New York, 1965.
99. "Micro Particles Classified Centrifugally," Bulletin SL-52, Harry W. Dietert Co., Detroit, Michigan, 1968.
100. Todd, W. F., Hagen, J. E., and Spaite, P. W., "Test Dust Preparation and Evaluation," U.S. Public Health Service, Division of Air Pollution, Cincinnati, Ohio, 1964.
101. Goetz, A., "Application of Molecular Filter Membranes to the Analyses of Aerosols," *American Journal of Public Health,* Vol. 43, No. 2, pp. 150-159, February 1953.
102. Katz, S. H., and Smith, G. W., "Determination of Suspended Matter in Gas by Collection on Filter Paper," U.S. Bureau of Mines Report of Investigations, No. 2378, 1922.
103. Roesler, J. F., "Application of Polyurethane Foam Filters for Respirable Dust Separation," *Journal of the APCA,* Vol. 16, No. 1, pp. 30-34, January 1966.
104. "Manual for Counting and Sizing Particles from Kraft Recovery Furnaces," Atmospheric Pollution Technical Bulletin No. 19, National Council for Air and Stream Improvement, New York, New York, July 1963.
105. Walker, C. G., "Counting and Sizing Particles from Kraft Recovery Furnaces," Masters Thesis Submitted to the Department of Environmental Engineering University of Florida, Gainesville, Florida, 1963.

Table 29A. Particulate Sampling Trains In Use. A. Low Flow Rate Systems Where Q Is Less Than 3.0 cfm.

Organization	Number	Operating Characteristics				Collection Devices						Sample Probe		Flow Measurement	
		Flow Range cfm	Pressure Drop in. Hg.	Maximum Temperature °F	Nozzle Size Inches	Primary		Secondary		Tertiary		Material Used	Heated Or Cooled	Dry Gas Meter	Orifice Meter
						Type	Location	Type	No. Stages	Type	Min. Temp. °F				
American Industrial Hygiene Association		0.3-1.5	2-10	400	1/4-1/2	—	—	GSI	3	MF	130	St. St.	X	X	—
Air Poll. Cont. Assoc.—Foundry Committee		0.5-2.5	3-6	1800	1/4-1/2	PT.	Ext.	DGF	1	Cond.	60	Iron	X	X	X
American Petroleum Institute		0.5-2.0	3-8	400	1/8-1/2	PT.	Int.	Cond.	1	—	—	St. St.	—	X	—
		0.5-2.0	4-10	750	1/8-1/2	GCT	Int.	Cond.	1	—	—	St. St.	—	X	—
American Society of Mechanical Engineers		0.5-2.0	5-10	1000	1/4-1/2	AT	Int.	—	—	—	—	St. St.	—	X	X
Bay Area Air Poll. Cont. Dist.	1	0.5-2.0	4-10	800	3/16-1/2	AT	Int.	GFF	1	GWB	60	St. St.	X	X	—
	2	0.5-2.0	3-8	250	3/16-1/2	GWF	Int.	—	—	GWB	60	St. St.	—	X	—
British Standards	893	0.5-4.0	5-10	750	1/4-1	AT	Int.	—	—	GWB	60	Mild Steel	—	—	X
	3405	0.5-3.0	3-10	1250	3/8-3/4	—	—	GWF	1	GWB	60	St. St.	—	—	X
Cooper-Rossano System		0.5-2.0	3-10	800	1/4-1/2	KNB	Ext.	GWB	4	MF,PT	130, 80	St. St.	X	X	X
Dade County PCA		0.5-1.5	5-10	1000	3/8-5/8	AT	Int.	GSI	4	—	—	St. St.	X	X	—
Industrial Gas Cleaning Institute		0.5-1.5	5-10	1000	1/4-1/2	AT	Int.	—	—	—	—	St. St.	X	—	X
Kaiser Steel		0.5-1.0	5-10	1800	1/4-1/2	AT	Ext.	GSI	2	—	—	Titanium	X	X	—
		1.0-3.0	1-2	1200	1/4-1	ESP	Ext.	—	—	GFF	80	—	X	X	—
Los Angeles County APCD	1	0.5-1.0	3-8	600	1/4-1/2	—	—	GSI	3	PT	80	Glass In Stainless Steel	—	X	—
	2	0.5-1.0	3-8	250	1/4-1/2	PT	Ext.	GSI	3	—	—	"	X	X	—
	3	0.5-1.0	3-8	1000	1/4-1/2	AT	Ext.	GSI	3	PT	80	"	X	X	—
	4	0.3-0.5	3-8	1000	1/4-1/2	Cycl.	Ext.	AT	—	GSI	60	"	X	X	—
	5	0.5-0.3	3-8	800	1/4-1/2	ESP	Ext.	GSI	3	PT	80	"	—	X	—
Nat. Council for Air & Stream Imp.	1	0.5-2.0	3-8	400	1/4-1/2	KNB	Ext.	GWB	4	—	—	St. St.	—	X	—
	2	0.5-2.0	5-10	800	1/4-1/2	AT	Int.	GWB	4	—	—	St. St.	X	X	—
Research-Cottrell, Inc.		0.5-2.0	5-10	1000	1/8-1/2	AT	Int.	Cond.	1	—	—	St. St.	—	X	—
		0.5-2.0	3-8	300	1/8-1/2	PT	Int.	Cond.	1	—	—	St. St.	—	X	—
U.S. Public Health Service		0.5-1.0	5-10	1200	1/4-1/2	Cycl.	Ext.	DGF	1	GSI(3)	60	Glass In St. St.	—	X	X
USSR (Warshavsky)		1.0-2.0	3-8	1000	—	GWF	Int.	KNB	1	—	—	Mild St.	—	WTM	—
USSR (Gas Pur. Inst.)		1.0-2.0	5-15	—	—	PF	Ext.	—	—	—	—	St. St.	—	—	X
Western Precip. Co.		0.5-2.0	5-10	1000	1/4-5/8	AT	Int.	GSI	4	—	—	St. St.	—	X	—
		0.5-2.0	3-8	300	1/4-5/8	PT	Int.	GSI	4	—	—	St. St.	—	X	—

Table 29A, Continued

Prime Mover			Isokinetic Sampling			Gas Condition			Sources Where Primarily Used									References
									Combustion Sources				Industrial Sources					
Vacuum Pump	Air Ejector	Water Ejector	Vel. Prior To Test	Pitot Tube In Parallel	Null Balance	Hot, Dry	Wet, Sat'd.	Org. Material	Coal Boiler	Oil Boiler	Wood Boiler	Refuse Incin.	Refinery Units	Metall. Oper.	Kilns & Drying	Pulp & Paper	Chemical Process	
X	—	—	X	—	—	—	X	—	—	X	X	—	—	—	X	X	X	[1]
X	—	—	—	X	—	X	—	—	—	—	—	—	—	X	X	—	—	[2]
—	—	X	X	—	—	X	—	—	X	—	—	—	X	—	—	—	—	[3]
—	—	X	X	—	—	X	—	—	X	—	—	—	X	—	—	—	—	[3]
X	X	X	—	X	—	X	—	—	X	—	—	X	—	X	—	—	—	[4]
X	—	—	X	X	—	X	—	—	X	X	X	X	X	X	X	—	—	[5]
X	—	—	X	X	—	—	X	—	—	—	—	—	—	—	X	X	X	[5]
—	X	—	—	X	—	X	—	—	X	X	—	—	—	X	—	—	—	[6]
—	X	—	—	X	—	X	—	—	X	X	—	—	—	X	X	—	—	[7]
—	—	X	—	X	—	—	X	X	X	X	X	X	—	—	X	X	X	—
X	—	—	—	X	—	X	—	—	X	X	—	X	—	—	X	X	X	[8]
X	—	—	X	X	—	X	X	—	X	—	—	—	X	X	X	X	—	[12]
X	—	—	—	X	—	X	—	—	X	—	—	X	—	X	X	—	—	[9]
X	—	—	X	—	—	X	—	—	X	—	—	—	—	X	—	—	—	[10]
X	—	—	—	X	—	—	X	X	X	X	X	X	X	—	X	X	X	[11]
X	—	—	—	X	—	X	—	X	—	X	X	X	X	X	X	—	—	[11]
X	—	—	—	X	—	X	—	X	X	—	X	X	—	—	X	—	—	[11]
X	—	—	—	X	—	X	—	—	X	—	—	—	—	X	X	—	—	[11]
X	—	—	X	X	—	—	X	X	—	—	X	—	—	—	X	X	X	[13]
X	—	—	X	X	—	X	X	—	—	—	X	—	—	—	X	X	X	[13]
—	X	—	X	—	—	X	—	—	X	—	—	—	X	X	X	—	—	[14]
—	X	—	X	—	—	X	—	—	X	—	—	—	—	X	X	—	—	[14]
X	—	—	—	X	—	X	X	X	X	—	X	X	X	—	X	X	X	[15] [16]
X	—	—	X	—	—	X	—	—	X	—	X	—	—	X	X	—	—	[17]
X	—	—	X	—	—	X	X	—	X	X	X	X	X	X	X	X	X	[75]
X	—	X	X	—	—	X	—	—	X	X	X	—	X	X	X	X	X	[18]
X	—	X	X	—	—	X	—	—	—	—	—	—	—	—	X	X	—	[18]

Table 29B. Particulate Sampling Trains in Use. B. High Flow Rate Systems Where Q Is Greater Than 3.0 cfm.

Organization	Number	Operating Characteristics				Collection Devices			Minimum Temperature °F	Material Collected grams
		Flow Range cfm	Pressure Drop in. Hg.	Maximum Temperature °F	Nozzle Size inches	Primary	Secondary	Tertiary		
American Society of Mechanical Engineers		10-80	0-1	1200	1/4-1-1/2	Cyclone	Filter Bag	—	150	5-100
Babcock and Wilcox		5-20	0-1	—	3/4-2	Cyclone	Filter Bag	—	150	5-20
Armour Research Foundation										
British Standards	1	4	0-3	1250	1/2-1-1/2	Cyclone	GFF	—	150	5-20
	2	—	0-3	750	1/2-1-1/2	Cyclone	GFF	—	150	1-5
Cooper-Rossano System		10-50	0-3	1200	1/2-2	Cyclone	Filter Bag	GFF	150	5-50
Incinerator Institute of America		5-20	0-1	2000	3/4-2	Cyclone	Filter Bag	—	150	5-20
Los Angeles County APCD		60-100	0-2	—	1/2-1-1/2	Cyclone	Filter Bag	—	150	10-100
Oregon State University		10-40	0-2	—	1/2-1-1/2	Cyclone	Screen	GWB	—	10-20

Table 29B, Continued

Location	Sample Probe		Flow Measurement			Prime Mover			Isokinetic Sampling			Reference
	Material Used	Heated Or Cooled	Cyclone Meter	Orifice Meter	Gas Meter	Centrifuge Blower	Vacuum Motor	Water Ejector	Vel. Prior To Test	Pitot Tube in Parallel	Null Balance	
External	St. St.	—	—	X	—	X	X	X	X	X	X	[4]
External	St. St.	X	X	—	—	X	X	—	X	—	—	[19] [20]
External	St. St.	X	X	X	—	X	—	—	—	X	—	[7]
External	St. St.	X	X	X	—	X	—	—	—	X	—	[7]
External	St. St.	X	—	X	—	X	X	—	—	X	—	—
External	St. St.	X	X	—	—	—	—	—	—	—	X	[21] [22]
External	St. St.	—	X	X	—	X	X	—	—	X	—	[11]
External	Aluminum	—	Gas	—	X	—	X	—	X	—	—	[23]

Table 30. Distribution of Particle Collection in a Sampling Train on Gases from a Furnace for Burning Insulated Copper Wire (2).

Collection Device	Material Collected in Train					
	Sample 1		Sample 2		Sample 3	
	grams	%(1)	grams	%(1)	grams	%(1)
Probe Washings	0.2050	28.6	0.3494	32.1	0.2378	28.9
Paper Thimble	0.4225	56.8	0.6088	56.2	0.5265	63.8
Wet Impingers	0.1149	15.6	0.1272	11.7	0.0594	7.3
Total Collected	0.7424	100.0	1.0854	100.0	0.8237	100.0

NOTES: 1. Percent by weight of the total material collected in the sampling train.
2. Courtesy of the Journal of the Air Pollution Control Association. Kaiser, E. R., "Control of Air Pollution from the Burning of Insulated Copper Wire," January 1963 [72].

Table 31. Minimum Particle Sizes Collected by Final Stage Devices.

Device	Minimum Size Particle Collected microns	Organization
1. Greenburg-Smith Impinger	1.0	Dade County, USPHS [8]
2. Glass-Fiber Filter	0.3	Bay Area APCD [5]
3. Paper Thimble	0.3	Los Angeles APCD [11]
4. Membrane Filter	*	Amer. Indus. Hyg. Assoc. [1]

* Depends on the filter porosity selected. Sizes range from 0.02 to 5.0 microns. Porosities of 0.45 and 1.0 microns are common.

Table 32. Techniques for Particle Size Determination.

Technique	Collection Mechanism	Devices Used	Particle Size Range-microns
Sieving	Physical barrier	Sieve trays, Vibrating screens	5.0-1000.0
Filtration	Physical barrier	Membrane filters	0.01-5.0
Impaction	Inertial	Cascade impactor Multiplate impactor Greenburg-Smith impinger	0.1-100.0
Elutriation	Gravitational Sedimentation	Centrifugal fractionation Single stage fractionation Multiple stage fractionation	5.0-100.0
Sedimentation	Centrifugal Sedimentation	Aerosol spectrometer	0.02-10.0
Electrical	Electrical Resistance	Particle counter	1.0-100.0
Thermal	Thermal Forces	Thermal precipitator	0.1-15.0
Optical	Light Scattering	Aerosol counter	0.2-50.0
Condensation	Nuclei Generation	Condensation nuclei counter	0.01-0.1

Courtesy of Chemical Engineering, vol. 75, no. 11, p. 155, Table 5 in article by C. E. Lapple, "Particle Size Analyses and Analyzers," May 20, 1968 [78].

Table 33. Microscopy in Particle Size Analysis.

Method	Size Range-microns
Optical	0.2-100.0
Ultra	0.005-1.0
Electron	0.002-15.0

Courtesy of: Chemical Engineering, vol. 75, no. 11, p. 155, Table 5 in article by C. E. Lapple, "Particle Size Analyses and Analyzers," May 20, 1968 [78].

Table 34. The Effect of Particulate Concentration on Sampling Period When Using a Cascade Impactor (90) (91).

Particulate Concentration gr/SDCF (1)	Sampling Period minutes (2)
>5.0	1/2
2.0	1
1.0	2
0.5	6
0.2	15
0.1	30
0.05	60

Notes: 1—At 60°F, 29.92 in. Hg, dry
2—Based on $Q_m = 0.5 - 1.0$ cfm
3—Courtesy of Mr. John C. Bosch (91)

Table 35. Properties of the Andersen Sampler.

Collection Stage	Jet Diameter (inches)	Jet Velocity (ft/sec)	Particle Size Range (microns*)
1	0.0465	3.54	>8.3
2	0.0360	5.89	5.0-10.5
3	0.0280	9.74	3.0-6.0
4	0.0210	17.31	1.7-3.0
5	0.0135	41.92	0.9-2.0
6	0.0100	76.40	0.5-1.0

Data Courtesy of: Journal of Bacteriology, Vol. 76, No. 10, A. A. Andersen, "New Sampler for the Collection, Sizing, and Enumeration of Viable Airborne Particles," November 1958 [92].
* Stokes equivalent diameter.

CHAPTER 9

GASEOUS SAMPLING

I. Introduction

Gaseous sampling in flue gases is simpler than particulate sampling, principally because the gas molecules are small enough to be governed by the random nature of Brownian motion; inertial effects became insignificant. This means that it is not necessary to sample at isokinetic conditions, and in continuous sampling it is necessary only to withdraw a sample from the flue at a known rate. The task of obtaining a representative sample therefore is considerably easier because the sampling rate can be independent of the velocity in the duct.

Sampling for gaseous materials in flue gases, however, involves several complications. First, particulate matter from the flue gas can react with the gases of interest or the absorbing solution, to constitute a possible interference. Particulates can also act to plug impinger tips and tubing, and interfere with sample withdrawal from the duct. Hence, it is often necessary to remove particulate materials from flue gas upstream of the collection system. Second, moisture from the flue gas may condense in the sampling train and dissolve some of the gaseous materials. Losses resulting from condensation may be avoided by several different techniques. Third, the sample probe and all tubing upstream of the collection system should be constructed of inert material which does not react chemically with, or absorb physically, the gaseous molecules to be analyzed. Fourth, the techniques used to collect the gaseous materials must be either 100 percent efficient or of known efficiency. Fifth, the analytical techniques used for the particular gaseous components of interest must be specific, accurate, sensitive, reproducible, and free from interfering substances.

The two major steps in gaseous sampling involve collection of a representative sample from the flue gas, and making an accurate analysis of the amount of constituent gas collected. A known volume of flue gas is withdrawn from the duct and collected prior to analysis, unless continuous monitoring is employed. The sample must contain a sufficient amount of the gas of interest to facilitate accurate analysis, and collection should be done without losses or interferences resulting from particulate materials, moisture condensation, or wall effects. The rate and duration of sampling is important in determining the amount of constituent material collected, depending on the collection technique used.

A. Sample Collection

It may be desirable to withdraw the sample from the flue at a rate proportional to the velocity in the duct at the point of sampling. This prevents selectively weighting times of gas flow rates at less than the mean velocity during the time of sampling at the expense of times of greater than mean velocity [14]. Proportional sampling may be necessary during periods where major fluctuations of $\pm$ 25 to 50 percent or greater occur. Where possible, process units should be operated at constant conditions to maintain relatively steady flow conditions during the test. These constant conditions will minimize the need for proportional sampling, and allow a constant sampling rate to be used. Adjustments in sampling rate may be difficult to make to maintain a constant sampling rate where an orifice meter is being used and operated at a pressure drop of greater than 14 inches of mercury.

One technique for proportional sampling involves making a study of the change in duct velocity with time during operating conditions similar to the ones to be tested, prior to the actual gas sampling period. This gives an estimate of the range in velocities. A mean velocity is selected and used with a specified flow rate so that both maximum and minimum velocities may be covered within the range of the sampling flow meter capacity and accuracy. The method is similar to that previously described for isokinetic particulate sampling, with the exception that the sampling flow rate need be proportional only to the velocity, and not necessarily be at the isokinetic rate. Normally, gas flow rates used in gaseous sampling are in the order of magnitude of 0.05 to 0.15 cubic feet per minute, while for particulate sampling they are 0.50 to 2.00 cubic feet per minute. It is necessary to determine the hydraulic characteristics of the sampling train prior to testing, however.

The basic relationship for relating meter flow rate to duct velocity has the following form:

$$Q_m = K \frac{U_s}{P_m} \quad (99)$$

$$K = (P_s)\left(\frac{T_m}{T_s}\right)\frac{(100 - MC_s)}{(100 - MC_m)} \quad (100)$$

where:

- Q_m = Flow rate at the meter in cubic feet per minute.
- U_s = Duct velocity in feet per minute.
- P_m = Absolute pressure at the *upstream* side of the flow meter in inches of Hg.
- K = Constant for the particular sampling train.

The changes of pressure drop across the orifice and upstream meter static pressure, with the sampling train in place exactly as it will be used in the field, are then determined by varying the flow rate. It is necessary to calibrate the flow meter, either an orifice flow meter or rotameter with a wet test meter, prior to setting up the sampling train. The ratio of flow rate with velocity change between the mean and some other condition is shown as follows:

$$\frac{(Q_m)_1}{(Q_m)_{Avg.}} = \frac{K\left(\frac{U_s}{P_m}\right)_1}{K\left(\frac{U_s}{P_m}\right)_{Avg.}} \quad (101)$$

$$(Q_m)_1 = \frac{(U_s/P_m)_1}{(U_s/P_m)_{Avg.}}(Q_m)_{Avg.} \quad (102)$$

If there is only a small change in absolute static pressure on the upstream side of the meter, the relationship is approximated by the following:

$$(Q_m)_1 = \frac{(U_s)_1}{(U_s)_{Avg.}}(Q_m)_{Avg.} \quad (103)$$

This condition normally applies with impingers when the flow rate is 0.10 cubic feet per minute (3 liters per minute) or less with a ¼ inch or larger nozzle. Variations of pressure drop across the orifice meter with flow rate at the meter are shown in Figure 84.

Sample Problem No. 11—Sample Flow Rates for Proportional Sampling

Problem:

Average, maximum, and minimum velocities in a duct were found to be 2,400, 3,000, and 1,800 feet per minute, respectively. Select average, maximum, and minimum flow rates as large as possible, yet within the capacity of the orifice flow meter shown in Figure 84. Plot the meter flow rate as a function of velocity in the duct when using a ¼ inch nozzle.

Expressions:

(1) Rough Estimation:

$$(Q_m)_1 = \frac{(U_s)_1}{(U_s)_{Avg.}}(Q_m)_{Avg.}$$

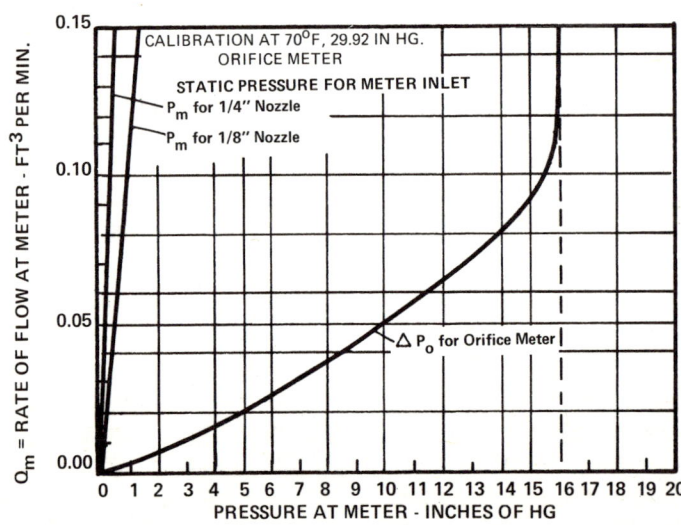

Fig. 84.

Solution:

(1) Selection of Average Flow Rate
Let $(Q_m)_{Avg.} = 0.075$ cfm
Maximum:

$$(Q_m)_{max} = \frac{(U_s)_{max}}{(U_s)_{Avg.}} (Q_m)_{Avg.}$$

$$(Q_m)_{max} = \frac{3{,}000}{2{,}400} (0.075) = (1.25)(0.075)$$

$$(Q_m)_{max} = 0.094 \text{ cfm}$$

$$(U_s)_{max} = 3{,}000 \text{ fpm}$$
$$(U_s)_{Avg.} = 2{,}400 \text{ fpm}$$
$$(U_s)_{min} = 1{,}800 \text{ fpm}$$

This is within the capacity of the orifice meter:

$$(Q_m)_{max} = 0.11 \text{ cfm from Figure 84.}$$

Minimum:

$$(Q_m)_{min} = \frac{(U_s)_{min}}{(U_s)_{Avg.}} (Q_m)_{Avg.}$$

$$(Q_m)_{min} = \frac{1{,}800}{2{,}400} (0.075) = (0.75)(0.075)$$

$$(Q_m)_{min} = 0.056 \text{ cfm}$$

(2) Plotting on Graph:

Shown in Figure 85 is the variation in sampling rate with duct velocity. This assumes that changes in meter static pressure are negligible over the range of operation.

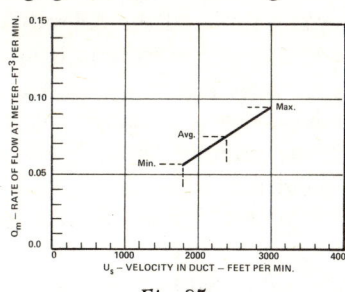

Fig. 85.

B. Sample Analysis

Chemical analysis for the constituents of interest follows completion of the collection step. The method used must be sensitive to low concentrations of the material and free from potential interfering substances, or else have their potential effects known. In tests where the material collected is not stable during extended storage periods, the sample should be analyzed as soon after collection as possible. An additional problem is involved in measuring for a small amount of a constituent gas present in a large background of a potential interfering substance. Measurement of a small amount of sulfur trioxide in a large background of sulfur dioxide in power boiler flue gases is one example of this problem. It should also be emphasized that potential errors in analysis, as listed above, may be of sufficient magnitude to make proportional sampling unnecessary. Wet chemical analytical techniques cannot be considered suitable substitutes for continuous monitoring unless the above problems can be overcome.

II. SAMPLE COLLECTION METHODS

Several methods are available for collection of gaseous constituents from flue gas streams. These include absorption into a liquid phase, collection in an evacuated container, collection in a flexible fabric bag, adsorption on a solid material, or freeze-out techniques. Each method is useful for particular applications, depending on the temperature and moisture content of the flue gas, the material being analyzed for, and the method of analysis used.

A. Liquid Absorption

Absorption into a liquid medium is perhaps the most common method for collecting gaseous components from sources. Flue gas is withdrawn from a duct at a known rate through a series of impingers containing a liquid into which the gas of interest is dissolved. Following the collection step, the liquid is removed from the impingers and analyzed. A typical sampling train is shown is Figure 86. The major components are the sample probe, impingers, drying system, flow measurement and control section, and the prime mover. Flow rates for gaseous sampling are normally much lower than with particulate sampling, being on the order of one to five liters per minute.

1. Sample Probe

Flue gas is first withdrawn from the duct through the sample probe. The probe should be constructed of a material such as glass or stainless steel which does not react with or physically adsorb the gases to be analyzed [1]. The probe opening should be pointed downstream to minimize the intake of particulate materials and water droplets from the gas stream. An alundum thimble may be placed either inside hot ducts or

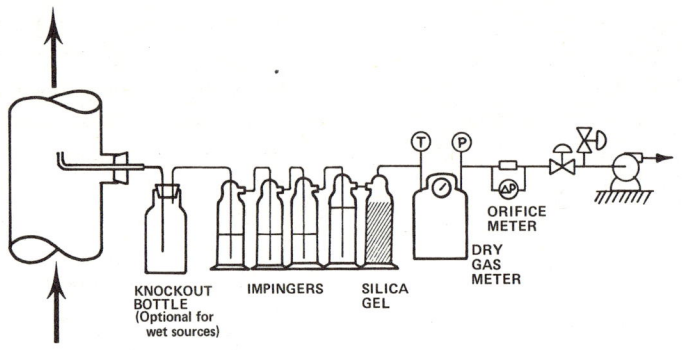

Fig. 86.

externally, with provision in the latter case for heating. The above techniques tend to prevent particulate plugging in the impingers. A third problem is associated with condensation of water, which may be alleviated by giving a downhill slope to the sampling line from the duct to the impingers. Any water which condenses then tends to end up in the impingers. It is also possible to heat the sample line, but this involves additional complexity and, if done electrically, may present a safety problem in wet weather.

An alternative technique employing placement of a knockout jar upstream of the impingers is useful for wet flue gases at saturation following scrubbers. It is especially useful for removing entrained droplets of salt-containing liquid which would otherwise interfere with the chemical test being used. An example is the removal of entrained black liquor droplets containing sodium sulfide in flue gases from a Kraft pulp mill recovery furnace, following a venturi recovery unit. The droplets are removed in the knockout bottle so as not to appear as hydrogen sulfide in the absorbing liquid. The tubes should be kept short enough so that flue gas does not bubble through the collected liquid and strip out gaseous materials, as the presence of carbon dioxide in flue gases causes many gases to be stripped from the collected liquid. Where possible, the condensate liquid should be analyzed for losses of gaseous materials.

2. Impingers

a. Parameters

Impingers are designed to provide intimate contact of the sample gas stream with the liquid phase so that the gaseous material of interest can be absorbed into the liquid for analysis. Calvert and Workman [2] [3] investigated the factors which influenced the efficiency of gas absorbers, and developed a method for making a rough estimate of their efficiency using mass transfer relationships. They found that diffusivity of the gaseous material of interest in the respective gaseous and liquid phases, retention time, the reciprocal of bubble size, and solubility of the gas of interest in the liquid phase were the most important factors.

The above factors influence the collection efficiency of the gaseous material in the absorbing liquid as follows: the diffusivities of the gas at the respective interfaces of the gaseous and liquid phases determine the ease with which the gas molecules pass from the gas stream to the absorbing liquid. It is important to consider which phase is controlling with regard to mass transfer. The degree of gas/liquid contact increases with retention time, and it is influenced by the sampling rate and the height of liquid in the impinger. The retention time is increased by a decrease in flow rate but has the limitation that excessively long sampling times may be required if the sampling rate is too low. Contact time increases with the height of liquid above the impinger tip, but is also limited by increases in pressure drop, and the desire not to have liquid carryover to the following impinger. The degree of gas/liquid contact increases as the bubble size decreases because the ratio of exposed surface area to volume increases. This ratio is normally increased by decreasing the size of opening in the impinger tip, but the technique is limited by the resultant increase in pressure drop across the impinger. The above parameters can be altered to increase the amount of gas-liquid contact, but there are definite limitations in the degree of change.

It has been observed by Calvert and Workman [2] that the degree of absorption of a gas into a liquid is low when its solubility in that liquid is low. This occurs because there is only a very small chemical driving force tending to draw the gas into the liquid. This concentration driving force may be greatly increased by using a liquid having a greater solubility for the gas. An example is the use of a sodium hydroxide solution to absorb the acidic carbon dioxide gas, instead of distilled water because of the much larger driving force. It should also be observed that polar substances dissolve more easily in polar solvents, and nonpolar substances dissolve more easily in nonpolar solvents. Therefore, most inorganic gases are best collected in aqueous solvents while organic gases are normally collected in organic solvents. The vapor pressure of the gas at the temperature of collection is important because it determines how easily the gas may be driven out of the solution. Gases with appreciable vapor pressures at ambient conditions should be collected in absorbing liquids in which they have high solubilities.

In addition to solubility, causing the gas to undergo a chemical reaction increases the driving force for collecting the gas in the absorbing liquid. An example is collection of hydrogen sulfide in a cadmium chloride solution to form a cadmium sulfide precipitate. It is necessary to provide a large excess of reacting chemical so that it does not become depleted during the sampling period.

The absorbing liquid used should be stable, nonfoaming, nonvolatile, and of high purity to avoid potential interferences. For volatile organic solvents it may be necessary to place the impingers in a cold water bath to minimize evaporation.

b. Types:

Several types of impingers are available for absorbing gases, as discussed by Hendrickson in *Air Pollution* [4]. These are used for sampling for a number of constituent gases in varying applications. Several types are shown in Figure 87.

(1) Midget Impingers:

Midget impingers are small impingers, with fine pointed tips, which normally contain between 10 and 25 milliliters of absorbing liquid. The size of the orifice is one millimeter in diameter, and the jet impinges against the bottom of the impinger five millimeters away from the tip of the orifice. Impingement against the flat surface increases the degree of gas/liquid contact by

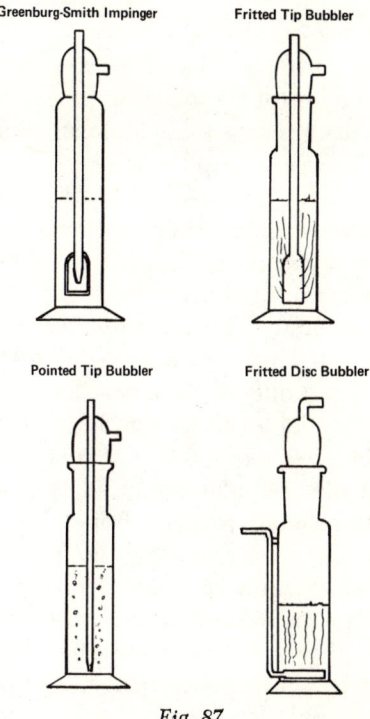

Fig. 87.

breaking up the larger bubbles into smaller ones. The devices are normally limited to a flow rate of about three liters per minute because the head loss becomes excessive at higher flow rates. Their efficiency is also limited by the small capacity and resultant low liquid height, which results in short retention times of gas in the liquid phase. Roberts and McKee [5] found that midget impingers were less efficient at comparable flow rates than either Greenburg-Smith impingers or gas washing bottles employing fritted glass tips. These impingers find limited application in source testing for the above reasons.

(2) Greenburg-Smith Impingers:

Greenburg-Smith impingers are similar in design to the midget impingers previously mentioned except that they are considerably larger and can handle sample flow rates up to about 30 liters per minute [1.0 cfm]. They have a total volume of 500 milliliters, and are normally filled with between 100 and 250 milliliters of liquid. The orifice diameter of the tip is 2.3 millimeters, and the gas impinges against a plate 5.0 millimeters away, which is connected to the tip. They provide a means for efficient collection of gaseous constituents with minimal pressure drop, at flow rates between one and 15 liters per minute.

(3) Fritted Glass Tip Bubblers:

Gas streams are caused to flow through a porous medium of fritted glass which tends to break up the gaseous phase into extremely small bubbles. The very small bubbles produced provide for a high degree of gas/liquid contact, and these devices are normally the most efficient types of gas absorbers. The frit is either a vertical cylindrical tip at the bottom of the shaft or facing horizontally upward from the bottom of the impinger. There are several different grades of frits of varying pore sizes which are commercially available. The finer the pore size the greater the degree of gas/liquid contact because of the smaller bubbles produced, but also the greater the resultant pressure drop.

These devices are normally used only over a range in flow rates of one to three liters per minute. The frits are also subject to plugging; therefore, the gas stream must be filtered upstream of the impinger to remove any particles. Fritted tip bubblers cannot be used where a precipitate is formed in the absorbing liquid because the frit will become plugged. It is often a good procedure to place a conventional impinger upstream as a first stage collector, and use the fritted bubbler as a final stage collector. Normally, two fritted bubblers in series provide an efficient means for collecting gaseous substances in liquids, even for such relatively insoluble gases as nitrogen dioxide. However, they should be thoroughly cleaned after sampling by several washings of the frit with the absorbing solution.

(4) Straight Tip Bubblers (Gas Washing Bottles):

Gas washing bottles normally come with a straight tip of about seven millimeters bore diameter. The straight tip is not normally used in gaseous sampling because the bubbles produced are too large to provide for efficient gas/liquid contact. Either pointed or fritted glass tips may be placed on the shaft instead of the straight tip to increase the efficiency of the scrubber. The gas washing bottles are of 250 or 500 milliliters capacity and are normally filled with 150 or 250 milliliters of absorbing solution, respectively.

Roberts and McKee [5] found that when placed with a fritted glass tip, a single gas washing bottle removed greater than 95 percent of the incoming ammonia gas at the one ppm level into distilled water, at flow rates up to ten liters per minute. The devices are less costly than Greenburg-Smith impingers and may be used for similar purposes. Three gas washing bottles in series with pointed tips, or two pointed tip gas washing bottles with a fritted tip bubbler as a final stage, can provide an efficient absorption train for gaseous constituents.

(5) Multistage Impingers:

Bergshoeff [6] describes the use of a six-stage vertical fritted disc scrubber for absorbing gases. Liquid is first placed in the six stages above the sintered discs and gas then drawn upward through the absorber. The sintered discs act as frits to produce small bubbles and provide efficient gas/liquid contact. Absorption efficiency for nitrogen dioxide at a flow of one liter per minute in Saltzman absorbing solution approached 100 percent. Little other information is available on these units.

c. Evaluation:

Four recent studies have been devoted to measure the respective collection efficiencies of the different types of absorbers. Roberts and McKee [5] investigated the comparative scrubbing efficiencies of three series combinations consisting respectively of gas washing bottles with fritted glass tips, Greenburg-Smith impingers, and midget impingers. They studied sulfur dioxide, chlorine, and ammonia at flow rates between 0.05 and 0.90 cfm at concentrations of about one ppm. They found that absorption and recovery of both chlorine in sodium hydroxide and sulfur dioxide in sodium tetrachloromercurate were complete for all conditions studied. Scrubbing efficiencies for ammonia varied between 70 and 100 percent of the material added to the dilution system. Fritted glass impingers were found to have the highest recoveries of ammonia for gas flow rates up to 0.5 cfm.

Three recent studies of impinger efficiency were made using radioactive tracers for gases at levels approximating those found in ambient air. Bostrom [7] found that absorption efficiencies of pointed tip impingers were greater than 97 per cent for hydrogen peroxide, sodium tetrachloromercurate, and distilled water at flow rates between one and six liters per minute. Bracewell and Hodgson [8] found that the absorption efficiency of sulfur dioxide was greater than 99 percent at a flow rate of 1.2 liters per minute for concentrations down to 0.1 ppm by volume. Bostrom [9] studied the collection efficiency of hydrogen sulfide in cadmium hydroxide solution, and found that fritted glass disc impingers were the most efficient over a flow range of one to six liters per minute. Straight tipped impingers were found to show rapid decreases in efficiency at flow rates exceeding three liters per minute.

A recent study by the National Council for Air and Stream Improvement shows the distribution of sulfur dioxide recovery between three impingers in series containing sodium tetrachloromercurate as a function of inlet sulfur dioxide concentration [10]. Nearly all the sulfur dioxide is absorbed in the first scrubber for concentrations below 150 ppm during a one hour sampling period at 0.1 cfm. The proportion collected in the first impinger was reduced to 50 percent at an inlet concentration of 750 ppm. The presence of significant quantities of material in the final absorption stage is an indication of incomplete collection in the earlier stages.

3. Flow Measurement and Control:

Sampling rates for gaseous sampling are normally between one and five liters per minute (0.03 to 0.20 cfm). It is normally desirable to use a critical flow orifice meter which limits the sampling rate to a certain maximum, such as three liters per minute. The orifice should be operated at a pressure drop of 14 to 16 inches of mercury to maintain nearly constant flow. This arrangement is particularly valuable for maintaining a constant rate of flow for variable downstream pressure, such as occurs with some vacuum pumps. It is advantageous to dry the gas upstream of the orifice meter with silica gel or Drierite to prevent condensation. The gas stream should be filtered with glass wool after the drying stage to minimize the possibility of plugging at the orifice. A needle valve-bleed arrangement is placed downstream of the orifice to regulate the flow rate. A rotameter may be used as an alternative, but the float is subject to oscillations caused by vacuum source fluctuations, which tends to make setting and observation of flow rate more difficult. This fluctuation may be offset by placing surge bottles downstream of the flow meter, immediately upstream of the prime mover. Under conditions where proportional sampling is necessary, it is useful to operate the flow meter so that the maximum flow rate is within its capacity and minimum flow rate within its accuracy.

4. Prime Mover:

A vacuum pump of suitable capacity which is capable of pulling at least sixteen inches of mercury is necessary, though a water, compressed air, or steam aspirator may also be used. These devices have been discussed in a previous section.

B. Solid Adsorption:

1. Principles:

Gaseous substances may be collected by adsorption onto the surfaces of a porous granular solid, and then desorbed off the solid phase following collection for analysis. These are several factors which affect the efficiency of adsorption, both on the part of the gas stream or adsorbate, and the solid phase or adsorbent [4]. The efficiency of adsorption of the solid bed is influenced by the type of material used, the total surface area per unit weight of solid, the degree of adsorption which has previously taken place, and any losses due to chemical reactions. The rate of flow through the adsorbent, the gas temperature, and the possible presence of other gases which may compete for the available adsorption sites, also influence the degree of adsorption.

The solid phase should be a granular solid with a large surface area per unit weight, relatively specific for the gases being adsorbed, and of low head loss through the bed. The temperature of the gas stream passing through the bed should be low enough to minimize revolatilization of the molecules once adsorbed. It is noted from the discussion by Hendrickson [4] that the adsorptive capacity of a solid for a gas rises dramatically as the critical temperature of the substance increases. Therefore, the method is particularly suitable for collecting organic and other vapors with critical temperatures in the ambient range and above.

Removal of the gas or vapor following collection is then made to facilitate analysis. This may be done by heating the bed to between 250°F and 350°F at the same time that an air stream is being passed over the bed. The offgas may then be analyzed by means of gas chromatography, mass spectrometry, or absorption into a liquid with subsequent chemical analysis. The bed may be stripped by passing steam over it, and condensing the vapors to a liquid. The condensed liquid may then be analyzed by gas chromatography or wet chemical analyses. A third method employs drawing a sufficient vacuum over the bed to cause the boiling point to be reached, whereupon the gases are driven off. A fourth technique involves selective removal by washing in a liquid solvent.

Adsorption onto a solid phase provides a potentially useful sampling method for gases present in flue gases in small quantities. It is possible to concentrate them greatly to facilitate collection of a sufficient amount of material for collection. In desorption it is necessary to utilize a method which provides for essentially complete removal of vapor from the solid bed without significant loss or alteration of composition. The bed should also be easily regenerated.

2. Applications:

Adsorption of vapors and gases to facilitate chemical analyses has proved particularly useful in certain applications, such as with gas chromatography, for ambient air analyses. It is also useful in source sampling applications. Turk [11] made use of activated carbon canisters to sample sulfur dioxide, carbon tetrachloride, benzene, and other gases. He found the collection efficiencies varied between 95.5 and 99.5 percent for the compounds tested, and that air oxidation of sulfur dioxide on the bed was insignificant. Vacuum desorption proved to be a useful method for removing certain constituents, but removal by steam stripping was far less time-consuming.

Feldstein [12] described the use of silica gel for collection of organic solvents such as esters, ketones, hydrocarbons, and chlorinated organic compounds. The gases were drawn through a three stage collection device with silica gel beds placed in each stage. The three stage system was used to overcome the effects of moisture adsorption in the first stage. Following collection, each stage was eluted with dimethyl sulfoxide to desorb the collected vapors prior to gas chromatographic analysis.

A third application is the use of detector tubes for gas analysis which employ an indicator that changes color upon reaction with a gas. Several of these are commercially available and may be used to estimate gaseous concentrations in ducts.

C. Freezing-Out:
1. Principles:

Freezing-out employs withdrawal of a gas stream from a duct through a series of cold traps at progressively lower temperatures to successively condense out gases and vapors. These condensed liquids may then be analyzed by gas chromatography, infrared spectrophotometry, mass spectrometry, or wet chemical analyses. Each cold trap is placed in a bath which is regulated to a certain temperature by addition of a cooling medium. Several possible cooling agents are listed in Table 36 [4].

The method provides a potentially useful method for trapping a large number of pollutants for analyses. It is dependent on the boiling points of the respective materials, temperatures of the cold traps, their design and capacity, the flow rate through the sampling train, and the moisture content of the flue gas. It may be desirable to use a water-cooled condenser ahead of the sampling train to remove most of the water vapor by collection in a container [4]. The gas is then passed through the series of freezeout traps to successively remove the gaseous and vapor phase constituents.

2. Applications:

Freezeout methods have been found particularly suitable for sampling organic constituents in flue gases from combustion sources such as refuse incinerators and automobile exhausts. A typical sampling train employing freezeout traps is shown in Figure 88. The gas is first drawn through an alundum thimble to remove the major portion of the particulate materials, and then through a water condenser to remove most of the water vapor present. It is then passed through a series of freezeout traps of progressively lower temperatures to remove most of the organic gaseous constituents present. The gas is then reheated and passed through an orifice or dry gas meter where the rate of flow is measured. The prime mover is either a vacuum pump or water aspirated ejector of suitable capacity to pull at a sampling rate of between 0.1 and 1.0 cfm. Following the collection step, the condensed water and the liquid from each of the freezeout traps are analyzed for the constituents of interest. It is important to keep the contents of each freezeout trap at the low temperature of collection to avoid volatilization.

Yocom and Hein [13] used a freezeout trapping method for collecting gaseous constituents from incinerator flue gases. They used a

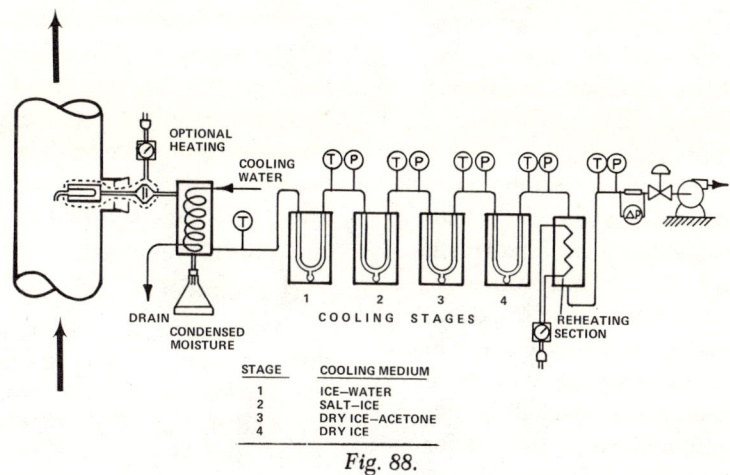

Fig. 88.

filter to remove particulates, two ice baths, a trap at a somewhat lower temperature, and finished with two dry ice traps, at a sampling rate of three cfm. Temperature and pressure recordings were taken following each collection stage during the test. Infrared spectrophotometric analysis were then made on the contents of each of the freezeout traps.

D. Grab Sampling:

1. Principles:

The technique involves in taking grab samples is to withdraw a known volume of flue gas into a container. The container is then taken to the laboratory and the contents analyzed either by absorption into a liquid medium followed by wet chemical analysis or by direct instrumental analysis with either gas chromatography, mass spectrometry, or infrared spectrometry. Major problems with the method are losses of gaseous constituents caused by moisture condensation, or by adsorption on, or reaction with, the wall materials of the container. Inert grease should be used on stopcocks or fittings to minimize losses.

2. Rigid Wall Containers:

Flue gas may be withdrawn into an evacuated rigid wall container such as a glass bottle or stainless steel cylinder. The size of sample taken depends on the capacity of the container. Pressure measurements are taken after evacuation and sampling to make the necessary volume corrections. Wall materials such as glass and stainless steel are normally inert to most gaseous constituents. The containers are often heated by means of wrapping with heating tapes or nichrome wires to prevent condensation during sampling [15]. The use of heated and evacuated five gallon jugs has been found suitable for sample collection from Kraft recovery furnace flue gases.

Martin [15] describes the use of 250 ml glass bombs for sample collection from Kraft mill sources. The container is open at both ends, and a sufficient volume of gas is purged through to obtain a representative sample. The container is closed following sampling, and taken to the laboratory for gas chromatographic analysis. A similar technique is described by the Los Angeles County Air Pollution Control District [14]. Leonard [15] uses heated two-neck five liter flasks for sample collection from Kraft mill sources.

It is also possible to use evacuated stainless steel cylinders for sample collection. The cylinder is first evacuated to a known absolute pressure, then repressurized with a stack gas sample. Portions of the gas may be removed and the cylinder then successively repressurized to make necessary dilutions for high gas concentrations. It is necessary to make note of absolute pressures in the cylinder both before and after any evacuation steps, to make to necessary volumetric corrections for concentration measurements. Cylinders should be heated to prevent losses caused by condensation.

Brief and Drinker [16] made use of evacuated two liter flasks for collecting samples from flue gases. Approximations of an integrated sample during a cyclic operation were made by taking constant rates of rise of the absolute pressure in the flask during the sampling period.

3. Flexible Bags:

Flexible fabric bags may be used to collect samples by pumping in the gas from the duct.

The method provides for simplified and rapid collection of a gaseous sample, but it may become necessary to cool the gas stream to prevent damage to the bag. Several materials such a mylar, polyethylene, and others may be used, but the wall material should be inert to the gases being sampled. The method is subject to losses caused by moisture condensation or adsorption on the walls of the bags. Two recent studies by Baker and Doerr [17] and Altshuller [18] document the storage properties of several gases in various bag materials. Normally, it is a good idea to analyze the samples immediately following collection, to minimize the changes for adsorption or reaction of the gases of interest with the wall material.

An additional problem is involved with possible alteration or dilution of the sample by passing through a vacuum pump upstream of the bag. This may be alleviated by placing the bag inside a sealed box. The bag is connected to the flue gas and the sample drawn into it by progressive evacuation of the outer portion of the sealed box. The method has proved successful for sample collection in incinerator flue gases [19].

III. WET CHEMICAL METHODS

Following the sample collection step, it is necessary to perform an analysis for the presence of the material or materials of interest. A large number of analytical techniques are available for many gaseous substances, and brief descriptions of many individual methods have been presented in several recent publications [20] [21] [22] [23] [24] [25] [26] [27]. These are classified into inorganic pollutants such as sulfur dioxide and nitrogen dioxide, plus organic gases such as ethylene and formaldehyde. A complete discussion of the many techniques available would be the subject of another book, so only selected examples will be presented.

The analytical method should meet several criteria with regard to the substance being measured. The method should be specific for the gas of interest and free of interfering substances, but these requirements are difficult to meet, since many methods depend on the nonspecific oxidation-reduction or acid-base reactions [26]. Selective pre-removal of interfering substances is one method of dealing with the problem. These may be removed either upstream of the collection system, by addition of counteracting substances to the absorbing medium, by filtration, or by chemical reaction before analysis. Reagents should be stable with regard to both time and temperature. The method should also be reproducible, accurate, and suitable for use in a variety of field conditions.

A. Sulfur Compounds:

1. Oxides of Sulfur:

Both sulfur dioxide and sulfur trioxide are often present in flue gases, particularly from combustion of sulfur-containing fuels such as coal and fuel oil. They may also be emitted from petroleum refinery, Kraft and Sulfite pulp mill, metallurgical, sulfuric acid making, and other industrial operations. The major portion of the sulfur oxides is normally present as sulfur dioxide with between one and ten percent of the total in the form of sulfur trioxide. The sampling procedures are complicated by the tendency of sulfur trioxide to form sulfuric acid in the presence of water, and difficulties in differentiating between SO_2 and SO_3 in aqueous media.

A summary of collection and analytical methods using wet impingement techniques is presented in Table 37. Sulfur trioxide is first selectively removed from the sample gas stream, and the sulfur dioxide then absorbed into an aqueous medium. Analyses for the individual constituents are then made for the individual constituents by any of several different methods. Individual references should be consulted for the exact procedure to be used. It is noted that prefiltration of the gas upstream of the sample collection system is often necessary to remove particulate materials, which is particularly necessary if sulfate-containing particles are present, to avoid interferences.

2. Sulfur Trioxide:

There are two methods currently available for selective separation of sulfur trioxide from sulfur dioxide in the sample gas stream. The first involves formation of sulfuric acid by reaction of sulfur trioxide with water from the flue gas. Subsequent condensation of the acid droplets without condensation of the water is then performed at a temperature below the acid dewpoint but above the water dewpoint. The acid droplets are then removed by means of a sintered glass filter following the condenser. A plot of acid dewpoint temperature as a function

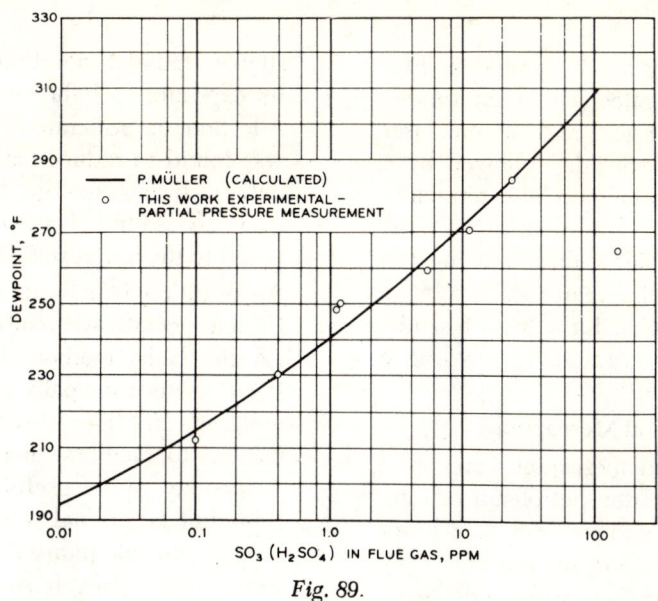

Fig. 89.

Courtesy of: American Society of Mechanical Engineers, PCT. 19.10-1968 (41).

of sulfur trioxide concentration is illustrated in Figure 89. Lisle and Sensenbaugh [31] devised a method for determining acid dewpoint temperature, making use of selective condensation of sulfur trioxide. Calibration was achieved by means of addition of known amounts of sulfuric acid to the system, and results agreed closely with thermodynamic considerations.

A second method involves absorption of sulfur trioxide into a solution of 80 percent isopropyl alcohol/20 percent water. The oxidation of sulfur dioxide is inhibited by the presence of the isopropyl alcohol [39]. The sulfur dioxide collected in the alcohol solution is then removed following the sampling by purging with air, and is collected in the sulfur dioxide scrubbers. The sulfur trioxide remaining in the isopropyl alcohol is then analyzed by one of several procedures.

Several methods are available for analysis of the sulfur trioxide from either the filter washings or the isopropyl alcohol solution. The first involves titration with standardized sodium hydroxide to either a bromthymol blue [31] or methyl red endpoint [40]. The second involves precipitation with barium chloride followed by turbidometric [36] or gravimetric analysis [34]. A recently developed method makes use of titration with barium chloride in the presence of Thorin indicator [35]. There is a definite need for standardization of existing methods to provide for uniform interpretation of data.

3. Sulfur Dioxide

Sulfur dioxide is absorbed into any of several aqueous solutions following collection of the sulfur trioxide. One technique involves scrubbing in a solution of hydrogen peroxide, whereupon the sulfur dioxide is oxidized to sulfate ion. The amount of sulfate ion present may then be analyzed by one of two methods, and results calculated as sulfur dioxide. One method involves titration of the peroxide solution with sodium hydroxide to a bromthymol blue or phenolphthalein end point. A second method involves addition of barium chloride to form a barium sulfate precipitate, which is then analyzed gravimetrically.

Sulfur dioxide may also be absorbed in a solution of sodium tetrachloromercurate [38]. A disulfitomercurate complex ion is formed, and this is analyzed colorimetrically by subsequent addition of formaldehyde and pararosaniline. The method is subject to interference from oxides of nitrogen in the flue gas, which may be overcome by addition of sulfamic acid prior to the color development step [42]. The presence of hydrogen sulfide in the flue gas causes the formation of a mercuric sulfide precipitate. This has been found to constitute an interference with the method upon storage, which can be minimized by analysis of the sample immediately following collection [43]. The method has the advantage, however, of being specific for sulfur dioxide and not subject to interference from most acidic or basic gases [44].

Additional methods for analysis of sulfur oxide gases have been presented by Chory [45], Grondona [46], and Bent [47].

Two typical sampling are illustrated in Figure 90. The first incorporates a filter for removing particles and sulfuric acid mist, a condenser and filter for the sulfur trioxide, followed by a series of scrubbers containing sodium tetrachloromercurate solution for sulfur drioxide. The second train includes a filter for removing the particulate and acid mist material, scrubbers containing isopropyl alcohol for sulfur trioxide, followed by scrubbers containing hydrogen peroxide for sulfur dioxide.

4. Hydrogen Sulfide and Mercaptans:

Hydrogen sulfide and mercaptans are often present in the flue gases from petroleum refining and Kraft pulp mill operations, and to a lesser extent from ferrous metallurgical sources. They are important because of their strong odor even at low concentrations. The gases are slightly acidic in nature, and are normally collected by drawing the flue gas through impingers containing an aqueous solution which is either alkaline or causes rapid precipitation of a sulfide salt. Jacobs [20] describes a number of methods which have been used for analysis of hydrogen sulfide either in flue gases or in the ambient air.

A summary of analytical methods for hydrogen sulfide and mercaptans is presented in Table 38. The methylene blue colorimetric method has been successfully used in ambient air analyses with two different absorbing solutions. The sample is first collected in either cadmium hydroxide or zinc acetate solution, and then analyzed colorimetrically at 670 mμ by addition of phenylene diamine dye, sulfuric acid, and ferric chloride. Recent work by Adams [55] has shown that deterioration of the samples occurs by exposure to light, oxidation, and on storage. Addition of Stractan to the absorbing solution was found to reduce losses of sulfide ion by oxidation during storage to less than 20 percent over an extended storage period, when compared to the initial sulfide concentration.

An additional method makes use of absorption in zinc acetate with subsequent analysis by the Lauth's violet method. The above methods have been used principally for hydrogen sulfide in ambient air. The absorbing solutions may be diluted to facilitate analyses of source samples. They may prove useful in source test applications but additional work is necessary to evaluate the possible interferences of compounds such as oxides of nitrogen, sulfur dioxide, and others.

Moore [54] described a method for measuring mercaptans by using a modification of the methylene blue technique. Flue gas is drawn through impingers containing mercuric acetate to collect the mercaptans. Following collection, the methylene blue dye solution is added and a red complex is formed, which is analyzed at 500 mμ. Hydrogen sulfide and sulfur dioxide did not interfere, but results showed an interference from oxides of nitrogen at levels found in flue gases from combustion sources.

Three methods have been developed which provide for simultaneous analysis of hydrogen sulfide and mercaptans in flue gases. Felicetta [52] devised a method for collecting hydrogen sulfide and mercaptans in a solution of sodium hydroxide. Following collection, an aliquot of the solution is titrated potentiometrically with

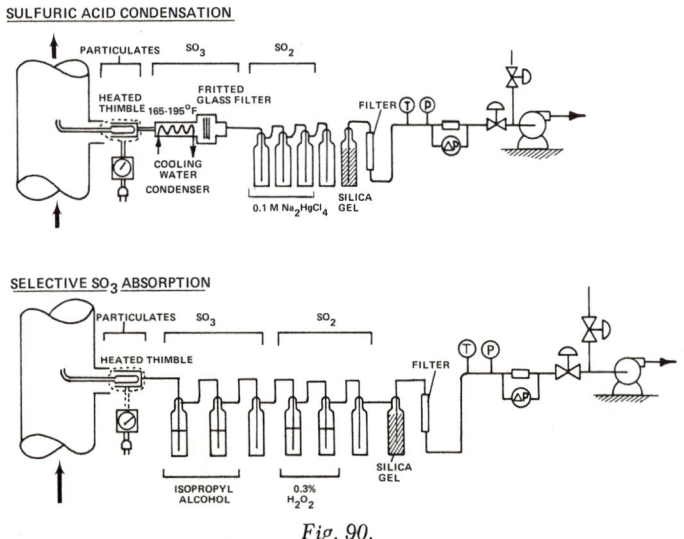

Fig. 90.

silver nitrate. The respective sulfide and mercaptide ion concentrations are determined from the respective inflection points of the titration curve. Ammonium hydroxide should be added to prevent the precipitation of silver oxide, which would interfere with the titration. The titrations should be performed rapidly to prevent oxidation of the mercaptide ion. In most instances, it is difficult to observe the mercaptide inflection point because hydrogen sulfide is normally present in the flue gas in much larger amounts than mercaptan compounds.

Harding [42] developed a method for collection of both hydrogen sulfide and mercaptans in cadmium chloride solution, where they are subsequently precipitated as their respective cadmium salts. Following collection, the sample is split into two well-mixed aliquots of equal volume. One aliquot is acidified to a pH of 1.1 to selectively dissolve the mercaptide precipitate. The sample is then filtered and then the precipitate is placed in a reaction flask, acidified, iodine added, and then titrated with sodium thiosulfate solution. This gives the hydrogen sulfide present. The other aliquot is then filtered without pH adjustment and subjected to the same analysis as the first to give the total sulfur present. Recent work by Megy [43] has shown the method to be suitable for hydrogen sulfide but that significant losses of mercaptan may occur.

A recent development employs techniques used in the two previous methods [53]. Hydrogen sulfide is first selectively absorbed in a solution of cadmium sulfate and boric acid, as methyl mercaptan has been found to pass through unreacted [56]. The mercaptan is then absorbed in subsequent impingers containing sodium hydroxide. Following collection, the hydrogen sulfide is analyzed in the cadmium sulfate by acidification and iodine thiosulfate titration. The mercaptan is analyzed in the caustic by potentiometric titration. The method provides a means for selective collection and analysis of the two constituents in flue gases.

Colombo [57] has described several methods for reduced sulfur gases, and possible limitations with their use.

5. Total Sulfur:

Bialkowsky and DeHaas [58] described a method for measuring the total sulfur present in flue gases. Gas is drawn at three liters per minute through a combustion furnace heated to at least 1400°F to oxidize the sulfur to SO_2 or SO_3. It may be necessary to add air or oxygen if there is not a sufficient amount present in the flue gas. It has been found that complete combustion occurs with five percent oxygen in the flue gas [10]. The gas then passes through a series of impingers containing three percent hydrogen peroxide, where the sulfur oxides are absorbed to form sulfate ion. The sulfate is then analyzed by either gravimetric or turbidometric means [59]. It is necessary to filter the gas stream upstream of the furnace if sulfate-containing particles are present, to eliminate possible interferences. The method provides a means for determining the sulfur present in the flue gas from a source, without attempting to differentiate between the individual compounds.

B. Nitrogen Compounds:

1. Oxides of Nitrogen:

Oxides of nitrogen are produced as byproducts from combustion processes, because of the tendency for the oxygen and nitrogen in air to react at elevated temperatures. Oxides of nitrogen are normally found in the flue gases from coal, oil, and gas-fired boilers, refuse incinerators, automobiles and buses, and nitric acid manufacturing processes. Nitric oxide is first formed in the combustion zone of boilers or incinerators, and is subsequently oxidized to nitrogen dioxide in the flue gas or in the ambient air. These two major oxides of nitrogen are of primary interest because of their potential direct health effects and their tendency to participate in photochemical reactions in the atmosphere.

Measurement of oxides of nitrogen consists of a collection step followed by analysis. The gases are soluble in water and are normally collected by grab sampling techniques. Nitric oxide is normally oxidized to nitrogen dioxide prior to or during the sample collection step. Nitrogen dioxide ionizes in solution to both nitrite and nitrate ions to complicate attempts at its analysis. Several methods which are currently available for oxides of nitrogen are listed in Table 39.

The method most commonly used for measuring oxides of nitrogen in flue gases is the phenoldisulfonic acid method [60]. Flue gas is drawn into a collection flask of known volume, which has been previously evacuated to the vapor pressure of the absorbing solution. The absorbing solution is a solution of hydrogen peroxide in 0.1 N sulfuric acid. The presence of hydrogen peroxide causes the nitric oxide and nitrogen dioxide to become oxidized to nitrate ion. Following collection, the liquid is then made alkaline by NaOH addition, evaporated to dryness, redissolved in phenoldisulfonic and sulfuric acids, treated with ammonia and diluted in water, filtered and then read in a spectrophotometer.

The method is specific and accurate for the oxides of nitrogen and can be used over a wide range of concentrations. The analytical procedure is complicated and time-consuming, however. Recent work by Mills [62] has shown a possible interference with the method caused by reaction with sulfur dioxide in tests run on oil-fired boilers. These losses occurred on storage when moisture droplets appeared on the walls of the sampling flask. These losses were eliminated by maintaining the flask at less than saturated conditions at all times prior to analysis, by heating it. An added precaution is to make the analysis as soon after sample collection as possible. It should be emphasized that the method provides a means for measuring total oxides of nitrogen and makes no attempt at differentiation. Results are reported in parts per million by volume as nitrogen dioxide.

The Griess-Saltzman method has been extensively used for measuring oxides of nitrogen in ambient air [63]. The color development reaction is sensitive to nitrite ion and therefore only nitrogen dioxide is measured directly. Nitric oxide may be measured by placing a scrubber containing potassium permanganate upstream of the collection flask. The method consists of collecting a sample of flue gas in a gas-tight syringe and injecting it into a previously partially evacuated sample collection flask containing the absorbing solution. The absorbing solution consists of a mixture of sulfanilic and acetic acids in distilled water, with an ethylenediamine dihydrochloride dye solution added as a color developing solution. The presence of nitrite ion produces a red azo dye on reaction with the absorbing solution. Measurement is at 550 mμ with a spectrophotometer.

Experience with the method to date in source test applications has been limited. The method provides, in theory, a means for differentiating between nitrogen dioxide and nitric oxide in flue gases. The concentration range of usefulness may be adjusted by changing the size of the syringe, the collection flask, and the amount of absorbing liquid used. It is necessary to correct for the portion of nitrogen dioxide which ionizes to nitrate ion in solution. Saltzman [63] found that 72 percent of the nitrogen dioxide present ionized to form nitrite ion and participated in the color development reaction. The other 28 percent ionized to form nitrate ion, which did not participate in the reaction to form the azo dye. Theoretically, equal amounts of both should be formed and it is therefore necessary to correct for this phenomenon. Patty and Petty [64] evaluated the method in field applications and found that the absorption efficiency of nitrogen dioxide was about 60 percent when using the syringe collection technique. They found no serious interference from hydrogen sulfide, but ozone was observed to oxidize nitrite to nitrate ion. Sulfur dioxide has also been found to interfere with the method. Additional work is necessary to ascertain the suitability of the method in source test applications.

Two other methods are available for analysis of oxides of nitrogen in flue gases. One is a method for total oxides of nitrogen where a gas sample is drawn into an evacuated flask containing hydrogen peroxide [61]. The solution is then titrated to a methyl red endpoint with standardized hydrogen peroxide. The method is nonspecific because sulfur dioxide, sulfur trioxide, ammonia, and nitric acid mist also react in the titration. It is useful in nitric acid manufacturing operations, but not in flue gases from coal and oil-fired boilers. Cholak and McNary [65] devised a method for measuring oxides of nitrogen by polarography. A flue gas sample was drawn into an evacuated flask containing hydrogen peroxide. The sample was then analyzed for nitrate ion by addition of lanthanum chloride and barium hydroxide.

2. Ammonia:

Ammonia is a pollutant which results from fuel combustion, decay of vegetation, and certain chemical processes. It is soluble in acidic solutions and may be measured by drawing through impingers containing dilute sulfuric acid. Following the collection step, the absorbing solution is made slightly alkaline and the ammonia then distilled off into a solution of boric acid and reabsorbed. A solution containing mercuric and potassium iodides is then added for colorimetric development. The degree of colorimetric development is then read with a spectrophotometer. The reader is referred to several other references on the subject [14] [20] [26] [59] [66] [67].

C. Other Gases:

1. Chlorine Compounds:

Chlorine compounds are sometimes emitted from certain chemical manufacturing operations such as caustic production, metallurgical plants, or bleachery operations in pulp mills. The four constituents most likely to occur are chlorine, hydrogen chloride, phosgene, and chlorine dioxide. These gases are toxic, and proper safety precautions should be used when working near sources where these can be emitted.

Two methods are currently used for analysis of chlorine. One method employs drawing flue gas through a series of impingers containing potassium iodide solution. Chlorine reacts with the iodide to liberate iodine, which then forms triiodide ion. Following collection, the absorbing solution is titrated with standardized sodium thiosulfate to a starch end point. Sulfur gases and any material which can react with the iodide ion will interfere, so the method is primarily useful only where chlorine is the main constituent being emitted. An alternative method employs drawing flue gas through impingers containing sodium hydroxide. Following sample collection, the absorbing solution is heated, neutralized with sulfuric acid, and analyzed colorimetrically at either 435 or 490 mμ by addition of ortho-toluidine color developing solution. The method is also suitable for analysis of chlorine dioxide [26].

Hydrogen chloride may be collected by drawing the flue gas through a solution of sodium hydroxide. The sample may then be analyzed by titration of the excess base with standardized sulfuric acid, or by measurement of the chloride ion concentration. This may be done either argentometrically or gravimetrically by precipitation with silver nitrate. The methods may suffer from interferences by sulfate or chloride ions present in particulate matter.

Crummett and McLean [68] have devised a method for analysis of phosgene by drawing flue gas into a solution containing distilled water and aniline. The reaction of aniline with phosgene produces diphenylurea, which is then analyzed colorimetrically at its absorbance maximum of 257 mμ. The amount of diphenylurea produced is proportional to the amount of phosgene present in the sample.

2. Carbon Monoxide:

Carbon monoxide is produced in combustion reactions by the incomplete oxidation of carbon to carbon dioxide. Normally, automobile exhaust provides the greatest source in populated areas. Flue gases from combustion sources such as oil, gas, or coal-fired boilers, refinery catalytic cracker regenerators, and Kraft pulp mill recovery furnaces may also produce significant amounts of carbon monoxide. Carbon monoxide is important as a pollutant because it inhibits oxygen transfer in the blood of humans.

Two methods are currently available for wet chemical analysis of carbon dioxide. The first is the Orsat analyzer, which has been described previously. The third absorption stage consists of a scrubber containing cuprous chloride solution. The method is useful for carbon monoxide concentrations greater than 0.2 percent by volume, or 2,000 parts per million. The difference in volume between readings taken before and after the carbon monoxide absorption step is taken as the amount of carbon monoxide present.

The second method was developed by Teague [69], and makes use of the reaction whereby carbon monoxide reacts with iodine pentoxide to liberate carbon dioxide and iodine gas.

$$5\,CO + I_2O_5 \rightarrow 5\,CO_2 + I_2 \qquad (104)$$

The flue gas is first drawn through a series of traps to remove potentially interfering substances [20]. A chromic acid washing stage removes olefinic hydrocarbons, aldehydes, and other organic materials. Potassium hydroxide removes acid spray and carbon monoxide, while silica gel removes water and sulfur gases. The gas then passes through a drying tube containing iodine pentoxide. A series of impingers containing potassium iodide are placed following the drying tube to absorb the iodine liberated. Following collection, the absorbing solution is titrated with standardized sodium thiosulfate solution to determine the amount of triiodide ion present. The amount of thiosulfate used is proportional to the carbon monoxide concentration. The method is complicated, subject to interferences if the traps become exhausted, and field experience to date is limited.

3. Organic Materials:

Organic gases and vapors can be emitted from numerous chemical processing, solvent, and degreasing operations, refuse incinerators, fuel oil and coal burning, petroleum refining, paper coating, and printing and rotogravure processes. They may be present as hydrocarbons, ketones, esters, ethers, organic acids, aldehydes, aromatics, and other compounds. They are particularly suitable for collection by freezeout techniques or solid adsorption, but may also be absorbed in impingers containing organic solvents.

Numerous methods are available for the analysis of organic compounds. A limited number of wet chemical techniques are available, as described in Jacobs [20] and Stern [26]. Los Angeles County has developed methods for analysis of organic acids, aldehydes and formaldehyde [14]. The most common techniques, however, employ instrumental methods such as gas chromatography, mass spectrometry, or infrared spectrophotometry.

IV. INSTRUMENTAL ANALYSES

Several instrumental methods are available for the analysis of gases. These techniques provide for speed and accuracy in analyzing for a wide range of gases and vapors. Their disadvantages are that the equipment used is expensive, complicated, requires highly trained personnel for successful operation, and is normally confined to the laboratory. Techniques employed include gas chromatography, ultraviolet and infrared spectrophotometry, hydrogen flame ionization, and mass spectrometry. General reviews of instrumental methods for gaseous analyses have recently been presented by Willard [70] and Lodding [71].

A. Gas Chromatography:

1. Principles:

Gas chromatography provides for separation of constituents in a gaseous mixture based on their differences in relative affinity for a given packing material in a column. This continuous process of repeated adsorption and desorption steps results in their passing through the column at different rates. They can then be analyzed separately at the column exit by means of a suitable detector. The main elements of a gas chromatograph are the carrier gas for transporting the sample, the sample injection system, the column for separating the constituent gases and vapors, a detector for measuring the amounts of each present, and a recorder. It may be necessary to maintain both the sample injection system and column at elevated temperatures to facilitate separations and eliminate losses of materials. Two recent discussions of the principles involved in gas chromatography have been presented by McNair and Bonelli [72] and Rushing [73]. A diagram of a typical gas chromatograph is shown in Figure 91.

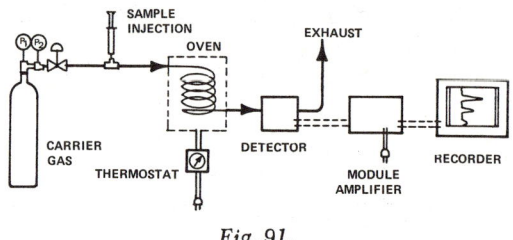

Fig. 91.

The major variables in gas chromatography are the sample handling procedure, the sample size, the column, and the detector used. Samples should be handled in inert and possibly heated containers to minimize losses. It may be necessary to concentrate samples prior to analysis to provide for a sufficient amount of material to be measured accurately. This may be done by freeze-out, solid adsorption, or absorption in an inert solvent or a chemically reactive solution. The size of the sample injected, via either syringe or sample loop, determines the amounts of given constituents being emitted from the column. The degree of separation achieved and suitability of a given column depends on its length, diameter, material, packing, and support. The detector used must be able to measure the constituents of interest and be sufficiently sensitive.

Detectors frequently used include thermal conductivity, electron capture, flame ionization, flame photometry, and thermistor. Thermal conductivity is theoretically sensitive to any gas but is limited in its sensitivity. Electron capture is useful for halogen-containing compounds, oxides of nitrogen, oxides of sulfur, and certain other gases, but is adversely affected by water. Flame ionization is useful for organic materials, while flame photometry is useful for phosphorus- and sulfur-containing substances.

There are several limitations involved in gas chromatography. First, it is not suitable for truly continuous monitoring since it requires taking a series of batch samples. Second, it is complicated and subject to numerous operational problems. Third, properties of materials are often measured which are not necessarily unique, and careful selection of column and detector is required. Fourth, separation of individual compounds on columns is often difficult and requires tedious experimentation. Fifth, losses often occur in the column which cause low results. Sixth, detectors are often insensitive to low concentrations of materials. In summary, the method provides a potentially useful method for analysis of a wide range of materials, but is very complicated in its operation.

2. Applications:

Gas chromatography has found frequent use in measuring constituents in flue gases. Grune [74] described the use of a thermal conductivity detector, and the column technology used in the separation and subsequent analysis of sulfur dioxide, methyl mercaptan, nitrogen dioxide, carbon dioxide, hydrogen sulfide, methane, and ammonia. Rushing [73] described the use of thermal conductivity detectors in the analysis of organic solvents. The lighter molecular weight constituents of a given hydrocarbon group were observed to pass through the column faster than the heavier ones, and were analyzed first. Ettre [75] described the use of thermal conductivity and flame ionization detectors, plus the column technology involved in the separation and analysis of complex hydrocarbon mixtures. Tuttle [76] used a thermal conductivity detector for analyzing hydrocarbons present in refuse incinerator flue gases. Samples were collected in heated bottles and passed through Drierite for water removal prior to analysis.

Gas chromatography has also been applied in the analysis of flue gases from industrial sources. Hayek [77] described the use of a thermal conductivity detector in the analysis of CO_2, CO, CH_4, and other gases from metallurgical sources. Adams [78] described the use of a bromine microcoulometric detector in the analysis of reduced sulfur gases in Kraft mill sources. Included were discussions of the column technology involved and the detector response. Blosser and Cooper [15] presented the results of a recent seminar on sample handling techniques and column technology in gas chromatographic analyses of Kraft mill sources. Walther and Amberg [79] used two gas chromatographs outfitted in a mobile laboratory to make a thorough study of reduced sulfur gas emissions from a Kraft pulp mill.

B. Spectrophotometry:

Both infrared and ultraviolet spectrophotometry have been utilized for analyses of gaseous constituents in flue gases. The principle involved is the selective absorption of light at a given wavelength by a particular substance. The degree of absorption is proportional to the concentration of the particular compound. These instruments may be adapted for continuous as well as single sample analyses.

1. Infrared Analyzers:

Two different type of infrared instruments are available. The nondispersive type relies on selective absorption of infrared light at a particular wavelength by a single compound. It makes a comparison in signals between a reference and a sample cell. It may be adapted for continuous monitoring of oxides of nitrogen, hydrocarbons, carbon monoxide, carbon dioxide, and ammonia by drawing gas through the sample cell.

Dispersive infrared analyzers make a wavelength scan from one to 16 microns on a single sample. It is necessary that the gas be dried prior to being placed in the sample cell, to prevent lens damage. Variable sensitivity may be attained by changes in the path length. Different constituents may be measured by their characteristic peaks in the infrared region. It is not adaptable for continuous monitoring.

Infrared techniques have been used in analyses of both auto exhaust and organic solvent emissions. Twiss [80] employed grab sampling of automobile exhaust gases followed by water removal with Drierite. The samples were then analyzed by dispersive infrared techniques to determine the hydrocarbons present. McPhee [81] used the characteristic carbon dioxide peak for analyses of organic solvents. The vapors were collected in a two-liter flask by a freeze-out technique, then passed through a combustion furnace where they were burned to carbon dioxide, and thence into a nondispersive infrared analyzer. The amount of carbon dioxide present was taken as being proportional to the amount of organic solvent material originally present in the flue gas.

2. Ultraviolet Analyzers:

Ultraviolet radiation may be used for analysis of gaseous constituents in a manner analogous to those employed with infrared techniques. Monitors have recently been developed for monitoring sulfur dioxide in flue gas based on its peak in the UV region. The monitor is placed in the duct and a beam of light at the given wavelength passed across. The degree of absorption is taken as being proportional to the sulfur dioxide concentration [82] [83]. Particulates and moisture may cloud the lenses and interfere with the method, however.

C. Flame Ionization:

Flame ionization is useful for detecting the amount of total organic constituents present in flue gas. The principle of operation is that a small current is detected when a gas stream containing carbon atoms is passed through a hydrogen flame where an electrical potential is applied across the flame. Oxidation of the carbon atoms produces the current, which is proportional to the number of carbon atoms being reacted. King [84] describes its use in analysis of auto exhaust. Gas is drawn from the source at a known rate, mixed with air and hydrogen, and passed through a combustion chamber. The current of the potential field across the flame is amplified, recorded, and used as being proportional to the hydrocarbon or total organic carbon concentration. It is useful in a limited number of sources for this application.

D. Mass Spectrometry:

Mass spectrometry makes use of the electronic charge to mass ratio for ions as a means of their identification. It operates by placing an ionic charge on gases in a mixture and then recording the mass spectrum as a function of ionic current (concentrations) at given mass-to-charge ratios (constituents). The method provides a potentially powerful tool for analysis of a wide number of constituents, and may be used as a gas chromatography detector. It is an extremely expensive instrument, and requires highly trained personnel for operation. Shepherd [85] describes the use of mass spectrometry in determining the types and amounts of individual constituents present in ambient air. King [84] describes the use of mass spectrometry in analyses of auto exhaust.

V. SAMPLE PROBLEM

Sample Problem No. 12. Gaseous Emissions.

Seventy-five milligrams of sulfur dioxide was collected from 2.0 ft³ of gas at meter conditions of 70°F, 27.8 inches of Hg, and 3.0 percent moisture by volume. Stack conditions during the test were 275°F, 29.70 inches of Hg, and 30.0 percent moisture by volume. Calculate the SO_2 concentration in ppm by volume and the emission rate in lbs per day for a gas flow rate of 70,000 cfm.

Solution:

$(V/m)_i$ = Volume of constituent gas per unit mass at stack conditions is ft³ per mg.

V/n = Volume occupied by a lb-mole of gas in ft³ per lb-mole.

MW_i = Molecular weight of constituent gas in lb per lb-mole.

m_i = Mass of constituent gas collected in milligrams.

T_s = Absolute flue gas temperature in °R.

T_m = Absolute meter temperature in °R.

P_s = Absolute static pressure of flue gas in inches of Hg.

P_m = Absolute static pressure of meter in inches of Hg.

MC_s = Moisture content of flue gas in percent by volume.

MC_m = Moisture content of saturated gas at meter in percent by volume.

V = Gas volume at meter conditions in ft³.

V_i = Volume of constituent gas collected in ft³.

V_s = Gas volume sampled at stack conditions in ft³.

Q_s = Volumetric stack flow rate in ft³ per minute.

C_i^T = Constituent gas concentration in parts per million by volume.

Expression for Volume per Unit Mass:

$$PV = nRT = \frac{m}{MW}RT$$

$$R = 21.85 \frac{\text{ft}^3\text{-in.Hg}}{\text{lb-mole-}°\text{R}} \quad (105)$$

$$\left(\frac{V}{m}\right)_i = \frac{RT_s}{MW_i P_s}$$

$$\left(\frac{V}{m} = \frac{\text{ft}^3}{\text{mg}}\right)_i = \frac{\left(21.85 \frac{\text{ft}^3\text{-in.Hg}}{\text{lb-mole-}°\text{R}}\right)}{\left(454 \frac{gm}{lb} \times 10^3 \frac{mg}{gm}\right)} \times$$

$$\frac{(T_s - °\text{R})}{(P_s\text{-in.Hg})(MW_i - \text{lb/lb mole})}$$

$$\left(\frac{V}{m} = \frac{\text{ft}^3}{\text{mg}}\right)_i = \frac{21.85 \times 10^{-3}}{454} \times$$

$$\frac{(T_s - °\text{R})}{(P_s\text{-in.Hg})(MW_i - \text{lb/lb mole})}$$

$$\left(\frac{V}{m} = \frac{\text{ft}^3}{\text{mg}}\right)_i = 48.2 \times 10^{-6} \times$$

$$\frac{(T_s - °\text{R})}{(P_s\text{-in.Hg})(MW_i - \text{lb/lb mole})} \quad (106)$$

Expression for Volume of Pollutant Gas:

$$(V_i - \text{ft}^3) = \left(\frac{V}{m} - \frac{\text{ft}^3}{\text{mg}}\right)_i (M_i - \text{mg}) \quad (107)$$

Expression for Gas Volume Sampled (Stack Conditions):

$$(V_s - \text{ft}^3) =$$
$$\frac{(T_s - °\text{R})}{(T_m - °\text{R})} \frac{(P_m\text{-in.Hg})}{(P_s\text{-in.Hg})} \frac{(100 - MC_m)}{(100 - MC_s)} \times$$
$$(V_m - \text{ft}^3) \quad (108)$$

Expression for Gas Concentration:

$$(C_i - \text{ppm}) = \frac{(V_i - \text{ft}^3)}{(V_s - \text{ft}^3)} \times 10^6 \quad (109)$$

Expression for Volume per Mole of Gas:

$$359.0 \text{ ft}^3 = 1.0 \text{ lb-mole at } 32°\text{F}$$

$$\left(\frac{V}{n} - \frac{\text{ft}^3}{\text{lb-mole}}\right)$$
$$= \left(359.0 \frac{\text{ft}^3}{\text{lb-mole}} \frac{(T_s - °\text{R})}{(492 °\text{R})}\right) \quad (110)$$

Expression for Emission Rate of Gas:

$$ER_i = \left(1440 \frac{\text{min}}{\text{day}}\right)\left(\times 10^{-6} \frac{\text{ft}^3_i/\text{ft}^3_{\text{gas}}}{\text{ppm}}\right) \times$$
$$\frac{(MW_i - \text{lb/lb-mole})}{(V/n - \text{ft}^3/\text{lb-mole})} \times$$
$$\left(Q_s - \frac{\text{ft}^3}{\text{min}}\right)(C_i - \text{ppm})$$

$$ER_i = (1{,}440 \times 10^{-6}) \times$$
$$\frac{(MW_i - \text{lb/lb-mole})}{(V/n - \text{ft}^3/\text{lb-mole})} \times$$
$$\left(Q_s - \frac{\text{ft}^3}{\text{min}}\right)(C_i - \text{ppm}) \quad (111)$$

Measured:

$T_s = 735°\text{R}$; $T_m = 530°\text{R}$; $MC_s = 30.0\%$ by vol
$P_s = 29.70$ in.Hg; $P_m = 27.8$ in.Hg;
$MC_m = 3.0\%$ by vol
$V = 2.0 \text{ ft}^3$; $Q_s = 70{,}000 \text{ ft}^3/\text{min}$;
$m_{SO_2} = 75.0$ mg

$359.0 \text{ ft}^3/\text{lb-mole at } 32°\text{F}, 29.92 \text{ in.Hg};$
$MW_{SO_2} = 64.0$

Amount of SO_2 Collected:

$$(V/m)_{SO_2} = 48.2 \times 10^{-6} \frac{1}{MW_i} \frac{T_s}{P_s}$$

$$(V/m)_{SO_2} = 48.2 \times 10^{-6} \frac{1}{64} \frac{735}{29.70}$$
$$= 18.6 \times 10^{-6} \text{ ft}^3/\text{mg}$$

$$V_{SO_2} = (V/m)_{SO_2}(m_{SO_2}) = (18.6 \times 10^{-6}) \times$$
$$(75.0) = 1{,}395. \times 10^{-6} \text{ ft}^3$$

Sample Volume:

$$V_s = \frac{T_s}{T_m} \frac{P_m}{P_s} \frac{(100 - MC_m)}{(100 - MC_s)} V$$

$$V_s = \frac{735}{530} \frac{27.8}{29.7} \frac{97.0}{70.0} (2.0) = (1.80)(2.0)$$

$$V_s = 3.60 \text{ ft}^3$$

Gas Concentration:

$$C_{SO_2} = \frac{V_{SO_2}}{V_s} \times 10^6 = \frac{1{,}395 \times 10^{-6}}{3.60} \times 10^6$$

$$C_{SO_2} = 387 \text{ ppm}$$

Emission Rate:

$$V/n = (359.0)\frac{T_s}{492} = (359.0)\frac{(735)}{(492)} =$$

$$V/n = 536 \text{ ft}^3/\text{lb-mole}$$

$$ER_{SO_2} = 1440 \times 10^{-6} \frac{(MW_{SO_2})}{(V/n)}(Q_s)(C_{SO_2})$$

$$ER_{SO_2} = 1440 \times 10^{-6} \frac{(64)}{(536)}(70{,}000)(387)$$

$$ER_{SO_2} = 4{,}660 \text{ lb/day}$$

REFERENCES

1. Altshuller, A. P., and Wartburg, A. F., "The Interaction of Ozone with Plastic and Metallic Materials in a Dynamic System, *International Journal of Air and Water Pollution*, Volume 4, Nos. 1/2, pp. 70-78, January—February 1961.
2. Calvert, S., and Workman, W., "The Efficiency of Small Gas Absorbers," *American Industrial Hygiene Association Journal*, Volume 22, No. 4, pp. 318-324, July—August 1961.
3. Calvert, S., and Workman, W., "Estimation of Efficiency for Bubbler-Type Gas Absorbers," *Talanta*, Volume 4, No. 2, pp. 89-100, April 1960.
4. Hendrickson, E. R., "Air Sampling," Chapter 11 in Stern, A. C., ed., *Air Pollution*, Volume 1, 1st ed., Academic Press, New York, New York, 1962.
5. Roberts, L. R., and McKee H. C., "Evaluation of Absorption Sampling Devices," *Journal of the APCA*, Volume 9, No. 1, pp. 51-53, May 1959.
6. Bergshoeff, G., "Improved Absorbers for Sampling Air Contaminants," *Air and Water Pollution*, Volume 10, No. 9, pp. 629-631, September 1966.
7. Bostrom, C. E., "The Absorption of Sulfur Dioxide at Low Concentrations Studied by an Isotopic Tracer Method," *Air and Water Pollution*, Volume 9, No. 6, pp. 333-341, June 1965.
8. Bracewell, J. M., and Hodgeson, A. E. M., "The Hydrogen Peroxide Method for Sulfur Dioxide in the Atmosphere—Efficiency at Low Concentrations by a Radioactive Tracer Technique," *Air and Water Pollution*, Volume 9, Nos. 7/8, pp. 431-438, August 1965.
9. Bostrom, Carl—Elis, "The Absorption of Low Concentrations of Hydrogen Sulfide in a Cd(OH)$_2$ Suspension as Studied by an Isotopic Tracer Method," *International Journal of Air and Water Pollution*, Volume 10, Nos. 6/7, pp. 435-441, June—July 1966.
10. "Laboratory Evaluation of Gas Sampling Procedure for Recovery Furnace Flue Gases," Atmospheric Pollution Technical Bulletin No. 11, National Council for Air and Stream Improvement, New York, New York, December 1959.
11. Turk, A., Sleik, H., and Messer, P., "Determination of Gaseous Air Pollution by Carbon Adsorption," *American Industrial Hygiene Association Quarterly*, Volume 13, No. 1, pp. 23-28, March 1952.
12. Feldstein, M., Balestrieri, S., and Levaggi, D., "The Use of Silica Gel in Source Testing," *American Industrial Hygiene Association Journal*, Volume 28, No. 4, pp. 381-385, July—August 1967.
13. Yocom, J. E., Hein, G. M., and Nelson, H. W., "A Study of the Effluents from Back-Yard Incinerators," *Journal of the APCA*, Volume 6, No. 2, pp. 84-89, August 1956.
14. Devorkin, H., Chass, R. L., Fudurich, A., and Kanter, C. V., ed., Holmes, R. G., "Source Testing Manual," County of Los Angeles, Air Pollution Control District, Los Angeles, California, 1965.
15. Blosser, R. O., and Cooper, H. B. H., ed., "Analytical Equipment and Monitoring Devices for Gases and Particulates," Proceedings of the West Coast Region Air Quality Control Workshop, Atmospheric Pollution Technical Bulletin No. 35, National Council for Air and Stream Improvement, New York, New York, March 6, 1968.
16. Brief, R. S., and Drinker, P. A., "Collection of Integrated Samples of Gaseous Effluents," *A.M.A. Archives of Industrial Health*, Volume 17, No. 6, pp. 654-658, June 1958.
17. Baker, R. A., and Doerr, R. C., "Methods of Sampling and Storage of Air Containing Vapors and Gases." *International Journal of Air and Water Pollution*, Volume 2, No. 2, pp. 142-158, October 1959.
18. Altshuller, A. P., Wartburg, A. F., Cohen, I. R., and Sleva, S. F., "Storage of Vapors and Gases in Plastic Bags," *International Journal of Air and Water Pollution*, Volume 6, No. 1, pp. 75-81, January—February 1962.
19. "Specifications for Incinerator Testing at Federal Facilities," U.S. Public Health Service, National Air Pollution Control Administration, Durham, North Carolina, October 1967.
20. Jacobs, M. B., *The Chemical Analysis of Air Pollutants*, Interscience Publishers, Inc., New York, New York, 1960.
21. Clayton, G. D., ed., "Stack Sampling," Chapter 9 in *Air Pollution Manual, Part 1—Evaluation*, American Industrial Hygiene Association, Detroit, Michigan, 1961.
22. "Selected Methods for the Measurements of Air Pollutants," U.S. Public Health Service Publication No. 999-AP-11, Cincinnati, Ohio, May 1965.
23. "Recommended Methods in Air Pollution Studies," California State Department of Public Health, Air and Industrial Hygiene Laboratory, Berkeley, California, 1962.
24. Ruch, W., *Chemical Detection of Gaseous Pollutants*, Ann Arbor Science Publishers, Inc., Ann Arbor, Michigan, 1967.
25. Magill, P. L., Holden, F. R., and Ackley, C., *Air Pollution Handbook*, McGraw-Hill Book Co., New York, New York, 1956.
26. Katz, M., "Analysis of Inorganic Gaseous Pollutants," Chapter 17 in Stern, A. C., ed., *Air Pollution*, Volume 2, 2nd ed., pp. 53-114, Academic Press, Inc., New York, New York, 1968.
27. Altshuller, A. P., Analysis of Organic Gaseous Pollutants," Chapter 18, in Stern, A. C., ed., *Air Pollution*, Volume 2, 2nd ed., pp. 115-145, Academic Press, Inc., New York, New York, 1968.
28. Nestell, R. J., and Anderson E., "Determination of Sulfur Dioxide and Sulfur Trioxide in Flue Gases,"

Journal of Industrial and Engineering Chemistry, Volume 8, No. 3, pp. 258-260, March 1916.

29. Goksoyr, H., and Ross, K., "The Determination of Sulfur Trioxide in Flue Gases," *Journal of the Institute of Fuel,* Volume 35, No. 255, pp. 177-179, April 1962.

30. Hissink, M., "A. Instrument for Determining Sulfur Dioxide in Flue Gases," *Journal of the Institute of Fuel,* Volume 36, No. 272, pp. 372-376, September 1963.

31. Lisle, E. S., and Sensenbaugh, J. D., "The Determination of Sulfur Trioxide and Acid Dew Point in Flue Gases," *Combustion,* Volume 36, No. 7, pp. 12-16, January 1965.

32. Radwanska, A., "Determination of SO_2 and SO_3 in Waste Gas," *Gazetta Woda i Technika Sanitrana* (Warsaw), Volume 41, No. 5, pp. 171-173, May 1967.

33. Wang, G. K. M., "An Instrument for Determining Sulfur Oxides in Flue Gases," *Combustion,* Volume 38, No. 11, pp. 46-49, May 1967.

34. Flint, D., "A Method for the Determination of Small Concentrations of Sulfur Trioxide in the Presence of Larger Concentrations of Sulfur Dioxide," *Journal of the Society of the Chemical Industry* (London), Volume 67, No. 1, pp. 2-5, January 1948.

35. Seidman, E. B., "Determination of Sulfur Oxides in Stack Gases," *Analytical Chemistry,* Volume 30, No. 10, pp. 1680-1682, October 1958.

36. Fiedler, R. S., Jackson, P. J., and Raask, E., "The Determination of Sulfur Trioxide and Dioxide in Flue Gases," *Journal of the Institute of Fuel* (London), Volume 33, No. 229, pp. 84-89, February 1960.

37. Nacovsky, W., "Determination of Sulfur Oxides in Flue Gases," *Combustion,* Volume 38, No. 7, pp. 35-38, January 1967.

38. West, P. W., and Gaeke, G. C., "Fixation of Sulfur Dioxide as Disulfitomercurate (II), Subsequent Colorimetric Estimation," *Analytical Chemistry,* Volume 28, No. 10, pp. 1816-1819, December 1956.

39. Corbett, P. F., "A Photometric Method for the Estimation of Sulfur Trioxide in the Presence of Sulfur Dioxide," *Journal of the Society of the Chemical Industry,* Volume 67, No. 6, pp. 227-230, June 1948.

40. "Atmospheric Emissions from Sulfuric Acid Manufacturing Processes," U.S. Public Health Service Publication No. 999-AP-13, Washington, D.C., 1965.

41. "Flue and Exhaust Gas Analyses," Performance Test Code 19.10 1968, American Society of Mechanical Engineers, New York, New York, 1968.

42. Harding, C. I., "A Méthod for Measuring the Concentration of Sulfur Compounds in Process Gas Streams," National Council for Air and Stream Improvement Atmospheric Pollution Technical Bulletin No. 28, New York, New York, December 9, 1965.

43. Blosser, R. O., Cooper, H. B. H., Owens, E. L., and Megy, J. A., "West Coast Research Center Progress Report," National Council for Air and Stream Improvement, Corvallis, Oregon, October 1967.

44. Hochheiser, S., "Methods of Measuring and Monitoring Atmospheric Sulfur Dioxide," U.S. Public Health Service Publication No. 999-AP-6, Cincinnati, Ohio, August 1964.

45. Chory, J. P., "Method for Quantitative Determination of SO_3 and SO_2 in Flue Gases," *Brennstoff-Wärme-Kraft,* Volume 14, No. 12, pp. 601-603, December 1962.

46. Grondona, A., and Marcucci, G. P., "An Automatic Analyzer for Determining Sulfur Trioxide in Flue Gases," *Termotechnica* (Milan), Volume 14, No. 4, pp. 185-188, April 1960.

47. Bent, R., Ladner, W., and Mullin, W., "A Method for the Estimation of Sulfur Trioxide," *Chemistry and Industry,* No. 11, pp. 461-462, March 18, 1967.

48. Jacobs, M. B., Braverman, M. M., and Hochheiser, S., "Ultramicrodetermination of Sulfides in Air," *Analytical Chemistry,* Volume 29, No. 9, pp. 1349-1351, September 1957.

49. Budd, M. S., and Bewick, H. A., "Photometric Determination of Sulfide and Reducible Sulfur in Alkalies," *Analytical Chemistry,* Volume 24, No. 10, pp. 1536-1540, October 1952.

50. "Laboratory Evaluation of Gas Sampling Procedure for Recovery Furnace Stack Gases," Atmospheric Pollution Technical Bulletin No. 11, National Council for Air and Stream Improvement, New York, New York, December 1959.

51. Murray, F. E., and Rayner, H. B., "Procedure for Sampling and Analysis of Hydrogen Sulfide in Kraft Stack Gases," *Tappi,* Volume 44, No. 3, pp. 219-221, March 1961.

52. Felicetta, V. F., Peniston, Q. P., and McCarthy, J. L., "Determination of Hydrogen Sulfide, Methyl Mercaptan, Dimethyl Sulfide, and Disulfide in Kraft Pulp Mill Process Streams," *Tappi,* Volume 36, No. 9, pp. 425-432, September 1953.

53. Blosser, R. O., Cooper, H. B. H., Owens, E. L., and Megy, J. A., "West Coast Research Center Progress Report," National Council for Air and Stream Improvement, Corvallis, Oregon, October 1968.

54. Moore, H., Helwig, H. L., and Graul, R. J., "A Spectrophotometric Method for the Determination of Mercaptans in Air," *American Industrial Hygiene Association Journal,* Volume 21, No. 6, pp. 466-470, December 1960.

55. Adams, D. F., "Analysis of Malodorous Sulfur-Containing Gases," *Tappi,* Volume 52, No. 1, pp. 53-58, January 1969.

56. Thoen, G. N., DeHaas, G. G., and Austin, R. R., "Instrumentation for Quantitative Measurement of Sulfur Compounds in Kraft Gases," *Tappi,* Volume 51, No. 6, pp. 246-248, June 1968.

57. Colombo, P., Corbetta, D., Pirotta, A., and Sartori, A. "Critical Discussion on the Analytical Methods for Mercaptan and Sulfur Compounds," *Tappi,* Volume 40, No. 6, pp. 490-498, June 1957.

58. Bialkowsky, H. W., and DeHaas, G. G., "A Catalytic Oxidation Procedure for Determining Sulfur Compounds in Kraft Mill Gases," *Tappi,* Volume 36, No. 7, pp. 330-336, July 1953.

59. "Standard Methods for the Examination of Water and Wastewater," 12th ed., American Public Health Association, New York, New York, 1965.

60. Beatty, R. L., Berger, L. B., and Shrenk, H. H., "Determination of Oxides of Nitrogen by the Phenolsulfonic Acid Method," U.S. Bureau of Mines Report of Investigations No. 3687, February 1943.

61. "Atmospheric Emissions from Nitric Acid Manufacturing Processes," U.S. Public Health Service Publication No. 999-AP-27, Washington, D.C., 1966.

62. Mills, J. L., Luedtke, K. D., Woolrich, P., and Perry, L., "Emissions of Oxides of Nitrogen from Stationary Sources in Los Angeles County," Los Angeles County Air Pollution Control District, Los Angeles, California, July 1961.

63. Saltzman, B. E., "Colorimetric Microdetermination of Nitrogen Dioxide in the Atmosphere," *Analytical*

Chemistry, Volume 26, No. 12, pp. 1949-1955, December 1954.

64. Patty, F. A., and Petty, G. M., "Nitrite Field Method for the Determination of Oxides of Nitrogen," *Journal of Industrial Hygiene and Toxicology,* Volume 25, No. 8, pp. 361-365, October 1963.

65. Cholak, J., and McNary, R., "Determination of the Oxides of Nitrogen in Air," *Journal of Industrial Hygiene and Toxicology,* Volume 25, No. 10, pp. 354-360, October 1943.

66. Bloomfield, B. D., "Source Testing," Chapter 28 in Stern, A. C., ed., *Air Pollution,* Volume 2 2nd ed., pp. 487-536, Academic Press, Inc., New York, New York, 1968.

67. *1965 Book of Industrial Standards,* part 23, Industrial Water; Atmospheric Analyses, American Society for Testing and Materials, Philadelphia, Pennsylvania, 1965.

68. Crummett, W. B., and McLean, J. D., "Ultraviolet Spectrophotometric Determination of Trace Quantities of Phosgene in Gases," *Analytical Chemistry,* Volume 37, No. 3, pp. 424-425, March 1965.

69. Teague, M. C., "The Determination of Carbon Monoxide in Air Contaminated with Motor Gases," *Journal of Industrial and Engineering Chemistry,* Volume 12, No. 10, pp. 964-969, October 1920.

70. Willard, H. H., Merritt, L. L., and Dean, J. A., "*Instrumental Methods of Analysis,* 3rd ed., D. Van Nostrand Co., Princeton, New Jersey, January 1964.

71. Lodding, W., Gas Effluent Analysis, Marcel Dekker, Inc., New York, New York, 1967.

72. McNair, H. M., and Bonelli, E. J., "Basic Gas Chromatography," Varian Aerograph Corp., Walnut Creek, California, 1967.

73. Rushing, D. E., "Gas Chromatography in Industrial Hygiene and Air Pollution Problems," *American Industrial Hygiene Association Journal,* Volume 19, No. 3, pp. 238-245, May—June 1958.

74. Grune, W. N., "Analysis by Gas Chromatography, *Industrial Water and Wastes,* Volume 7, No. 2, pp. 29-36, March—April 1962 and Volume 2, No. 3, pp. 72-74, May—June 1962.

75. Ettre, L. S., "Application of Gas Chromatographic Methods for Air Pollution Studies," *Journal of the APCA,* Volume 11, No. 1, pp. 34-41, January 1961.

76. Tuttle, W. N., and Feldstein, M., "Gas Chromatographic Analysis of Incinerator Effluents," *Journal of the APCA,* Volume 10, No. 6, pp. 427-429, December 1960.

77. Hayek, H., Mandl, M., and Doubek, J., "The Use of Gas Chromatography in Sampling Metallurgical Fumes," *Hutnicke Listy,* Volume 21, No. 8, pp. 532-538, August 1966.

78. Adams, D. F., and Koppe, R. K., "Direct GLC Coulometric Analysis of Kraft Mill Gases," *Journal of the APCA,* Volume 17, No. 3, pp. 161-165, March 1967.

79. Walther, J. E., and Amberg, H. R., "Experience with a Mobile Laboratory in Source Sampling Kraft Mill Emissions," *Tappi,* Volume 51, No. 11, pp. 126A-129A, November 1968.

80. Twiss, J. B., Teague, D. M., Bozek, J. W., and Sink, M. V., "Application of Infrared Spectroscopy to Exhaust Gas Analysis," *Journal of the APCA,* Volume 5, No. 2, pp. 75-83, August 1955.

81. MacPhee, R. D., and Kuramoto, M., "Methods of Organic Solvent Analyses Used by the Air Pollution Control District," Paper 68-7, Presented at the 61st Annual Meeting of APCA, St. Paul, Minnesota, June 24, 1968.

82. Bailey Meter Company; Instrument Division, Wickliffe, Ohio, 1968.

83. Barringer Research, Ltd., 304 Carlingview Drive, Rexdale, Ontario, Canada.

84. King, W. J., Wilson, K., and Schwartz, D. J., "Analysis of Automotive Exhaust Gas," *Journal of the APCA,* Volume 12, No. 1, pp. 5-21, January 1963.

85. Shepherd, M., Rock, S. M., Howard, R., and Stormes, J., "Isolation, Identification, and Estimation of Gaseous Pollutants in Air," *Analytical Chemistry,* Volume 23, No. 10, pp. 1431-1440, October 1951.

Table 36. Cooling Agents Used for Freezeout Traps.

Cooling Agent	Temperature		
	°C	°F	°R
Ice-Water	0°	32	492
Ice-Salt	−21	−6	454
Dry Ice	−79	−110	350
Liquid Air	−147	−233	227
Liquid Nitrogen	−183	−298	162

Courtesy of: Academic Press, Inc., New York, N. Y., Hendrickson, E.R., "Air Sampling" 1962 [4].

Table 37. Analytical Methods for Sulfur Dioxide and Sulfur Trioxide.

Method	Sulfur Trioxide			Sulfur Dioxide		Reference
	Separation	Collection	Analysis	Collection	Analysis	
1	Humidification	Filtration	NaOH Titration	Scrub in Na_2CO_3	HCl Titration	[28]
2	Condensation	Filtration	NaOH Titration	Scrub in H_2O_2	NaOH Titration	[29] [30] [31] [32] [33]
3	Solubility*	Scrub in i-C_3H_7OH	$BaSO_4$ Precipitation, Gravimetry	Scrub in H_2O_2	$BaCl_2$ Precip.	[34]
4	Solubility*	Scrub in i-C_3H_7OH	$BaCl_2$ Titration, Thorin Indicator	Scrub in H_2O_2 (Total)	$BaCl_2$ Precip.	[35]
5	Solubility*	Scrub in i-C_3H_7OH	$BaSO_4$ Turbidity	Scrub in I_2 Soln.	$Na_2S_2O_3$ Titration	[36] [37]
6	Condensation	Filtration	NaOH Titration	Scrub in Na_2HgCl_4	West Gaeke Colorimetric	[38]
7	Solubility	Filtration	NaOH Titration	Scrub in Na_2HgCl_4	West Gaeke Colorimetric	[38]
8	Condensation	Filtration	NaOH Titration	Scrub in NaOH	$BaCl_2$ Precip.	[14]

Note: * Two parallel trains. Gives total sulfur oxides. Sulfur dioxide is determined by difference.

Table 38. Analytical Methods for Hydrogen Sulfide and Mercaptans.

Method	Hydrogen Sulfide		Methyl Mercaptan*		Reference
	Absorbing Solution	Analytical Method	Absorbing Solution	Analytical Method	
1	$Cd(OH)_2$	Colorimetric (Meth. Blue at 670 mμ)	—	—	[48]
2	$Zn(Ac)_2$	Colorimetric (Meth. Blue at 670 mμ)	—	—	[49] [50]
3	$Zn(Ac)_2$	Colorimetric (Lauth's Violet)	—	—	[51]
4	NaOH	Potentiometric Titration $AgNO_3$	NaOH	Potentiometric Titration $AgNO_3$	[52]
5	$CdCl_2$	Titration I_2-$Na_2S_2O_3$	$CdCl_2$	Titration I_2-$Na_2S_2O_3$	[42]
6	$CdSO_4$	Titration I_2-$Na_2S_2O_3$	NaOH	Potentiometric Titration with $AgNO_3$	[53]
7	—	—	$HgAc_2$	Colorimetric Methylene Blue at 500 mμ	[54]

Note: * Includes higher mercaptans in addition to the methyl constituent.

Table 39. Analytical Methods for Oxides of Nitrogen.

Method	Nitric Oxide			Nitrogen Dioxide			Reference
	Name	Treatment	Location	Collection Liquid	Ion Detected	Analytical Method	Limitations	
1	Phenoldisulfonic Acid	Oxidation	In Flask	H_2O_2-H_2SO_4	NO_3^-	Colorimetric	Not less than 5 ppm SO_2 interference	[60] [61]
2	Griess-Saltzman	Oxidation	$KMnO_4$ Scrubber	Acetic plus Sulfanilic Acid	NO_2^-	Colorimetric	SO_2 Interference not more than 500 ppm	[61] [63]
3	Hydrogen Peroxide	Oxidation	In Flask	H_2O_2	NO_3^-	NaOH Titration	SO_2, SO_3, NH_3 Interference (Nonspecific)	[61]
4	Polarographic	Oxidation	In Flask	H_2O_2	NO_3^-	Polarography	—	[65]

CHAPTER 10

CONTINUOUS MONITORING

I. INTRODUCTION

Continuous monitoring of source gas streams is useful in providing a record of emission levels for one or more constituents over an extended time period. It is useful in process control, particularly as a means of indicating irregular or upset conditions for process or combustion units. It can provide a continuous record of losses to the atmosphere as a means of computing materials inventories. Its main purpose in air pollution work is to provide a continuous record of emissions for one or more gaseous or particulate materials over a prolonged time interval. This makes it possible to avoid reliance on a single or limited number of batch samples as being representative of actual emissions. Maximum and minimum values may then be observed for individual sources and related to long term average conditions.

Information obtained regarding emission levels of particular constituents may be used for several purposes. First, it is possible to ascertain whether a source is in compliance with existing air pollution regulations. Second, the information obtained may be used by an industrial plant as protection against potential lawsuits or nuisance complaints regarding particular "episodes." Third, sudden changes in emission levels may be used as indicators of process unit or control equipment malfunctions, to show the need for correction. Fourth, studies may be made by variations in process equipment operation to determine optimum conditions for minimum air pollutant emissions. Fifth, the information may be used as the basis for evaluating control devices and programs relative to possible future need for additional improvements [1].

The wealth of information obtained with continuous monitoring is often accompanied by an increase in the complexity of the sampling equipment used as compared to batch sampling techniques. The added complexity makes the equipment more expensive, and often requires the services of highly trained personnel for successful operation. It is desirable to keep this increase in equipment complexity, expense, and skilled personnel to a minimum. The objective is a reliable and accurate system which can operate for an extended period with a minimum of attention and maintenance. Other requirements are that results be reproducible and that the device be sensitive and respond rapidly to changes in concentration. The device should also be able to monitor accurately over the entire range of concentrations to be expected. It is necessary to calibrate monitoring instruments at regular intervals to assure their continued accuracy. Routine servicing and preventive maintenance is also necessary to assure trouble-free operation for extended periods.

II. PARTICULATE MONITORING

Several different methods are potentially useful for monitoring particulate concentrations in source gases [1]. One method employs attenuation of light or heat energy by particles in the gas stream. The degree of absorption of energy by the particles is then taken as being proportional to the particulate concentration. Sound or nuclear radiation may be employed as alternative sources of energy employing the same technique. A second technique employs withdrawal of the flue gas through a tape for a given time interval to collect the particles. The degree of darkening of the tape is used as the measure of particulate concentration on a continuous or semi-continuous basis. A third method employs use of ion-specific membrane electrodes for continuous measurement of indicator ions by withdrawal of the flue gas through a measuring cell containing an aqueous solution. A fourth method consists of collection and direct gravimetric weighing of particles at known flue gas withdrawal rates on a semi-continuous basis. A fifth method involves measurement of some physical parameter of the particles and relating it to the mass concentration [2].

A. Photometric Detectors:

1. Operation:

A commonly used method for providing an indicator of particulate emissions from combustion and other sources provides for passing a light beam across a duct. Increase in the degree of absorption of radiant energy is taken as being proportional to increase in the particulate concentration in the flue gas. The device is illustrated in Figure 92, and consists of a light beam produced by a projector, and passed across the duct to a receiver, which is either a photoelectric or heat energy cell. The reading is converted to an electrical signal, transmitted to a control unit, amplified, and then printed out on a recorder [3].

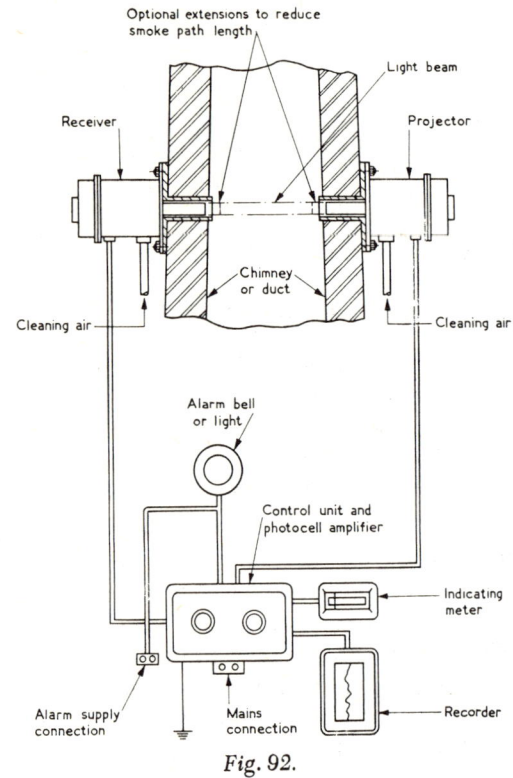

Fig. 92.

Courtesy of: British Standards Institution, B.S. 2811 (3).

The degree of sensitivity of the device is influenced by the path length across the duct, the intensity of the light beam, plus the particle concentration and size distribution within the duct. The sensitivity for observing low particulate concentration is increased by increasing the path length, but the degree of extension is limited by the greater degree of obscuration of the light beam by a greater number of particles. The normal range in path lengths is from three to fifteen feet [1]. The optical path length across a duct may be reduced if necessary by adding extension tubes to the inner ports of the duct. It is not necessary to account for isokinetic sampling conditions with this technique.

An alternative technique for mounting employs external location of the photocell and light beam where gas is withdrawn from the duct at a high rate and passed through a cell of constant length. The method makes for easier cleaning of lenses and may be adapted for use with wet, saturated plumes, which would not otherwise be suitable for measurement with the technique.

The intensity of the light beam must be maintained at a constant level to obtain consistent results. This requires that the optical lens surfaces facing the duct be kept clean at all times. This may be achieved by passing an air stream over the lens into the flue gas in order to prevent deposition of particles and condensation on its surface. Addition of a honeycomb mesh near the lens surface, with the surfaces of the mesh parallel to the light beam and perpendicular to the direction of gas flow, has been found to minimize migration of particles from the flue gas onto the lens [4]. The heat from the lamp and gas stream tends to minimize condensation of moisture from the flue gas onto the lens surfaces. It has been found in practice, however, that wiping the lenses with a clean cloth at least once every eight hours tends to prevent buildup of dirt on the lenses over a prolonged period.

The distribution of particles by size in a duct must be relatively constant if the mass concentration of particles is to be related to the degree of energy absorption. This occurs because small particles have a greater surface area per unit mass than large particles, and therefore tend to selectively absorb more energy. The technique is therefore more sensitive to changes in the numbers of small particles than for larger ones. Thus, it is necessary to make a calibration of mass concentration of particles as a function of the degree of light energy transmitted (following installation of the device) for each individual situation. Procedures have been developed by Hurley and Bailey [5] on a coal-fired boiler, and by Gansler [6] on a Kraft pulp mill recovery furnace. The technique used is to make a batch particulate test at a point adjacent to the unit when the optical density reading is approximately constant. The particle concentration is then related to the instrument reading as an indication of mass concentration. This procedure has met with limited success in other installations, however, and may require a large number of individual tests in order to obtain a reliable correlation.

The photometric method of measuring particulate concentrations can provide a useful indicator of particulate emissions from combustion sources in terms of operating conditions. It is useful for monitoring transmittance of a plume in terms of Ringelmann number in certain circumstances, but is subject to limitations regarding changes in particle size distribution. It has been used with some success on the flue gases from coal and oil-fired boilers, Kraft recovery furnaces, and steel mill blast furnaces. It cannot be used directly on wet, saturated plumes, but may be modified for use on flue gases following scrubbers.

2. Power Boilers:

Use of photometric smoke density meter is particularly suitable for coal- and oil-fired power boilers because it provides an index of combustion efficiency. The sudden increase in light absorption is an indicator of incomplete combustion which requires rapid correction. Sootblowing operations often cause rapid rises in optical density of the plume for short periods. Lucas and Snowsill [7] described the successful use of a photometric smoke density meter on a coal-fired boiler. They evaluated response of the instrument in terms of particle size, and described corrections in response to account for differences in duct size. Collins and Steele [8] designed an optical system to allow reading different ranges of light absorption across a duct, depending on particulate concentration and distance involved. Different ranges may be set by changing the optical wedge or the amplifier used.

3. Recovery Furnaces:

The photometric smoke density technique has also been successfully applied to measuring particulate from Kraft recovery furnaces following the electrostatic precipitators. Gansler [6] successfully used the device for determining the particulate concentrations and observing precipitator malfunctions. The scale reading for the bolometer was calibrated against particulate concentrations measured by drawing gas samples through a series of impingers. The liquid was analyzed for sodium ion and the results calculated as sodium sulfate and compared to the instrument reading. The following results were obtained in the instrument calibration check.

The instrument reading was found to have an output signal with a semilogarithmic relationship to the particulate concentration, the opposite of what would be expected from Beer's Law. This required changing the output signal from a logarithmic to a linear scale by use of an eccentric cam and restriction of the range, which provided for more accurate reading of high concentrations when precipitator malfunctions occurred.

Cooper and Haskell [9] also used a bolometer for measuring particulate emissions. They were able to obtain a correlation between particulate loss and instrument reading. McDonald [10] also used the device for measuring particulate losses from a Kraft recovery furnace.

4. Metallurgical Operations:

Heimke [11] describes the steps taken in mounting a light-photometer on an open hearth furnace vent. The unit should be able to withstand the high temperatures involved and be mounted securely. No operating data has been included.

5. Wet Plumes:

Direct placement of a light photometric unit in a wet stack, such as following a scrubber, is impossible because the condensed water vapor prevents transmission of the light beam. Therefore, it is necessary to locate the detection cell outside of the duct and draw flue gas through the unit. The gas stream must be heated above its dew point to prevent condensation so that the particulate matter present may be detected. The system may present problems, however, by deposition of entrained water droplets and particles on the walls of the tubing or cell. A system for measuring the light absorbed by particles in a wet stack following a scrubber was designed by Yocom [12]. It employs withdrawal of the flue gas through a heated line and into a detection cell 12 inches long and four inches in diameter. An electrical resistance heater was used to maintain the entire system at above the dew point of the flue gas.

B. Tape Samplers:

1. Operation:

Tape samplers operate on the principle of withdrawal of flue gas from a duct at isokinetic conditions. The gas stream is then cooled and passed through a filter paper for a given time period. The paper strip is then moved to the next spot and the sampling resumed. The amount of discoloration of the spot within a given time interval is used as the measure of particulate concentration. It may become necessary to cool the flue gas stream by dilution upstream of the tape. This is done in order to prevent heat damage to the paper, and condensation of water vapor from the flue gas on cooling. It also provides for dilution of the initial concentration so that the spot developed is not so dark that sufficient accuracy is impossible within a given sampling period. Otherwise, extremely short sampling periods must be used.

It is necessary to run periodic batch particulate samples at the same time that a particular spot is being developed to allow a correlation to be obtained between the degree of darkening of the paper and the mass particulate concentration. Dark particles are particularly suitable for analysis by spot discoloration with a tape sampler. Other particles, particularly those white in color, may not be easily measured by spot discoloration, and other techniques must be employed. Also, the method is only semicontinuous as a given spot must be developed for a given time interval of between 10 and 60 minutes duration. It may then be read following the sampling step. Existing commercially available instruments may be used for source monitoring with certain modifications.

The tape sampler has been found to be particularly suitable for monitoring flue gases from combustion and metallurgical processes. This occurs because the particles emitted have a dark color, and may be easily detected by spot discoloration. Price and Klemperer [13] used two tape samplers operating simultaneously to monitor the respective inlet and outlet gas streams of an electrostatic precipitator, to evaluate its performance. For the same sampling rates and exposure times, the ratio of the degree of darkening of the spots could be used as an indicator of performance of the precipitator. Changes in sampling rate could be used to facilitate greater accuracy in sampling.

2. Combustion Sources:

Tape samplers are useful in measuring particulate emissions from coal and oil-fired power boilers. Emission of dark gray carbon particles is an indicator of incomplete combustion, and these may be easily detected with the instrument. Hemeon, Haines, and Ide [14] used a tape sampler to measure particulate emissions from a coal-fired power boiler. The degree of darkening was found to depend principally on the number of small particles present, and was relatively independent of the large particles. The authors built a system which employed a thirty-to-one dilution of the flue gas with prefiltered air. A small aliquot of the diluted gas was then drawn into the tape sampler for measurement of the particulate material present. The flue gas stream was diluted to cool the gas stream to near ambient conditions without condensing the water vapor present. A similar system has been developed by Gruber and Schumann [15] for measuring particulate emissions from coal-fired power boilers.

3. Metallurgical Sources:

The tape sampler is also useful for measuring red iron oxide dust emissions from metallurgical sources such as steel mill blast- and open hearth-furnaces. Because of the hot flue gases present, it is necessary to provide a sufficient amount of dilution air to prevent paper damage and condensation. McShane and Bulba [16] [17] [18] developed the AISI tape sampler for automatic particulate monitoring of a basic oxygen furnace. The tape sampler operated through four consecutive 5.5 minute samples during an oxygen lance period. Gas was withdrawn from the stack, diluted, and then passed through the tape sampler. It was not necessary to sample at isokinetic conditions because the mean particle size was less than one micron. The device was calibrated by taking a series of batch particulate samples coinciding with the tape sampling, and subsequently weighing the filters. Beta-ray attenuation change was used as a measure of particulate concentration. A correlation between the attenuation change and the mass particulate concentration of iron oxide was obtained where the recorder chart reading increased from 5 to 100 as the mass concentration increased from 0.01 to 0.45 grains per standard dry cubic foot. The device operated successfully in the field during a four-week trial period.

Two additional systems used for sampling metallurgical sources have been developed by Groutsch [19] and Donoso [20]. It may be desirable to use either spot discoloration or an alternative technique such as beta-ray attenuation as the measurement technique [4]. Rozsa [21] described the use of a direct reading spectrograph to record beryllium oxide emissions from a source. Flue gas is withdrawn from the duct through a filter paper for 75 seconds, the spot moved and the spectrographic reading then made. The lower limit of sensitivity for the unit was found to be 0.5 micrograms of beryllium per cubic meter of flue gas.

C. Chemical Determinations:

1. Operation:

Chemical measurement of a particular substance present in the particulate matter provides a means for determining the mass particulate concentration in a flue gas stream. The devices normally operate on the principle of continuously drawing a sample from the source through a detection cell, which also has a continuous liquid flow. The concentration of the constituent of interest in the liquid passing the cell is measured by electrodes or other devices and placed on a continuous readout. It is then possible to relate the liquid concentration to the source gas concentration.

It is usually necessary to determine the weight relationship between the constituent being measured and the total particulate concentration in the flue gas. The substance being measured is usually only a portion of the total particulate present, thus, it is necessary to determine this weight ratio in order to obtain the total particulate concentration. This can be done by taking a number of batch particulate samples at the same time that the monitor is being operated. Correlation of the instrument reading with the mass particulate concentration during the test period provides the necessary information for determining this ratio.

Systems used for continuous particulate monitoring in flue gases must meet several requirements. First, it is normally necessary to sample at the isokinetic rate to obtain a representative sample. Suitable for maintaining isokinetic sampling conditions on a continuous basis are not yet available. Taking a number of traverses and locating the probe at the point of approximate average velocity and setting the rate at the approximate average condition is one way to compensate for this phenomenon. Second, the velocity in the duct should be monitored at the point of sampling to obtain a measure of total flow rate. Third, the ratio of liquid-to-gas flow rates should be such that the device always functions within its range of operation. This problem may require some field experimentation.

Fourth, the sample probe should be kept clean to prevent plugging, either by continuous flushing with water into the cell or by periodic blow-back. Fifth, the detection cell should be designed so as to be an efficient scrubber for the particles present in the gas stream. It should also be designed so that changes in concentration are rapidly seen with a minimum delay. Sixth, the device should be sensitive to low concentrations, be able to cover the range of concentrations to be expected, and to give both accurate and reproducible results. Seventh, the water used should have a low background level of the material being measured.

Three different chemical techniques have proved suitable for monitoring particulate concentrations in flue gases. These include conductivity measuring probes, ion specific electrodes, and colorimetric techniques. It will be necessary to develop other techniques for additional sources which are not feasibly measured with existing techniques.

2. Conductivity Probes:

Leonard [22] developed a continuous particulate monitor for a Kraft mill recovery furnace which employed continuous withdrawal of flue gas through a specially designed conductivity probe. The instrument provided an indicator of the presence of sodium sulfate, the primary constituent of the particulate material being emitted from the recovery furnace. The conductivity of the solution in the cell increased from one to 35 micromhos as the particulate concentration increased from zero to one grain per standard dry cubic foot (60°F, 29.92 in. Hg). The sample probe and pitot tube were placed at a location in the duct which had been previously determined as the point of average velocity.

Construction details of the unit are as follows: the water stream is added near the nozzle at the end of the probe, and acts to wash the probe by causing liquid flow into the cell along with the gas. It also provides for liquid change in the cell. The water used is deionized boiler feed water, which has a low background level of sodium sulfate and other soluble salts, and which has also been chlorinated to inhibit slime growth. The gas flow rate of 0.5 cfm is maintained through the cell while the boiler feed water flow rate is maintained at 1.0 liters per minute. A sealed pot, six inches in diameter and ten inches high, is used as the detection cell. It is divided into two stages by a baffled screen to provide for complete mixing of gas and liquid. A pitot tube is mounted adjacent to the probe and is used to monitor the velocity on a continuous basis. The outputs from both the conductivity cell and the pitot tube are mounted on the instrument panel in the control room to provide a continuous readout. This provides for observation of furnace and precipitator performance by operating personnel. A diagram of the system is shown in Figure 93.

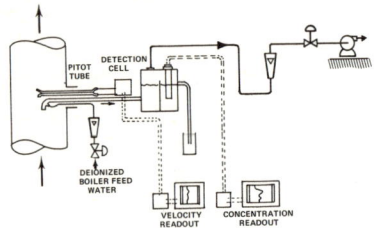

Fig. 93.

Courtesy of: Mr. James S. Leonard, Weyerhaeuser Company, Springfield, Oregon (22).

The device has proved successful in more than two years of continuous operation. It has been useful for continuously monitoring particulate concentrations and emission rates, and in providing a means for observing precipitator performance to spot malfunctions. The device has a certain degree of problems with interfering substances, however. Carbon dioxide from the flue gas may contribute to the conductivity, but is usually present at an essentially constant level. The presence of sulfur dioxide in the source will cause an additional conductivity to the solution, not caused by particulates. It is not normally present in significant amounts following the precipitator, but does constitute a potential interference. This problem gives rise to the next technique for particulate monitoring.

3. Ion Specific Electrodes:

Ion specific electrodes are membrane-type units which are used for detecting the presence of particular ions. They are commercially available for measuring a number of different ions, including the following: sodium, potassium, calcium, fluoride, chloride, sulfide, sulfate, nitrate, phosphate, plus others. They are normally specific for a particular ion but may be subject to changes with temperature and pH, which may require use of carefully buffered solutions. An electrode should be carefully checked for potential interferences from other substances prior to use, to evaluate its performance.

Leonard, Tretter, and Taylor [23] [24] have developed sodium ion electrodes for monitoring particulate emissions from Kraft recovery furnace flue gases. The system was used for measuring sodium ion as an indicator of sodium sulfate particles. The method also eliminated the potential interferences caused by sulfur dioxide in the flue gas, as described in the previous section. The design was the same as for the conductivity monitor except that a different electrode was substituted. Field trials have shown the system to be successful in monitoring particulate emissions from a Kraft recovery furnace. It is necessary to maintain an ammonia buffer for the solution to maintain the pH of the liquid between 8.5 and 9.5. This provides for optimum sensitivity of the electrode. It is also desirable to use chlorinated water so as to minimize the possibility for slime growth to occur. The system is illustrated in Figure 94.

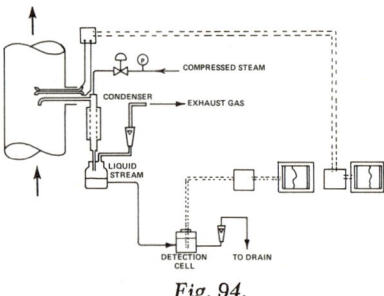

Fig. 94.

Courtesy of: Mr. Vincent J. Tretter, Georgia Pacific Corp., Portland, Oregon, 1970 (23).

Two recent articles have dealt with the potential usefulness of fluoride ion-specific electrodes for monitoring emissions from phosphate fertilizer, aluminum, and steel-producing plants. Warner [25] calibrated a lanthanum fluoride electrode over a range of fluoride ion concentrations in distilled water and in sodium chloride solution. The electrode operated successfully over a range in fluoride ion concentrations from 0.1 molar to 10^{-6} molar in both solutions, without needing corrections for ionic strength. Elfers and Decker [26] also evaluated the fluoride ion electrode and found it to be stable in the pH range from four to eight. It may be necessary to use a large amount of liquid per unit amount of gas to assure sensitivity when using the electrode as a continuous fluoride monitor.

4. Colorimetric Methods:

Instruments used for ambient air monitoring of constituents employing colorimetric reactions, may be adapted for particulate monitoring of certain ions. Necessary modification of the instrument would include a much larger liquid flow rate per unit gas flow rate to assure remaining within the operating range of the instrument. The basic operation of the instrument employs concurrent liquid and gas flow into a column or other contacting device for a sufficient period to allow a colorimetric reaction to take place. The colored liquid then flows to the detection cell, which compares the difference in intensity between the colored solution and the blank reagent. This difference in color is taken as being proportional to the concentration of the particular constituent. The only potential application to date has been a fluoride monitor developed by Adams [27] for ambient air work.

D. Additional Techniques:

Several additional methods have potential application as source particulate monitoring techniques. They are more complex, and would require considerably more sophisticated equipment and highly trained personnel for successful operation than those previously discussed. Detailed descriptions of the techniques used have been made by Mitchell and Engdahl [28] and by Schutz [2]. Techniques described include beta-ray and sound attenuation, measurements of electric charge and dielectric constants of particles, and use of a gravimetric balance. Problems may arise in using each of the techniques because of variations in velocity, temperature, and moisture content of the flue gas, and variations in size and physical properties of the individual particles [28].

1. Beta-Ray Attenuation:

Beta-radiation, or high velocity electrons, is used to measure particle densities in gas streams, fluids, or on paper sheets by bombardment. The particles present tend to absorb the beta-rays, and the concentration is taken as the difference in radiation intensity reaching the receiver from the blank and from the sample. The difference in signals increases as the particle concentration increases. Halliday [29] presents a review of the necessary nuclear theory for describing the action of beta-radiation.

The application of beta-radiation for particulate monitoring in source gases has seen limited application, where Breitling [30] was one of the first to use the technique. McShane and Bulba [17] used the technique for measuring iron oxide deposition with a tape sampler on a basic oxygen furnace in a steel mill. A series of four 5.5 minute samples were taken during a lancing period. The spot was exposed to the beta-radiation detector following each sample. The difference in attenuation between the discolored spot and the blank paper was used as the measure of particle concentrations.

Izmailov [31] developed a similar system for particulate monitoring on a continuous basis. Flue gas was continuously withdrawn from a source, and passed through a piece of slowly moving filter paper. Following exposure and filtration the paper passed through the detection cell where the degree of beta-ray attenuation was continuously monitored. The degree of beta-ray absorption increased with the particulate concentration on the filter paper. The system could be calibrated by taking batch particulate samples and relating instrument reading during the sampling period to the mass particulate concentration.

2. Sound Attenuation:

Passage of ultrasonic sound waves through a gas stream could be used for determining the particle concentration in a duct, in a manner similar to beta-radiation [32]. The degree of attenuation of the sound waves caused by the presence of particles could be used as a measure of particulate concentration [33]. Experience to date indicates, however, that the method has serious drawbacks. It is influenced by both the number of particles and their respective sizes. It would also require measurement of the small difference in attenuation between two very large signals (blank and sample), which greatly increases the chance for experimental error [28].

3. Electrical Properties:

It is possible to obtain an indication of particles present by their electrical properties, such as surface charge or dielectric constant. These techniques would be suitable for sources where inorganic particles are collected in the dry state. Metallurgical sources and coal-fired power boilers would appear to be the most suitable for these techniques.

Grindell [34] designed an instrument based on the principle of electrostatic precipitation, where the particle concentration was taken as being proportional to the exposed surface area of the particles. The gas stream is withdrawn from the duct, and drawn past a corona discharge wire, which causes the particles to become charged. The gases are then drawn through an electrical field where the particles are caused to collect on a plate of opposite charge. The particles release their charge on striking the plate to produce an electric current which is proportional to the number of particles. The charge accepted by a particle is approximately proportional to its surface area, and this is used as the measure of particle concentration. The relationship of surface area to mass for a particular case would have to be determined by experimental measurement. The collector plates are periodically cleaned to assure continued operation. The system would not be suitable for soot or sticky organic particles. It has not been used in the field to date, so there is no information available regarding actual experience.

Barkov [35] designed an instrument which measures the difference in dielectric constant for a flue gas stream before and after removal of the particles. The gas stream is withdrawn from the source through a cell where its dielectric constant is measured. The particles are then removed by passage through a filter and the dielectric constant of the cleaned gas stream is measured. The difference between the two is used as the measure of particle concentration. The filter must be cleaned periodically and the problem could amount to measuring the small difference between large numbers. The dielectric constant of the gas stream may also be so low as to be at the lower limit of sensitivity for the instrument. There is no field experience to date with the instrument, at least to the knowledge of the authors.

4. Gravimetric Analysis:

Duwel [36] describes the use of a modified analytical balance for semicontinuous gravimetric analysis of the particulate material present in flue gases. It is still in the development stage, and additional work remains to be done before it is available for field use.

E. Particle Size Analysis:

Measurement of particle size in flue gases is one of the major unsolved problems in source sampling. It is complicated by differences in the nature and size-ranges of particles present, and by high temperatures and moisture contents of flue gases. Obtaining a representative sample is a difficult task. A description of the different types of instruments used for particle size analysis has been presented by Lapple [37]. Peterson and Paulus [38] developed an instrument for analysis of particle size in ambient air over the size range from 0.001 up to 10.0 microns in size. It employed a combination of condensation, light scattering, and electrostatic collection to measure the particle size ranges. To date, it has not yet been applied to source monitoring.

III. GASEOUS MONITORING

A. Introduction:

Continuous analysis for gaseous constituents in flue gas streams is less complicated than particulate sampling because it is not necessary to sample at isokinetic conditions. It is then possible to withdraw a sample at a constant rate independent of flow rate in the stack. It is normally useful to have a pitot tube (placed adjacent to the probe) which is connected to a pressure transmitter and recorder. This provides a means for making a continuous record of velocity to facilitate computation of mass emission rates.

1. Sampling System:

It is important to measure the concentration of the constituent of interest as it actually exists in the stack. This requires that there be no significant losses through the sample line between the duct and the instrument, and that if losses occur, their extent should be a known quantity. Tubing and connections used should be of inert material so as not to be reactive with the gas to be measured. Particulate materials are often filtered just after entering the sample line to minimize the chances for line plugging and instrument malfunction. It is useful to point the probe downstream to minimize the number of particles pulled into the sampling line.

The presence of water vapor in flue gases makes possible losses of other gaseous constituents in the sample line by solubility in the condensed water. This may be overcome either by heating the entire line or by selectively removing the water without removal of the other gaseous constituents. It is possible to heat the sample line, but this is complicated and requires additional equipment. Heating the sample line is particularly suitable where water does not interfere or build up in the detection cell. The second method employs use of selective condensation or scrubbing of the gas stream in a suitable liquid. Its usefulness depends on the solubility of the gas being measured in the liquid.

Other features may also be necessary in the sampling system. For high pressure sources it may be useful to place a pressure reducing valve in the line to reduce the pressure to near-ambient conditions. A flow control valve is added following the detector to regulate the flow through the system. A prime mover such as a vacuum pump or water aspirator provides the means for maintaining gas flow through the sample line. Brown [39] presents additional information on the design of sampling systems.

2. Instrumentation:

Several techniques have been used to measure gas concentrations in flue gas streams on either a continuous or semicontinuous basis. The operating characteristics and usefulness of several continuous analyzers are presented in Table 41. These include wet chemical analyses, spectrophotometry, coulometry, flame ionization, mass spectrometry, paramagnetic effects, and gas chromatography. Brown [39] lists the types of instrumentation available as process analyzers; he lists their operating characteristics, limitations, suppliers and costs. Additional papers by Naumann [40], and Thoenes and Guse [41] describe the operation of several types of continuous analyzers for sulfur dioxide and other gases.

Instruments used for monitoring should be specific, accurate, and reproducible for the gases being measured, and subject to a minimum of interferences from other substances. The instrument should be able to operate for extended periods with a minimum of servicing and maintenance. It is necessary to calibrate the instrument periodically to assure reliability of results.

B. Wet Chemical Analyses:

The principle of analysis of flue gas constituents by wet chemical means involves withdrawal of the sample from the duct, and absorbing the substance in a liquid stream. The liquid stream is passed through the detection cell to give a reading. The instrument reading is then related to flue gas concentration in terms of the respective liquid and gas flow rates to the instrument. The two methods used to-date include conductivity and colorimetric analyses.

1. Conductivity:

Conductivity of a water stream has been used as a means of indicating the presence of sulfur oxides in flue gases. It involves continuous withdrawal of flue gas followed by contact with a water stream to absorb the sulfur dioxide and sulfur trioxide present. The solution is then monitored for conductivity as a measure of sulfur oxides concentration in the flue gas. It is necessary to remove any particulate materials which contribute to the conductivity upstream of the gas-liquid contact phase. This is particularly true for sulfate-containing particles. Otherwise, these will be indicated as total sulfur oxides in the conductivity cell, as previously described by Leonard [22] for Kraft recovery furnaces. The method is not specific for other materials which may contribute to the conductivity. The method is mainly suitable in flue gas streams where sulfur oxides are the major gaseous pollutants present, and where there is a minimum amount of sulfate-particulate present. These include coal- and oil-fired power boilers, sulfite process pulp mills, certain metallurgical operations, and refinery process flue gases.

Miller, Brown, and Abrams [42] described the use of conductivity cells as continuous monitors for the digester blowpit and sulfurous acid-making process vent stacks. In both cases, sulfur dioxide comprised nearly all the emission of materials which would add to conductivity. There was practically no particulate matter present in either source, but both flue gases were saturated with water vapor. The conductivity monitors for these two sources were then essentially functions of the sulfur dioxide concentrations present.

Construction details of the instrument used were as follows. Flue gas was withdrawn from the source at 3.0 liters per minute by means of a vacuum pump, with no effort made to prevent condensation. It was then contacted with de-ionized water from a constant head tank at 1.0 liter per minute. The liquid and gas streams were passed vertically downward and concurrently through a packed column two inches in diameter and 26 inches long. The gas stream was then vented off to the flow control and measurement section and through a vacuum pump while the liquid passed to the conductivity cell. Conductivity cells could be used to record concentrations between either zero to 1,000 ppm as SO_2, or zero to 10,000 ppm as SO_2. Readout from the cells was transferred to a conductivity monitor and a 0-10 millivolt recorder. The systems have proved successful in more than one year of continuous operation.

Nacovsky [43] described a similar system for simultaneous measurement of both sulfur dioxide and sulfur trioxide in flue gases from an oil-fired power boiler. The system is similar to the one previously described except that sulfur trioxide is separated from sulfur dioxide in the flue gas on a cyclic basis. The gas stream is removed from the source by first passing through a ceramic filter to remove the particulate material present. It is passed through a condenser maintained at 180°F to remove the sulfur trioxide by condensation as sulfuric acid mist. The sulfur dioxide is then absorbed in a scrubber containing hydrogen peroxide solution. In the next step, the sulfuric acid mist is flushed from the glass disc with distilled water to a reservoir. Then both the liquid used to flush the condenser (SO_3) and liquid from the scrubber (SO_2) are passed through their respective conductivity cells. The resultant signals are measured as indicators of the respective sulfur trioxide and sulfur dioxide concentrations on a semi-continuous basis, where the cycle lasts from 10 to 15 minutes. The system was found successful in a limited number of trials. It is a rather complex piece of equipment, and condenser malfunctions would limit the accuracy of sulfur trioxide measurement.

2. Colorimetry:

Colorimetric techniques may be used to measure concentrations of gaseous constituents in flue gas streams in a manner similar to instrumentation employed in ambient air monitoring. Concentrations are usually much higher in source flows, so that considerably larger relative liquid flows per unit of gas flow are required to maintain readings within the operating range of the instrument. Potential interfering substances present in the flue gas should be carefully checked because they are usually present in much greater relative concentrations than for ambient air. The reaction time must also be sufficiently short. To date, colorimetric wet chemical tests have found limited application in source monitoring.

Laxton and Jackson [44] developed a continuous analyzer for sulfur trioxide in the flue gases from coal- and oil-fired power boilers. The method made use of the reaction between sulfate ion and barium chloranilate to form an acid chloranilate ion, which gave a red color with an absorbance maximum at 535 mμ. Flue gas was drawn into the instrument after passing through a glass wool filter in the duct to remove particulate materials. It was then contacted with a solution of 80 percent isopropanol and 20 percent distilled water (both by volume), and passed through a thimble to cause intimate contact of gas and liquid. The gas stream was then withdrawn at a constant rate to a vacuum pump while the liquid stream was passed through the detection cell. The signal was sent to a 0-5 mv recorder, which allowed observing the sulfur trioxide concentration on a continuous basis. The range of the instrument was between 0.2 and 100 ppm as SO_3 by volume, depending on the setting. The color development time was about ten minutes, and once developed, it was stable for up to twelve hours. Concentrations of up to 3,000 ppm of sulfur dioxide did not interfere with the instrument. Operation was found to be successful in extensive field trials.

Two additional studies have been made using colorimetric techniques. Instrumentation has been developed by Adley and Skillern [45] for monitoring nitrogen dioxide, using the modified Saltzman method. The range of operation could be adjusted by modification in the gas flow rate between 5 and 400 milliliters per minute, and in the liquid flow rate between 1.5 and 20 milliliters per minute. To date, it has seen little use in source monitoring for nitrogen dioxide. Adams [27] developed a system for monitoring total fluoride ion from either ambient air or source gas streams, as previously described.

C. Coulometric Titration:

1. Operation:

Coulometric titration is a useful method for measuring the concentrations in flue gases of reduced sulfur gases and organic compounds containing unsaturated carbon-carbon bonds. It operates on the principle that a current being generated to produce bromine gas is proportional to the concentration of reactive gases passing through the detection cell. The technique is used for monitoring sulfur dioxide, hydrogen sulfide, mercaptans, organic sulfides and disulfides, thiophenols, and is also sensitive to olefinic hydrocarbons, terpenes, acrolein, and nitrogen dioxide.

Shaffer, Briglio, and Brockman [46] did the initial development work on application of a negative feedback amplification system employing two electrodes for monitoring the above gases. Subsequent work by Austin, Turner, and Percy [47], using two electrodes, met with only moderate success because the instrument had a definite upper limit with regard to allowable concentrations of reactive gases in the detection cell. This occurred because of plugging caused by elemental sulfur buildup in the detector.

Present operation of the system employs use of 16 percent hydrobromic acid solution as the electrolyte, with sensing, reference, and generating electrodes placed on a common shaft. The cell generates a fixed level of bromine gas. This level depends on the instrument setting so that different ranges in concentrations may be obtained. Flue gas is drawn into the cell and the presence of reactive gases causes consumption of the bromine gas being generated. This forces the cell to generate more bromine in attempting to maintain chemical equilibrium, and the current required increases as the total concentration of reactive gases increases. The current is converted to a potential-equivalent and recorded on a 0-100 millivolt recorder.

The current depends on the total concentration of all reactive gases passing through the cell and is not specific for any individual constituent. The current resulting from each gaseous compound passing through the cell depends on both the valences of the sulfur atoms relative to bromine, and on their concentrations. Therefore, it is necessary to calibrate each gas individually to determine its response to the cell. Analytical selectivity between the different constituents is obtained by a series of prescrubbing solutions placed upstream of the cell [48].

2. Applications:

Coulometric titration has been applied as a source-monitoring method for reduced sulfur gases in Kraft pulp mills on the flue gases from recovery furnaces, smelt tanks, and lime kilns. Thoen [49] used the instrument for sampling these sources on a batch basis by sample collection in 12 liter flasks. Following collection, the gas stream was withdrawn from the flask into the instrument. The total sulfur level present was first measured and the gas stream was then analyzed for its individual constituents by use of the selective prescrubbing solutions. Sulfur dioxide was first removed by drawing the gas through a solution of three percent potassium acid phthalate placed in an impinger upstream of the cell. The difference in signals between total sulfur and total sulfur minus sulfur dioxide was used as the measure of sulfur dioxide concentration. The actual sulfur dioxide concentration for a given instrument reading was then determined by referral to a previous calibration with sulfur dioxide. The process was then repeated for hydrogen sulfide, mercaptans, and organic sulfides and disulfides. Selective prescrubbing solutions to facilitate measurement of these gases are listed in Table 42.

Blank readings were obtained following the sample collection by disconnecting the sample line from the cell and allowing ambient air to purge through the instrument. The concentrations for each of the gases were computed based on previous calibrations for instrument readings. The mercaptans and organic sulfur compounds were computed as their respective methyl constituents, as these are normally the major ones present.

Development of a continuous monitoring system for reduced sulfur gases in Kraft mill flue gases is highly desirable for two reasons. First, the batch sampling procedure allows the obtaining of information during only brief time intervals. Second, the sample in the collection flask becomes diluted with air as the flue gas is withdrawn. The presence of particulate matter and flue gases saturated at 30 percent moisture by volume tends to complicate the situation. The prevention of losses by dissolving of gases in water is a major problem.

Blosser, Cooper, and Megy [50] developed a continuous monitoring system for reduced sulfur compounds in Kraft mill flue gases using the bromine coulometric titration system. The system employed withdrawal of the flue gas from a duct through a glass wool filter for particulate removal in hot flue gases not subject to condensation. A straight tube without a filter was used for flue gases following scrubbers and other saturated sources with low particulate concentrations. The gas was then drawn at 250 ml per minute into a scrubber of 40 ml capacity where a solution of potassium acid phthalate was being passed through at a rate of 0.5 ml per minute.

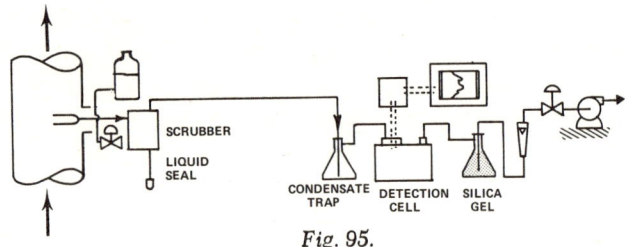

Fig. 95.

Courtesy of: Mr. Joseph A. Megy, National Council for Air and Stream Improvement, Inc., Corvallis, Oregon, 1970 (50).

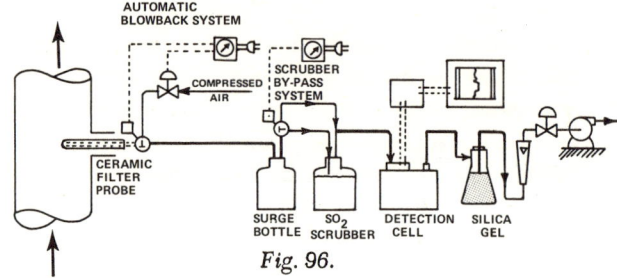

Fig. 96.

Courtesy of: Mr. Gary N. Thoen, Weyerhaeuser Company, Longview, Washington, 1970 (51).

This caused removal of sulfur dioxide from the flue gas, along with essentially all the remaining particulate matter. Most of the moisture was also condensed by cooling the flue gas to ambient conditions. This resulted in a total reduced sulfur loss of 0.1 to 0.2 ppm hydrogen sulfide at the one ppm level.

The flue gas containing the reduced sulfur gases, minus the sulfur dioxide, was then passed to the detection cell for analysis. It was thus possible to observe the total reduced sulfur concentration for a source on a continuous basis. Breakdowns for the individual constituents present could be run periodically by use of the previously mentioned prescrubbing solutions. A diagram of the sampling train used is shown in Figure 95. The system requires replenishment of the $KHC_8H_4O_4$ scrubbing solution approximately once per week, with change of the cell electrolyte, and calibration checks approximately once per month, depending on source concentrations.

Modification of the system is necessary when monitoring sulfur dioxide with the other reduced sulfur gases. It employs a heated sample line to prevent condensation of the water vapor, and two instruments placed in parallel. Flue gas is withdrawn from the duct through a filter and into the heated sample line and divided into two streams. One stream passes through a scrubber where the sulfur dioxide is selectively removed with potassium acid phthalate as previously described. Then both streams pass in separate tubes through a combustion furnace where the reduced sulfur gases are oxidized to sulfur dioxide by heating to greater than 1500°F. They then pass through two parallel coulometric titration units where the signals are recorded. The differential between the two is used as a measure of sulfur dioxide concentration.

During extended operation, the cell, which monitors total reduced sulfur including sulfur dioxide, will condense water from the flue gas because it has not been previously removed. This may require more frequent change of the electrolyte solution. A diagram of the system is shown in Figure 96 [51].

An alternative procedure employs two scrubbing systems using hydrogen peroxide solution for scrubbing the sulfur oxides in each gas stream. The liquid solutions in each cell are then passed through two parallel conductivity cells for respective measurement of total sulfur, as well as total sulfur minus the sulfur dioxide present. With this method, any sulfur trioxide formed during combustion of the reduced sulfur will be indicated and the problem with water condensation is not present.

Markovs [52] described the use of a coulometric titration for measuring reduced sulfur gas concentrations in the feed gas to an ammonia plant. The system employs a regulator to reduce the line pressure to near ambient level to facilitate passage through the instrument. Thus a prime mover is not necessary in the sample line and the system works on positive pressure so that dilution by air leakage does not occur. Carbonyl sulfide in the flue gas cannot be directly measured with the unit. It is converted to hydrogen sulfide upstream of the detection cell by passage through a catalytic converter, following addition of hydrogen gas. The carbonyl sulfide is then measured by first bypassing the converter and recording the signal, then passing the gas stream through it and measuring the difference. The method is not feasible in process streams containing oxygen because of the danger of explosion with hydrogen.

Coulometric titration units have also been used to monitor the reduced sulfur concentration in natural gas pipelines. The system employs pressure regulators and a series of orifices to reduce the line pressure to near atmospheric. The systems have been designed for continuous monitoring applications.

Coulometric titration units are also sensitive

to unsaturated hydrocarbons, including olefins such as ethylene and propylene, aromatic structures (such as terpenes), and acrolein. Altshuller and Sleva [53] evaluated the system for measuring olefinic hydrocarbons in automobile exhaust. Their findings showed that the results compared favorably with those obtained with suitable colorimetric methods. They also found the detection unit to be subject to several "interferences," including nitrogen dioxide, aromatic hydrocarbons, sulfur dioxide, hydrogen sulfide, and acrolein.

Studies of the relative response of interfering substances yielded the following results. Nitric oxide did not appear to react with the bromine electrolyte to any appreciable extent. Nitrogen dioxide, on the other hand, constituted a significant negative interference with the monitoring of olefins. This probably occurs because the nitrous acid formed in solution interferes with bromine generation. Its presence in flue gas samples in significant quantities would cause results for reduced sulfur concentrations to be low. This is particularly a problem in flue gases from coal- and oil-fired boilers when attempts are made to monitor sulfur dioxide. Tests showed that placing a tube containing Ascarite upstream of the detection cell removed approximately 90 percent of the nitrogen dioxide without removing significant amounts of olefins. It also removes sulfur dioxide and hydrogen sulfide to provide for measurement of olefins without major positive interferences.

Relative response of the instrument to several different compounds was tested by comparison to trans-2-butene. The degree of response depended on the ease of oxidation by bromine, of the chemical bond being attacked. Hydrogen sulfide was found to give by far the greatest response to any given concentration. It was also noted that certain of the compounds exhibited increasing response as their concentrations in a gas stream decreased, such as acrolein and thiophene. The relative response of several of the gases is listed in Table 43.

D. Spectrophotometry:

Instruments employing the passage of light of a given wavelength through a sample volume are used for measuring gaseous concentrations in flue gas streams, where the particular wavelength is absorbed by the constituent being measured. Instruments operating in the respective ultraviolet, visible, and infrared regions of the spectrum have been used with varying degrees of success. Continuous analyzers employ passage of the gas stream through a sample cell into which light of a given wavelength is directed. Light of the same wavelength is also passed through a reference cell and the degree of absorption of light by the sensitive component is taken as being proportional to its concentration.

1. Nondispersive Infrared Analyzers:

Nondispersive infrared analyzers operate in the manner described above where light energy of a narrow band of infrared wavelengths is passed through both reference and sample cells. Gases which absorb energy in the infrared wavelength region of the instrument may then be detected. It is desired to select a wavelength where only the constituent of interest absorbs the infrared radiation. If other substances absorb at the same wavelength, their interference can be removed by placing a "filter" in the reference cell light path. It consists of a gas mixture of the interfering substance at about the same concentration as it appears in the gas stream. Unfortunately, a considerable amount of overlapping of absorbed radiation occurs between the different components in flue gases.

Nondispersive infrared analyzers have been found useful in measuring hydrocarbons, carbon monoxide, carbon dioxide, sulfur dioxide, nitric oxide, nitrogen dioxide, water vapor, ammonia, and others [54] [55]. The technique has been found to be particularly suitable for monitoring automobile exhaust gases. Some difficulties have been observed with light absorption by interfering substances, however, when attempting to measure hydrocarbon components. Water vapor has been found to be a significant interference with measurement of nitric oxide. It is necessary to maintain a heated sample line to prevent losses of nitrogen dioxide by removal with condensed water vapor. The instrument must be kept above the dew point of the flue gas so that water condensation inside the sample cell does not occur. It is also necessary to filter the gas stream upstream of the cell to prevent deposition of particles in the sample cell.

Bent, Ladner, and Mullin [56] developed a method for continuous infrared analysis of sulfur trioxide. Flue gas is drawn through a column of oxalic acid, where sulfur trioxide reacts to liberate carbon monoxide and carbon dioxide.

These are then measured by nondispersive infrared spectrophotometry.

A nondispersive infrared analyzer used for measurement of carbon dioxide is shown in Figure 97 [55].

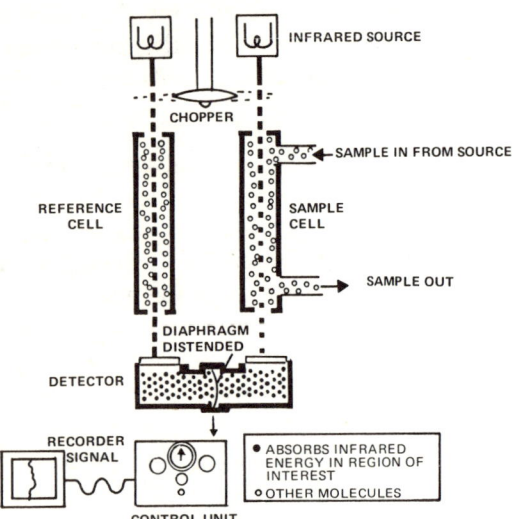

Fig. 97.

Courtesy of: Bechman Instrument Co., Fullerton, California.

2. Dispersive Infrared Analyzers:

Dispersive infrared analyzers are not continuous in operation because a single gas sample is placed in the instrument. The wavelength is then varied between one and sixteen microns, and the absorbance spectrum noted. The presence and concentration of certain gases is thus indicated by peaks at their characteristic wavelengths. The sensitivity for minimum concentrations of an instrument can be varied by changes in path length. Water must be removed from the flue gas prior to being placed in the cell to avoid condensation on the cell windows, which are constructed of sodium chloride. Shuck [57] used a long path unit for determining the concentrations of different hydrocarbons in studies on the irradiation of automobile exhaust.

3. Visible Spectrophotometry:

The absorption of nitrogen dioxide at 385 mμ in the visible range has been developed as a continuous monitoring method for automobile exhaust. At this wavelength, it is relatively free of effects of interfering substances. Operation of the instrument was similar to nondispersive infrared analyzers except that the wavelength is different. Nicksic and Harkins [58] developed the instrument for analyses of automobile exhaust gases by first drawing the gas through a drying trap to remove water vapor. The gas was then drawn through the instrument to measure nitrogen dioxide. The degree of absorbance was taken as being proportional to nitrogen dioxide concentration.

A modification to the method was made by oxidation of nitric oxide to nitrogen dioxide by means of pressurization using a compressor. On conversion of the nitric oxide, the gas was again expanded to atmospheric pressure before entering the sample cell. This allowed measurement of both the nitric oxide and nitrogen dioxide concentrations in the gas stream to be determined. The system was found to be successful for both automobile [58] and diesel exhaust gases [59].

It was noted that the major portion of the oxides of nitrogen being emitted from combustion sources was present as nitric oxide. It was thus necessary that the conversion of nitric oxide be essentially quantitative. Subsequent work by Sweeney [60] indicated the conversion efficiency to be essentially 100 percent by pressurization of a gas flow rate of three liters per minute to 80 psi in a compressor for five minutes. The pressure was then released to the atmospheric level, and the gas caused to flow through the instrument. The rate of reaction was found to increase with the square of nitric oxide concentration, and linearly with the oxygen concentration. It was then possible to compute the rate of oxidation of nitric oxide to nitrogen dioxide at varying initial nitric oxide concentrations for atmospheric pressure. Results are as listed in Table 44. They show that the rate of oxidation of nitric oxide decreased rapidly as its concentration decreased, particularly upon reaching the atmosphere.

To date, the main use of this technique has been on automobile exhausts. Conversion is essentially quantitative but side reactions involving nitrogen dioxide, once formed, may create a potential interference. Condensation in the sample line also may be the source of losses of NO_2. Additional work is necessary to provide for a fully suitable technique for measuring oxides of nitrogen in flue gases.

E. Other Methods:

Several other techniques have been used for continuous monitoring applications with varying degrees of success. These include gas chromatography, flame ionization, mass spectrometry, and paramagnetism. The first and third are normally semicontinuous techniques because a single sample is analyzed at a given time.

1. Gas Chromatography:

Gas chromatography provides for separation of individual components in a gaseous sample, based on their relative affinity for a column packing material. Each is then measured separately, by means of a suitable detector, after exiting from the column. Several types of chromatographic detectors and numerous types of columns are available, but a thorough discussion of these subjects is sufficient material for another book. The column should be selected so that it can provide adequate separation between the gases of interest in the sample. The detector should be sensitive for the gases being measured over the expected concentration range. By its nature, gas chromatography is a batch technique for each individual sample.

Gas chromatographs have been used as semicontinuous monitors in the petroleum and other industries in process streams where gas concentrations are fairly high. Walther and Amberg [61], however, developed a gas chromatograph for monitoring hydrogen sulfide, methyl mercaptan, and sulfur dioxide in the exit gases from a Kraft recovery furnace at low concentrations. Gas was withdrawn continuously from the duct through a cylindrical carborundum filter probe to selectively remove particulate matter. The gas stream was then drawn through a steam-jacket-heated line at a flow rate of two liters per minute, using a vacuum pump as a prime mover, and causing a portion of the flow to be sent through a sample loop. Samples were injected into the column at intervals of once every ten minutes by elution with helium carrier gas. A stripping column was added for water, and a polyglycol column used to separate the sulfur gases from the liquid. A thermal conductivity detector was used for measuring the respective gas concentrations of H_2S, CH_3SH, and SO_2. Full scale readings were at 1000 ppm and the minimum sensitivity was about ten ppm. The device proved successful during an initial three month trial period, but severe maintenance problems limited its usefulness during a later trial period.

Walther and Amberg [62] [63] also developed a mobile laboratory employing two gas chromatographs to facilitate studies of sulfur and other gas emissions from Kraft mill sources. Gas was pulled continuously from the duct through a ceramic filter and thence through a heated sample line to the trailer. Part of the gas stream was filtered, passed through a sample loop and into a stripping column, followed by passage through one of two dual columns employing polyglycol on Chromosorb T. Following separation, the gases passed in series through thermal conductivity and flame ionization detectors. This provided for detection of sulfur dioxide and hydrogen sulfide in the first, and for the organic sulfur gases in the second. Minimum sensitivity of the thermal conductivity detector was set at two ppm H_2S, and was about 0.2 ppm for the organic sulfur gases. Losses in the sample line and column became significant at concentrations below one ppm. The system proved successful in several field studies.

Gas chromatography provides a potentially useful method for measuring a number of compounds in flue gases. The needs for heated chemically inert sample lines, combined with complexity of the instrument, maintenance requirements, plus the necessity for using highly skilled personnel to achieve successful operation, limit the usefulness of this method in the field on a large scale.

2. Flame Ionization:

Hydrogen flame ionization analyzers operate on the principle that a small electric current is generated when gases containing carbon atoms are oxidized to carbon dioxide in a hydrogen flame when a potential is applied across the flame [54]. The output current is then proportional to the number of carbon atoms passing through the flame; measurement of this current provides an indication of concentration for organic compounds in a gas stream. The flame ionization units do not respond to water, carbon monoxide, carbon dioxide, sulfur dioxide, hydrogen sulfide, oxygen, and nitrogen to any measurable degree. Thus, this method may be used to measure the concentrations of organic compounds such as hydrocarbons and organic sulfur gases, in flue gas streams.

Hydrogen flame ionization has seen frequent use as a detector in gas chromatography applications, as previously described by Walther and Amberg [62] in Kraft mill studies. It has also been used for measuring hydrocarbon concentrations in automobile exhaust gases, as reported by King [54]. It can be adapted as a continuous monitor for total organic gases present in flue gas streams, because of its insensitivity to interfering substances. It is particularly suitable for monitoring hydrocarbons from automobile exhaust on a continuous basis. The major disadvantage of the method is that it is necessary to maintain a constant flow of hydrogen gas into the detector system.

3. Combustibles Analysis:

Combustibles gas analyzers operate on the principle that the electrical resistance of a wire placed in a flue gas changes as its temperature changes. The unit is first balanced for zero combustibles at the flue gas temperature, and resulting changes in temperature of the wire are then caused by the presence of gases which become oxidized upon contact with the wire. The temperature of the wire increases with the amount of combustibles present, changing the resistance and hence the current, by unbalancing the Wheatstone bridge circuit. The current is measured and used as the indicator of combustibles concentration in the flue gas stream.

The combustibles analyzers in use are non-specific, since they are sensitive to hydrogen, hydrogen sulfide, carbon monoxide, and organic compounds such as hydrocarbons. They are frequently used as indicators of combustion efficiency on power boilers, incinerators, recovery furnaces, and other combustion processes. The range may be set to observe different concentration ranges from 0-1 to 0-100 percent by volume.

Feldstein [64] described the use of a combustibles analyzer for monitoring hydrocarbon emissions from a refuse incinerator. Flue gas was withdrawn from a duct through an Ascarite column to remove carbon dioxide. The gas stream was then drawn through the instrument at 200 ml per minute and a reading established, which included both carbon monoxide and hydrocarbons. The gas stream was then drawn through a tube containing activated carbon ahead of the instrument to remove the hydrocarbon ethane and those higher in molecular weight. Separate analyses by gas chromatography and infrared radiation showed that methane and carbon monoxide passed through the bed, but that the higher molecular weight hydrocarbons were absorbed. The difference in instrument readings was used as the indicator of hydrocarbon concentrations for ethane and heavier carbon-hydrogen compounds.

4. Paramagnetism:

Paramagnetic techniques are used to measure oxygen because its magnetic susceptibility is much greater than any of the other constituents in flue gases. The method consists of drawing flue gas past a dumbbell suspended from a quartz fiber between two poles of a magnetic field. The presence of oxygen causes the dumbbell to change its angle by rotation, which is measured by an attached mirror. The degree of deflection is proportional to the oxygen concentration in the flue gas. The relative magnetic susceptibility of oxygen is much greater than those of nitrogen, carbon dioxide, or carbon monoxide. The only other gases having a significant value are nitric oxide and nitrogen dioxide, but normally these are not present in sufficient quantities to affect the oxygen reading by more than 0.2 percent oxygen by volume. The relative magnetic susceptibilities of several gases are listed in Table 45.

Tipping [65] described the use of an oxygen analyzer for flue gases in a chemical plant. Gas is withdrawn from the duct and washed in water to cool the gas stream, condense the water vapor, and remove sulfur dioxide to inhibit corrosion. The gas is then drawn through the analyzer and the oxygen level measured and recorded.

5. Mass Spectrometry:

Mass spectrometry measures the ratio of atomic mass-to-atomic charge for gas molecules, as a means of both quantitative and qualitative identification. A gas sample is introduced to the detection cell and subjected to bombardment by electrons in an electromagnetic field, which causes the gas molecules to become ionized. Analytical selectivity is obtained because the tendency of a gas molecule to be deflected decreases as its molecular weight increases. The presence of gasous molecules in the cell causes a dispersion of mass-to-charge spectra across the detector by differences in deflection in a manner analogous to dispersion of a light beam into its component wavelengths by a prism. The signal across the detector is varied as a means of changing the atomic mass-to-charge ratio. This gives qualitative identification of individual components. The amplitude of the signal at each particular ratio is used as the means of quantitative determination of the respective concentrations of individual components. The above procedure applies for analysis of the types and amounts of materials present in a gas sample on a batch basis.

The system may also be adapted for continuous monitoring by setting at one mass-to-charge ratio and drawing a flue gas sample continuously through the instrument. Changes in magnitude of the readout can then be used as indicators of concentration. Campau and Neerman [66] developed a system for measuring nitric oxide in auto exhaust. Gas from the exhaust was drawn through a condenser to remove water, and then through a filter for particulate removal. A small bleed was then passed through the instrument and the nitric oxide concentration recorded.

Mass spectrometry provides a method for measuring nearly any gaseous component in flue gas streams. It is necessary that the two constituents being measured not have the same atomic mass-to-charge ratio. The instrument is expensive, complicated, requires highly trained personnel, and can be used only in the laboratory at the present time.

F. Calibration Techniques:

Continuous gas monitoring instruments are normally calibrated by passage of a gas stream of a known concentration through the detection cell. It is important that the gas concentration be accurately prepared, to facilitate obtaining correct results when performing continuous monitoring with the instrument. Several devices are available for making gas mixtures of known concentrations, including rotating syringes, motor-driven syringes, rigid wall containers, flexible fabric bags, and permeation tubes. All of these employ withdrawal of a concentrated gas mixture into an air stream for dilution prior to passage through the instrument. Multiple dilution stages may be added for high inlet concentrations, but involve added complexity in the system and multiply the chances for error caused by flow rate fluctuations.

1. Rotating Syringe:

Rossano and Cooper [67] [68] described the use of a rotating syringe for preparing known concentrations of nitrogen dioxide for calibrating an instrument. The system employed injecting varying amounts of purified nitrogen dioxide into a 100 cc tuberculin syringe. A thermometer capillary acting as a limiting flow orifice was placed over the tip of the syringe. The syringe was then placed in a vertical direction. An air blast was directed against the rotor on the syringe plunger, causing the plunger to rotate and act as a piston. Constant pressure was maintained inside the syringe because of the frictionless seal between syringe plunger and casing, causing gas to be compressed and withdrawn from the syringe at a constant rate. Known concentrations were then prepared by adding a dilution air stream at a constant rate, as shown

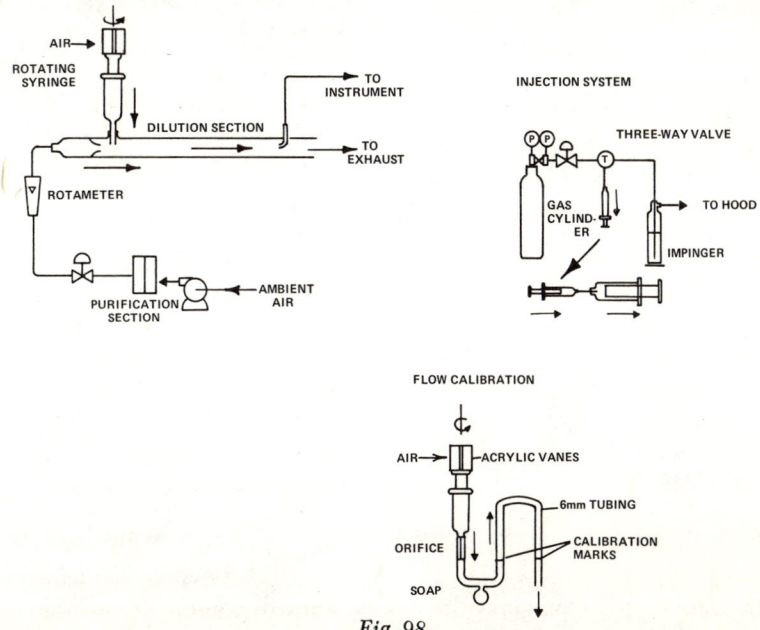

Fig. 98.

Courtesy of: *Journal of the Air Pollution Control Association*, Article by Rossano and Cooper, November 1963 (67).

in Figure 98. The readings obtained on the instrument were checked by means of a parallel wet chemical analysis, once a stable reading was achieved.

Megy [50] employed a similar system for calibrating a coulometric titrator with hydrogen sulfide, methyl mercaptan, and sulfur dioxide. Varying amounts of purified gases were withdrawn from cylinders into syringes ranging in size from 0.1 to 10.0 ml in size. The small syringe contents were then injected into a large 100 cc syringe through a rubber septum placed on the tip. The septum was removed following injection and the gas diluted with air to a given known volume. The capillary was attached to the tip and the syringe placed vertically as before. The flow rate was first tested by placing a 1.0 ml bubble tube to the end of the capillary, observing the time required to traverse the volume, diluting the gas stream with air and feeding to the instrument. The technique provided for making gas concentrations from one to 3,000 parts per million by volume.

Known concentrations of organic vapors were obtained by placing the liquid in a flask, sealing it and allowing it to stand overnight. The vapor reached an equilibrium concentration based on its characteristic vapor pressure at room temperature. A known volume was withdrawn from the flask and placed in the large syringe as before, and the concentration may then be computed for the diluted gas stream.

Concentration in the gas stream fed to the instrument was computed as follows:

$$C_L = \frac{V_s}{V_L} \times 10^6 \quad (112)$$

$$C_I = \frac{C_L Q_L}{Q_I} = \frac{V_s Q_L}{V_L Q_I} \times 10^6 \quad (113)$$

where

C_L = Concentration in large syringe in ppm by volume.
C_I = Concentration fed to instrument in ppm by volume.
V_s = Volume of small syringe in ml.
V_L = Volume of large syringe in ml.
Q_L = Flow rate through capillary in ml per minute.
Q_I = Flow rate of dilution air in ml per minute.

Concentrations were varied by changing the volume of gas added from the small syringe, or the orifice flow rate by use of different capillaries, or by changing the flow rate of dilution air. This technique provided for flexible operation over a wide range of concentrations.

2. Motor-Driven Syringes:

Kuczynski [69] described the use of a motor-driven syringe for preparing known concentrations of sulfur dioxide. The method was similar to the rotating syringe except that the plunger is slowly pushed inward by a revolving motor. It was necessary to use a gas-tight syringe to minimize leakage between the plunger and casing of the syringe. It was a suitable method, but required more complicated and expensive equipment for successful operation than the rotating syringe method.

3. Stainless Steel Cylinders:

Duckworth [70] described the use of prepared mixtures of sulfur dioxide for calibrating instruments in the field. The method employed evacuation of the cylinder, followed by addition of a known amount of sulfur dioxide, and subsequent repressurization of the cylinder with air or nitrogen to a given concentration. In the field, gas was bled from the cylinder at a known rate at about one liter per minute into a pre-purified stream of dilution air which could be varied between flow rates of 10 to 68 liters per minute. A small aliquot from the gas stream was bled to the instrument for calibration purposes, and parallel wet chemical tests were made, once a stable reading was obtained on the instrument.

The technique was suitable for field use, but a sensitive pressure regulator had to be used for accurate control of cylinder flow rate, and the method also has certain other inherent problems. The stainless steel cylinder was heavy and often difficult to carry to remote locations. Interaction of the test gas with stainless steel would create losses in concentration, thus making use of the method impossible, and dilution with nitrogen was necessary if the gas reacted with air. It was also necessary to use inert tubing such as teflon to minimize the chances for wall losses by interaction. A schematic diagram of a typical system is shown in Figure 99.

4. Flexible Wall Containers:

It is possible to make up gas mixtures of known volume in flexible bags, such as those

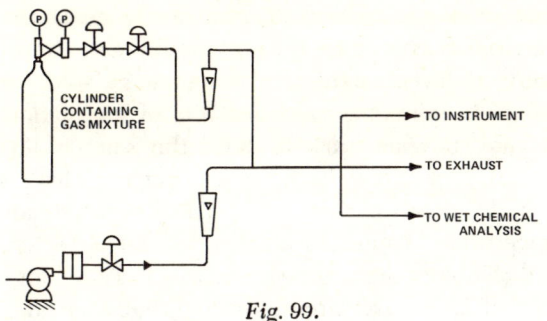

Fig. 99.

Courtesy of: *Journal of the Air Pollution Control Association*, Article by S. Duckworth, September 1963 (70).

made from Mylar, where accurate estimation of volume requires addition of a metered volume of air and a known volume of the test gas. The problems with the technique are that gas may interact with the walls during any prolonged storage, and losses may result if a pump is placed between the bag and the instrument. Both of these problems tend to cause the actual gas concentration to be less than that calculated, either due to losses by chemical reaction or by dilution.

Jutze and Lewis [71] placed the bag inside of a closed container and then added air at a known rate to the volume in the box not occupied by the bag. This tended to force gas out of the bag at the rate air was being blown in, thus eliminating the need for a pump. The same concentrations of sulfur dioxide or nitrogen dioxide were made up in two separate bags. One was taken to the field for instrument calibration while one was kept in the laboratory to be used as a check on actual concentration. The technique proved suitable for calibrating air monitoring instruments and involved construction of relatively inexpensive equipment.

5. Permeation Tubes:

Thomas and Amtower [72] used permeation tubes for field calibration of air monitoring instruments. Their principle of operation was that purified gas diffused through a membrane in a capsule at a known rate. When added to a dilution air stream, known concentrations of the gas could be prepared. The devices have proved successful in tests by several agencies for generating known concentrations of nitric oxide, nitrogen dioxide, sulfur dioxide, hydrogen sulfide, and other gaseous constituents. Gas concentrations may be varied by changes in the dilution air flow rate.

IV. DESIGN OF SYSTEMS

Continuous monitoring systems are designed to withdraw a representative sample at a known rate, and accurately measure the amount of one or more constituents present. Several elements are necessary for a system to operate successfully, which include the following: a system for continuous flow measurement, a sample line which brings gas from the source to the instrument, devices for removal of particulate or moisture if necessary, a detection and data readout system, a means for controlling flow rate, and a prime mover. Each system must usually be designed individually for a particular situation, depending on the type of material being measured, and the physical facilities at hand.

A. Continuous Flow Measurement:

The flow measuring system normally employs an S-type pitot tube placed adjacent to the sample probe. The pressure differential across the pitot tube is normally 0.0 to 0.5, or 0.0 to 1.0 inches of water, which may then be transferred to a pneumatic signal by being connected to a pressure transmitter and a pneumatic recorder. The pressure differential reading may also be converted to an electric current by means of a pressure transducer and the signal sent to a millivolt recorder. The recorders for both flow measurement and concentration in the detection cell are often placed on the process unit control panel. This provides a means for operating personnel to observe these recorders, as an aid for process control. The above system is particularly useful in corrosive atmospheres which are often found in the field.

For most sources, it is necessary to employ a blowback system to prevent pluggage of the impact side of the pitot tube by particulate matter. This normally consists of allowing compressed air to purge through the tube for approximately 30 to 60 seconds once every four to eight hours, where the exact interval depends on the particulate concentration of the source being monitored. Saturated sources may require periodic blowback through both tubes because entrained water may tend to condense in both upstream and downstream tubes. The blowback system may be activated either manually from the control panel, or automatically on a cyclic basis. It is important that the three-way or solenoid valves used prevent air surges toward the pressure sensing devices during the blowback periods.

It is necessary to relate velocity at a given point to gas flow rate in the duct. This is normally done by taking a number of velocity traverses across the duct at several process operating conditions to establish the profile. One point is then selected as the location for the pitot tube, which should be as near the average as possible. The velocities measured at the point selected are then related to the total gas flow rates at the different operating conditions. Emission rates may then be computed from knowledge of the gas flow rate and the concentration of the substance as indicated by the detector cell.

It may also be desired to measure temperature on a continuous basis by means of a thermocouple placed in the flue gas. It is important to protect the wires by placing the entire unit in an enclosed pipe with only the tip exposed, where occasional cleaning of the probe tip may be needed to assure continued accuracy. Static pressure in the duct may also be measured continuously by a system similar to that described for velocity head differential, but this does not normally change significantly in most ducts. Devices are also available to measure the dew point of gas streams on a continuous basis as a means of moisture content determination, but the presence of particulates and corrosive gases may limit the effectiveness of these devices. To date, they have not been evaluated in source test applications to any extent.

With the above techniques, it is possible to provide for continuous velocity measurement and flow rate determination for flue gas streams; a typical system is illustrated in Figure 100.

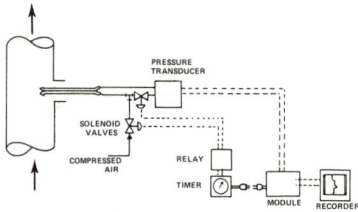

Fig. 100.

B. Sample Collection:

Gas is withdrawn at a known rate from the duct through a sample line to the detection cell. It is normally a good idea to make this section as short as possible to minimize possible losses. The tubing should also be of inert material such as 316 stainless steel or teflon, to minimize the possibility of interaction. It may be necessary to heat the sample line to prevent condensation of water. This is usually done by using electrically or steam-traced line from the source to the detection cell. Tubing diameter should be large enough so that load losses are minimized, but also small enough so that retention times are not excessive. Normally, tubing diameters of 3/16 to 1/2 inch (internal) are sufficient for this purpose.

Monitoring of either gaseous or particulate materials involves certain special considerations. For particulate monitoring systems, the line must be heated when condensation is a possible problem if the particles are to be measured in the dry state, and the number of bends or constrictions should be minimized. However, for measurement in detectors containing liquids, such as ion-specific or conductivity electrodes, this is not necessary. It is necessary to provide for overflow and removal of water in such cases where continuous liquid flows through the detector are employed. No devices should be added upstream of the detector which result in removal of materials from the gas stream, including filters and similar devices.

Gaseous monitoring systems normally entail particulate removal upstream of the detection cell. This may be accomplished by means of filtration, electrostatic precipitation, or other means. These devices require frequent cleaning by blowback or other means. A cyclic blowback system actuated for five minutes once per hour is often sufficient for a ceramic filter placed in a duct at the end of a probe. It is necessary to prevent air blowback into the detection cell, and also to provide for a minimum of air dilution. A filter of 0.3 micron porosity, immediately upstream of the detection cell, has been found suitable in several applications, but would have to be heated to prevent condensation if this posed a potential problem. A preselective scrubbing system may also be employed to selectively remove particulates and moisture. The technique cannot be used, however, if significant losses of the gas being measured occur. A combination of proper pH and low liquid flow rate may minimize possible losses, however. Heated sampling lines should be used in applications where water can cause losses by absorption in the line. The detector cell should be able to accommodate water vapor, however, and possible interferences must be compensated for.

C. Detection Cells:

The detector used for measuring the concentration of either gaseous or particulate matter should be capable of accurate measurement without interferences or losses. Types of detectors have been described in a previous section. Water poses a potential problem either as an interference or in operation. It may require heating of the detection cell, removal by condensation, or use of a continuous liquid flow under such circumstances. Calibration of the detection cell occurs at a constant rate of flow and operating conditions. Gaseous calibration is normally handled by preparation of known gas concentrations by use of the rotating syringe or other suitable techniques. Calibration of particulate sampling devices is normally formed by taking a batch sample in parallel with the continuous system, in an adjacent location. The mass concentration obtained is then related to the average instrument reading.

D. Sample Flow Rate:

Following the detection cell, the gas stream is filtered and dried if necessary, and then caused to flow through the flow measuring device. The cleaning steps are to protect the flow meter from plugging or corrosion. A metering valve is also placed in the line to control the rate of flow. Suction is provided by a prime mover at the downstream end of the train, and may be a vacuum pump or water aspirator. The flow rate for gaseous systems is normally two liters per minute or less, and is normally at a constant rate. The flow rate for particulate systems is normally 0.5 to 1.0 cfm because of the requirement for isokinetic sampling. It is often set at the isokinetic rate at average velocity and may be adjusted manually. A suitable automatic compensation for sampling rate in terms of velocity for maintaining isokinetic conditions, is not yet commercially available, unfortunately.

REFERENCES

1. Purdom, P. W., "Source Monitoring," Chapter 29 in Stern, A. C., ed., *Air Pollution*, Volume 2, 2nd ed., Academic Press, Inc., New York, New York, 1968.
2. Schutz, A., "Possible Methods for Recording Dust Measurements and Dust Content," *Staub*, Volume 26, No. 10, pp. 409-415, October 1966.
3. "Smoke Density Indicators and Recorders," British Standard 2811, British Standards Institution, London, England, 1957.
4. Crosse, P. A., Lucs, D. H., and Showsil, W. L., "Instruments for Recording the Dust Nuisance Emitted by Chimneys, "*Journal of Scientific Instruments*, Volume 38, No. 1, pp. 12-17, January 1961.
5. Hurley, T. F., and Bailey, D., "The Correlation of Optical Density with the Concentration and Composition of the Smoke Emitted from a Lancashire Boiler," *Journal of the Institute of Fuel*, Volume 31, No. 11, pp. 534-538, November 1958.
6. Gansler, N. R., "The Use of a Bolometer for Continuous Measurement of Particulate Losses from Kraft Recovery Furnaces," Presented at the Annual Meeting of the Pacific Northwest International Section of the Air Pollution Control Association, Vancouver, British Columbia, November 22, 1968.
7. Lucas, D. H., and Snowsill, W. L., "Some Developments in Dust Pollution Measurement," *Atmospheric Environment*, Volume 1, No. 5, pp. 619-636, November 1967.
8. Collins, K. E., and Steele, D. J., "High Sensitivity Recording Optical Density Meter," *Journal of Scientific Instruments*, Volume 38, No. 5, pp. 186-190, May 1961.
9. Cooper, S. R., and Haskell, G. F., "Cutting Chemical Ash Losses from a Kraft Recovery Furnace," *Paper Trade Journal*, Volume 151, No. 13, pp. 58-59, March 27, 1967.
10. McDonald, W., "BCFP Monitor Cuts Recovery Boiler Stack Losses 50%," *Pulp and Paper*, Volume 40, No. 17, pp. 39-41, August 19, 1966.
11. Heimke, W., "Mounting Devices for Radiation Pyrometers and Smoke Density Meters in Iron and Glass Foundries," *Siemens Zeitshrift*, Volume 41, No. 9, pp. 785-790, September 1967.
12. Yocom, J. E., "Problems in Judging Plume Opacity— A Simple Device for Measuring Opacity of Wet Plumes," *Journal of the APCA*, Volume 13, No. 1, pp. 36-39, January 1963.
13. Price, T. D., and Klemperer, H., "Evaluation of the Multiple Spot Discoloration Tester," *Air Repair*, Volume 4, No. 4, pp. 213-217, February 1955.
14. Hemeon, W. C. L., Haines, G. F., and Ide, H. M., "Determination of Haze and Smoke Concentrations by Filter Paper Samplers," *Air Repair*, Volume 3, No. 1, pp. 22-28, August 1953.
15. Gruber, C. W., and Schumann, C. E., "Soiling Potential—A New Method for Measuring Smoke Emission," *Journal of the APCA*, Volume 16, No. 5, pp. 272-275, May 1966.
16. McShane, W. P., and Bulba, E., "Automatic Stack Monitoring of a Basic Oxygen Furnace," Paper 67-120, Presented at the 60th Annual Meeting of the Air Pollution Control Association, Cleveland, Ohio, June 14, 1967.
17. McShane, W. P., and Bulba, E., "Automatic Stack Monitoring of a Basic Oxygen Furnace," *Journal of the APCA*, Volume 18, No. 4, pp. 214-215, April 1968.
18. Bulba, E., and Silverman, L., "A Mass Recording Stack Monitoring System for Particulates," Paper 65-141, Presented at the 58th Annual Meeting of the Air Pollution Control Association, Toronto, Canada, June 20, 1965.
19. Groutsch, E. R., "Automatic Dust Sampler," *Pro-

ceedings of the Australasian Institute of Mining and Metallurgy, Volume 16, No. 191, pp. 165-189, September 1959.

20. Donoso, J. J., "An Improved Automatic Smoke Sampler," *Transactions of the American Institute of Mining and Metallurgical Engineers,* Volume 188, pp. 610-612, 1950.
21. Rozsa, J. T., Stone, J., Uguccini, O. W., and Kupel, P. E., "Determination of Beryllium in Air," *Applied Spectroscopy,* Volume 19, No. 1, pp. 7-10, January 1965.
22. Leonard, J. S., "Continuous Kraft Mill Emission Monitoring," in Blosser, R. O., and Copper, H. B. H., ed., "Analytical Equipment and Monitoring Devices for Gases and Particulates," National Council for Air and Stream Improvement Atmospheric Pollution Technical Bulletin No. 35, New York, New York, March 6, 1968.
23. Leonard, J. S., Tretter, V. J., and Taylor, C. E., "Continuous Particulate Monitoring by Specific Ion Electrode Devices," Presented at the Sixth Tappi Air and Water Conference, Jacksonville, Florida, April 29, 1969.
24. Tretter, V. J., "Use of Continuous Monitors of Soda Loss and Malodorous Sulfur Loss in Process Control," *Tappi,* 52, (12), 2324-2326, December 1969.
25. Warner, T. B., "Lanthanum Fluoride Electrode Response in Water and in Sodium Chloride," *Analytical Chemistry,* Volume 41, No. 3, pp. 527-529, March 1969.
26. Elfers, L. A., and Decker, C. E., "Determination of Fluoride in Air and Stack Gas Samples by Use of an Ion Specific Electrode," *Analytical Chemistry,* Volume 40, No. 11, pp. 1658-1661, September 1968.
27. Adams, D. F., Koppe, R., and Dana, A., "An Automatic Atmospheric Fluoride Analyzer with Application to Other Pollutants," *Journal of the APCA,* Volume 9, No. 3, pp. 160-168, November 1959.
28. Mitchell, R., and Engdahl, R., "A Survey of Improved Methods for the Measurement of Particulate Concentrations in Flowing Gas Streams," TA-5 Committee Informative Report No. 1, *Journal of the APCA,* Volume 13, No. 11, pp. 558-562, November 1963.
29. Halliday, D., Introductory Nuclear Physics, John Wiley and Sons, Inc., New York, New York, 1950.
30. Breitling, K., "The Beta Method, A New Stack Sampling Procedure," *Staub,* Volume 20, No. 10, pp. 364-365, October 1960.
31. Izmailov, G. A., "Measuring the Gravimetric Concentration of Dust in the Air Using Beta-Radiation," *Zavodskaya Laboratoriya,* Volume 27, No. 1, pp. 40-43, January 1961.
23. Knudsen, V. O., Wilson, J. V., and Anderson, M. S., "The Attenuation of Audible Sound in Fog and Smoke", *Journal of the Accoustical Society of America,* Volume 20, No. 6, pp. 849, November 1948.
33. Pohlman, R., and Walters, K., "Measurement of Aerosol Concentration by Determination of High Frequency Sound Extinction," *Chemie Ingenieur Technik* (Frankfurt), Volume 31, No. 1, pp. 31-34, January 1959.
34. Grindell, D. H., "An Electrostatic Dust Monitor," *Proceedings of the Institution of Electrical Engineers,* Volume 107, Part A, No. 34, pp. 353-365, August 1960.
35. Barkov, N. N., "Electronic Analyzer Type EG-59P of the Amount of Dust Gases," *Proborostronic,* Volume 3, No. 3, pp. 23-25, March 1961.
36. Duwel, L., "Latest State of Development of Control Instruments for the Continuous Monitoring of Dust Emissions," *Staub,* Volume 28, No. 3, pp. 42-53, March 1969.
37. Lapple, C. E., "Particle Size Analysis and Analyzers," *Chemical Engineering,* Volume 75, No. 11, pp. 149-156, May 20, 1968.
38. Peterson, C. M., and Paulus, H. J., "Continuous Monitoring of Aerosols over the 0.001 to 10 Micron Spectrum," *American Industrial Hygiene Association Journal,* Volume 29, No. 2, pp. 111-119, March-April 1968.
39. Brown, J. E., "Onstream Process Analyzers," *Chemical Engineering,* Volume 75, No. 10, pp. 164-176, May 6, 1968.
40. Naumann, A., "Methods of Automatic Flue Gas Analyzers," *Glasstechnische Berichte,* Volume 24, No. 9, pp. 222-229, September 1952.
41. Thoenes, H. W., and Guse, W., "Latest State of Development of Instruments for the Continuous Monitoring of Gas Emissions," *Staub,* Volume 28, No. 3, pp. 53-63, March 1968.
42. Miller, A. M., Brown, J., and Abrams, R., "Applied Techniques of Analyses for Stack Emissions," Presented at the West Coast Regional Meeting of the National Council for Air and Stream Improvement, Portland, Oregon, October 2, 1968.
43. Nacovsky ,W., "Determination of Sulfur Oxides in Flue Gases," *Combustion,* Volume 38, No. 7, pp. 35-38, January 1967.
44. Laxton, J. W., and Jackson, P. J., "Automatic Monitor for Recording Sulfur Trioxide in Flue Gas," *Journal of the Institute of Fuel,* Volume 37, No. 276, pp. 12-17, January 1964.
45. Adley, F. E., and Skillern, C. P., "A Portable Multi-Range NO_2 Gas Monitor," *American Industrial Hygiene Association Journal,* Volume 19, No. 3, pp. 233-237, May—June 1958.
46. Shaffer, P. A., Briglio, A., Brockman, J., "Instrument for Automatic Continuous Titration," *Analytical Chemistry,* Volume 20, No. 11, pp. 1008-1014, November 1948.
47. Austin, R., Turner, B., and Percy, C., "A Continuous Sulfur Monitor," *Instruments,* Volume 22, p. 488, 1949.
48. "The Barton Model 286 Sulfur Titrator," ITT Barton Instrument Co., Monterey Park, California, 1967.
49. Thoen, G. N., DeHaas, G. G., and Austin, R. R., "Instrumentation for Quantitative Measurement of Sulfur Compounds in Kraft Gases," *Tappi,* Volume 51, No. 6, pp. 246-248, June 1968.
50. Blosser, R. O., Cooper, H. B. H., and Megy, J. A., "Gaseous Emissions—Automatic Techniques—Electrolytic Titration," Atmospheric Pollution Technical Bulletin No. 38, National Council for Air and Stream Improvement, New York, New York, 1968.
51. Thoen, G. N., DeHaas, G. G., and Austin, R. R., "Continuous Measurement of Sulfur Compounds and Their Relationship to Operating Kraft Mill Black Liquor Furnaces," *Tappi,* Volume 52, No. 9, pp. 1485-1487, August 1969.
52. Markovs, J., Lee, M., and Nasser, B. E., "Determining Impurities in Absorption Process Streams," *Chemical Engineering Progress,* Volume 65, No. 5, pp. 68-74, May 1969.
53. Altshuller, A. P., and Sleva, S. F., "Vapor Phase Determination of Olefins by a Coulometric Method," *Analytical Chemistry,* Volume 34, No. 3, pp. 418-422, March 1962.
54. King, W. J., Wilson, K. W., and Swartz, D. J.,

"Analysis of Automotive Exhaust Gas," *Journal of the APCA*, Volume 12, No. 1, pp. 5-21, January 1963.
55. Christman, R. F., "Chemistry of Air Pollution," Chapter 5 in Rossano, A. T., ed., *Air Pollution Control—Guidebook for Management*, Environmental Science Services Corp., Stamford, Connecticut, 1969.
56. Bent, R., Ladner, W., and Mullin, W., "A Method for the Estimation of Sulfur Trioxide," *Chemistry and Industry*, No. 11, pp. 461-462, March 18, 1967.
57. Shuck, E. A., Ford, H. W., and Stephens, E. R., "Air Pollution Aspects of Irradiated Automobile Exhaust as Related to Fuel Consumption," Report No. 26, Air Pollution Foundation, San Marino, California, October 1958.
58. Nicksic, S. W., and Harkins, J., "Spectrophotometric Determination of Nitric Oxide in Auto Exhaust," *Analytical Chemistry*, Volume 34, No. 8, pp. 985-988, July 1962.
59. Harkins, J., and Goodwine, J. K., "Oxides of Nitrogen in Diesel Exhaust," *Journal of the APCA*, Volume 14, No. 1, pp. 34-38, January 1964.
60. Sweeney, M. P., Swartz, D. J., Rost, G. A., MacPhee, R., and Chao, J., "Continuous Measurement of Oxides of Nitrogen in Auto Exhaust," *Journal of the APCA*, Volume 14, No. 7, pp. 249-254, July 1964.
61. Walther, J. E., and Amberg, H. R., "Continuous Monitoring of Kraft Mill Stack Gases with a Process Gas Chromatograph," *Tappi*, Volume 50, No. 10, pp. 108A-110A, October 1967.
62. Walther, J. E., and Amberg, H. R., "Experience with a Mobile Laboratory in Source Sampling Kraft Mill Emissions," *Tappi*, Volume 51, No. 11, pp. 126A-129A, November 1968.
63. Walther, J. E., and Amberg, H. R., "A Positive Air Quality Control Program at a New Kraft Mill," *Journal of the APCA*, 20, (1), 19-26, January 1970.
64. Feldstein, M., "Studies on the Analysis of Hydrocarbons from Incinerator Effluents with a Combustible Gas Indicator," *American Industrial Hygiene Association Journal*, Volume 22, No. 4, pp. 286-291, August 1961.
65. Tipping, F., "Oxygen Analysis in Chemical Plant," *Chemical Process Engineering*, Volume 49, No. 10, pp. 82-84, October 1968.
66. Campau, R. M., and Neerman, J. C., "Continuous Mass Spectrophotometric Determination of Nitric Oxide in Auto Exhaust," in Vehicle Emissions, Part 2, Technical Progress Series, Volume 12, Society of Automotive Engineers, Inc., New York, New York, 1967.
67. Rossano, A. T., and Cooper, H. B. H., "Procedure for Calibrating a Continuous NO_2 Analyzer," *Journal of the APCA*, Volume 13, No. 11, pp. 518-523, November 1963.
68. Cooper, H. B. H., "Source Testing Procedures," Chapter 11 in *Air Pollution Control—Guidebook for Management*, Environmental Science Services Corp., Stamford, Connecticut, 1969.
69. Kuczynski, E. R., "Simplified Method of Calibrating SO_2 Gas Analyzers in the Parts per Million Range," *Journal of the APCA*, Volume 13, No. 9, pp. 435-436, September 1963.
70. Duckworth, S., Levaggi, D., and Lim, J., "Field Dynamic Calibration of SO_2 Recording Instruments," *Journal of the APCA*, Volume 13, No. 9, pp. 429-434, September 1963.
71. Jutze, G. A. and Lewis, R. L., "A Method for Checking Instrument Performance at Remote Sampling Sites," *Journal of the APCA*, Volume 15, No. 7, pp. 323-326, July 1965.
72. Thomas, M. D., and Amtower, R. E., "Gas Dilution Apparatus for Preparing Reproducible Gas Mixtures in any Desired Concentration and Complexity," *Journal of the APCA*, Volume 16, No. 1, pp. 618-623, November 1966.

Table 40. Calibration of a Photometric Meter for Measuring Particulate Concentrations on a Kraft Recovery Furnace.

Scale Reading %	Concentration gr/SCF (1)
0.0	0.00
30.0	0.08
50.0	0.15
65.0	0.60
80.0	1.20

Note: 1. Concentrations for wet gas at 60°F, 29.9 in. Hg.
2. Courtesy of: Mr. Neil Gansler, Longview Fibre Corp., Longview, Washington [6].

Table 41. Operating Characteristics of Continuous Gaseous Analyzers.

Technique	Principle	Operation	Selectivity	Gases	Interferences
Wet Chemical	Colorimetric	Continuous	Specific	NO_2, SO_2, SO_3	O_3, Sulfate Particulates
	Conductivity	Continuous	Nonspecific	SO_2, SO_3	Inorganic Particulates, HCl, NH_3, H_2SO_4, CO_2, Cl_2
Spectrophotometry	Dispersive Infrared	Noncontinuous	Specific	Any with IR peak	—
	Nondispersive Infrared	Continuous	Specific	CO, CO_2, SO_2	Water, CO_2, CO
	Ultraviolet	Continuous	Specific or Nonspecific	NO, NO_2 SO_2	—
	Visible	Continuous		NO_2	—
Coulometry	Bromine Oxidation	Continuous	Nonspecific	Olefins, SO_2, H_2S, CS_2, RSR, RSSR	NO, NO_2, Olefins
Combustion	Combustibles	Continuous	Nonspecific	CO, H_2, Organic Compounds	—
	Flame Ionization	Continuous	Nonspecific	Organic Compounds	—
Mass Spectrometry	Atomic Mass to Charge Ratio	Continuous or Noncontinuous	Specific	Any Gas	Others of same Atomic Weight
Paramagnetic	Paramagnetic Relationship	Continuous	Specific	Oxygen	—
Gas Chromatography	Detectors	Noncontinuous	Specific		
	Thermal Conductivity			Any Gas	High Concentrations needed
	Flame Ionization			Organic Compounds	—
	Flame Photometry			Sulfur & Phosphorus Gases	—
	Electron Capture			NO, NO_2, SO_2, H_2S, SO_3, Halogen Compounds	Water
	Coulometry			SO_2, H_2S, Olefins RSR, RSSR	NO, NO_2, Olefins

Table 42. Selective Prescrubbing Solutions for Sulfur Gas Analyses with Coulometric Titration.

Scrubbing Solution	Concentration % by Wt.	Gases Removed
$KHC_8H_4O_4$	5	SO_2
$CdSO_4$-H_3BO_3	1-2	H_2S + SO_2
NaOH	10	RSH + H_2S + SO_2
$AgNO_3$	0.5	RSR + RSH + H_2S + SO_2
Blank	—	RSSR + Others

Table 43. Response of a Coulometric Titration to Different Gases.

Gas	Concentration ppm by volume	Response Relative to SO_2 — %	Concentration as SO_2 ppm by volume
Hydrogen Sulfide (2)	400	330.0	1320
Sulfur Dioxide (1)	400	100.0	400
trans-2-Butene (1)	400	100.0	400
Acrolein (1)	400	37.0	148
Thiophene (1)	400	98.0	392
Phenol (1)	400	1.0	4
n-Butyl Sulfide (1)	400	6.0	24
Methyl Mercaptan (2)	400	300.0	1200
Dimethyl Sulfide (2)	400	75.0	300
Dimethyl Disulfide (2)	400	300.0	1200

Notes: 1. Data from Altshuller and Sleva, Analytical Chemistry, 34, 3, 420, 1962 [53].
2. Data from Blosser, Cooper, and Megy, NCASI Technical Bulletin No. 38 [50].

Table 44. Oxidation of Nitric Oxide to Nitrogen Dioxide.

NO Concentration, ppm by volume	Conversion Rate, ppm NO_2/minute
1000	210.
100	2.1
10	0.021
1	0.00021

Notes: 1. From: Harkins & Goodwine, Journal of the APCA, 14, 1, 37, 1964 [53].
2. Data at 29.92 in. Hg, 70°F for dry air.

Table 45. Relative Magnetic Susceptibilities of Flue Gas Constituents.

Gas	Magnetic Suscept. %
Oxygen	+100.0
Nitric Oxide	+45.0
Nitrogen Dioxide	+3.0
Nitrogen	−0.36
Carbon Dioxide	−0.63
Carbon Monoxide	−0.35
Hydrogen	−0.12
Methane	−0.37

From: A. O. Beckman, Bulletin No. 108A; The Hays Corp. Bulletin, 56-829-56.
Courtesy of: Journal of the Air Pollution Control Association, King, W. J., et al, "Analysis of Automotive Exhaust Gas," January 1963 [54].

CHAPTER 11

SPECIAL APPLICATIONS

I. INTRODUCTION

Material in this chapter is related to measuring techniques which indirectly involve gaseous and particulate emissions from sources not previously discussed, plus those from mobile sources, and plume measurements. Measuring techniques described include determination of odor threshold levels of flue gases, measurement of acid deposition onto metal surfaces, the presence of radioactive materials, and bacterial emissions to the atmosphere. Techniques involved in the measurement of gaseous and particulate emissions from mobile sources, such as automobiles and jet aircraft, are described. Tracing the path of and determining concentrations of materials in plumes is a combination of techniques employed in both source and ambient air measurements.

II. ADDITIONAL MEASURING TECHNIQUES

Several types of source measurements do not fit neatly into either gaseous or particulate sampling. These include measurement of odor threshold levels of flue gases, acid deposition onto metal surfaces, plus the emissions of radioactive materials or bacterial cells. Each involves the use of techniques not previously presented in the text.

A. Odor Threshold Levels:

1. Problems

A problem in both source and ambient measurements is the determination of odor threshold levels, with the character, intensity, threshold, and degree of objectionability all being important factors. Variations occur when different odorous gases are mixed with each other. A particularly important factor is that response between human subjects varies not only with background odor levels but also with the components present, the degree and duration of exposure, and the degree of sensitivity of the subject. Wilby [1] recently performed a study of variations in response between 35 untrained subjects on exposure to 18 different odorous sulfur gases. He found that response varied somewhat between subjects, but much more so with certain compounds than for others.

Prediction of potential odor response in the community adjacent to a source of odorous substances is thus a difficult task. Variations between individuals of varying sensitivities is an important factor. Evaluation is made particularly difficult by problems of choosing a representative group of subjects and by the possible effects of prior exposure. People who have been exposed to an odor in the vicinity of its source often become deadened to low concentrations and are not reliable subjects. Accurate evaluation then involves removal of the sample to a relatively remote location where the subjects have not been previously exposed to the odor. Unfortunately, this may result in altering the character of the sample during transit by chemical reactions, or by adsorption of gaseous constituents onto the walls of the container. Atmospheric humidity may also have an effect on the response of subjects, as well as the possible presence of particulate matter.

2. Methods

Exposing human subjects to varying levels of flue gas diluted with "purified" air is used to determine odor threshold levels for three reasons. First, it provides for direct measurement of the odor threshold level regardless of the constituents present. Second, it is much less costly and time-consuming than a detailed chemical analysis. Third, it is not really possible to perform a complete identification of the odorous gases present. Work by Cederlof [2] on the flue gases from a Kraft pulp mill showed that it was not possible to determine the odor level of flue gas samples based strictly on chemical measurements.

Different quantitative methods for odor measurement have been described by Mills [3]. These include chemical matching using trained observers, and both static and dynamic dilution techniques. Following collection, the flue gas samples are diluted and submitted to a panel of either trained or untrained observers. Fox and Gex [4] devised a means for expressing the

187

relative strength of an odor based on successive dilutions in glass syringes. It is based on the dilution required to bring the original sample to the most dilute level where the odor is still perceptible. The concept of an "odor unit" is introduced, where one odor unit is defined as the odor being barely perceptible at its concentration in the duct. The number of odor units increases as the necessary dilution to odor threshold increases.

Odor Concentration:

$$C = \frac{100}{V_s} \times 1 \text{ odor unit/ft}^3 \quad (114)$$

Odor Emission Rate:

$$OER = 60\,(C)(Q_s) \quad (115)$$

where:

- OER = Odor emission rate in odor units per hour.
- C = Odor level of original sample for odor units per cubic foot.
- DF = Dilution factor.
- V_s = Volume in milliliters of original sample present in the most dilute sample where the odor is present.
- V_o = Initial volume of sample gas withdrawn from the duct in milliliters.
- 100 = Total volume of diluted sample in milliliters.
- Q_s = Rate of flow of flue gas in cubic feet per minute at stack conditions.

Sample Problem Number 13. *Odor Threshold Level and Emission Rate.*

Problem:

A sample of 100 ml of flue gas from a chemical plant exhaust gas is drawn into a glass syringe and taken to a laboratory. Evaluation of the strength of the odor by a panel indicates that three consecutive dilutions of 10 to 1 (90 ml clean air and 10 ml odorous gas) are required to reduce the concentration of the flue gas to its odor threshold level by 50 percent of the subjects. Gas flow rate at the source was found to be 20,000 cfm at stack conditions. Compute the odor concentration in odor units, and the odor emission rate.

Formulae:

1. Dilution Factor for a Single Dilution Step:

$$DF_i = \left(\frac{\text{Total Volume of Gas in Syringe}}{\text{Volume of Odorous Gas in Syringe}}\right)_i$$

i = For a single step (116)

2. Total Dilution Factor:

$$(DF_T) = \sum_i^n (DF)_i = (DF_1) \times (DF_2) \times \text{--} \times (DF_n) \quad (117)$$

n = Number of dilution stages

3. Volume of Original Sample in Most Dilute Sample:

$$V_s = \frac{1}{(DF_T)}(V_o - \text{ml}) = \text{ml} \quad (118)$$

4. Odor Concentration:

$$C = \frac{100 \text{ ml}}{(V_s - \text{ml})} \times 1.0 \text{ o.u./ft}^3 = \text{Odor Units/ft}^3$$

(114) Repeated

o. u. = Odor Units

5. Odor Emission Rate:

$$OER = (60)(C)(Q_s) = \text{Odor Units/Hour}$$

(115) Repeated

Solution:

1. Terminology:

- n = 3 dilution stages
- DF_i = $(90 + 10)/10 = 100/10 = 10/1$
- V_o = 100 ml
- Q_s = 20,000 ft^3/min

2. Dilution Factor:

$DF_1 = DF_2 = DF_3 = 100/10 = 10/1$

$DF_T = (DF_1)(DF_2)(DF_3) = \left(\frac{10}{1}\right)\left(\frac{10}{1}\right)\left(\frac{10}{1}\right)$

$DF_T = \frac{1{,}000}{1}$

3. Volume of Original Sample in Most Dilute Sample:

$V_s = \frac{1}{(DF_T)} V_o = \frac{1}{(1000)}(100) = \frac{100}{1000}$

$V_s = 0.10$ ml

4. Odor Concentration:

$$C = \frac{100}{V_s} \times \left(1.0 \, \frac{\text{o.u.}}{\text{ft}^3}\right)$$
$$= \frac{100}{0.10}(1.0) = 1,000$$
$$C = 1,000 \text{ Odor Units/ft}^3$$

5. Odor Emission Rate:

$$OER = (60)(C)(Q_s)$$
$$= (60)(1,000)(20,000)$$
$$OER = 1.20 \times 10^9 \text{ Odor Units/hour}$$

A method frequently used for measuring odor thresholds for flue gases involves progressive dilutions using 100 ml glass syringes. Gas samples are collected in a 250 ml gas collection tubes, then taken to the laboratory and a known volume displaced into a 100 cc syringe by mercury displacement. The gas sample in the syringe is diluted to 100 cc with air and progressive dilutions then made. The syringe volume is then injected into the nostrils of several subjects on an odor panel, and estimates made of the odor. The process is repeated with varying volumes of sample until the odor threshold is determined. Odor threshold is defined as the dilution at which 50 percent of the panel can just barely detect the odor.

The method is specified by both the American Society for Testing and Materials [5] and the Los Angeles County Air Pollution Control District [6]. It has been successfully used by Benforado [7] to evaluate the efficiency of afterburners used to reduce emissions of malodorous gases. Samples were collected from both inlet and outlet gas streams of a direct flame incinerator. Progressive dilutions were applied to each of the samples to allow establishing the odor threshold levels. It is important to make accurate dilutions with clean syringes in order to obtain reliable results. Following each test, syringes should be cleaned thoroughly with soap and water. Condensation of moisture from the duct in the gas collection tube can be minimized by sample collection into a 10 ml syringe followed by injection into the collection tube [6]. It is normally useful to work from low to high concentrations to avoid deadening the olfactory senses of the subjects. The method is illustrated in Figure 101.

Dynamic dilution has also been used for producing given concentrations of flue gas samples in air, as recently described by Nader [8]. Flue gas was bled into a stream of air which had been previously passed through activated carbon and other purification media. Normally, the air flow rate would be set at a level where there was a low concentration of odorous gases. The air flow rate was progressively reduced until the odor was just barely perceptible to a panel of subjects, who may be either trained or untrained. A face piece was attached to the unit to allow a portion of the diluted gas stream to pass through it.

Hemeon [9] designed a two-stage dilution system for measuring the odor threshold of flue gases. Considerably larger air flows (30-70 cfm) were used, than in the previously cited work, to minimize the possibility of adsorption of gases and vapors on the walls. Gas samples were collected in flexible Mylar bags, then drawn into the first air stream. A portion of this air stream was bled into a second air stream for additional dilution. Face pieces attached to both air streams provided for odor threshold evaluation over a wide range of concentrations. The system was larger and more complicated than Nader's, however, and required much more space. The author described the use of a scale for increasing odor intensities ranging from zero to five.

Leonardos [10] described the use of an odor chamber for measuring the odor threshold of 53 different odorous gases. The gas mixture was produced by injecting known amounts of gases into the chamber and having a panel of trained observers evaluate the resulting odor threshold levels.

Nickol [11] evaluated the rank odor method for measuring odor threshold levels of distillery offgases. Gas was drawn into a sampling train where the moisture was first removed by entrainment separation, and the odorous gases then collected in a freezeout trap. Known amounts of the material in the freezeout traps were placed in varying amounts of water and then exposed to subjects. Evaluation of the inlet and outlet gases from the vent scrubber showed it reduced the odor threshold by 98 percent. The method provided only a relative comparison because it involved dilution in water. There was also the possibility that odorous constituents might react in solution and be changed in character, leading to erroneous results.

Turk [12] described an additional technique where odor standards were made up in known

Collection:

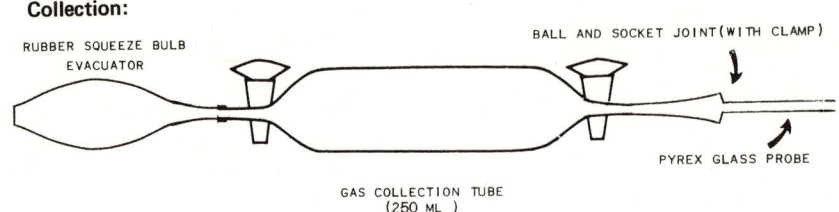

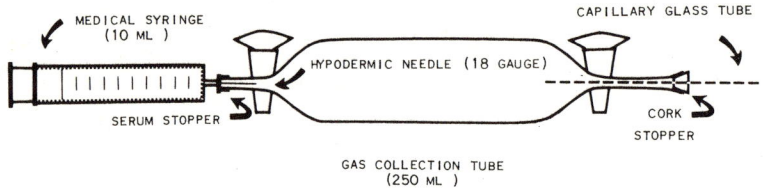

Dilution:

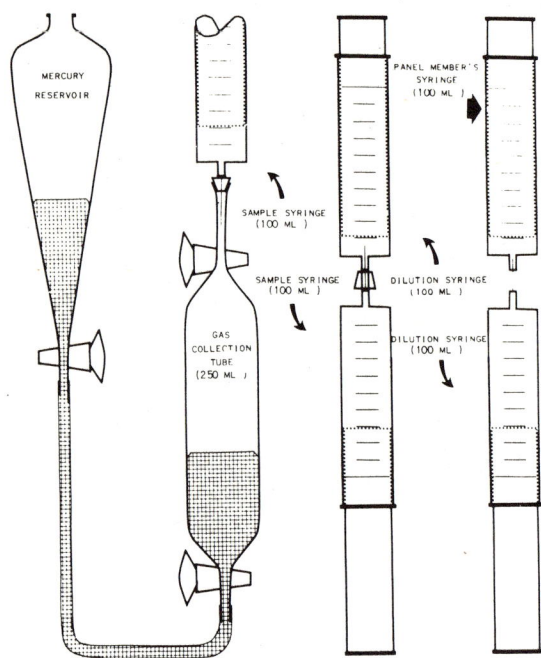

Fig. 101.
Courtesy of: Los Angeles County Air Pollution Control District (6).

concentrations. Trained subjects were exposed to these standards, and then allowed to smell unknown samples for comparison. A means for "matching" odors was thus provided, but a special test program would have to be established for each source, a time-consuming and expensive procedure. Turk [13] has also recently compiled a review of the concepts and methodology involved in measurement of odor threshold levels in flue gas streams.

B. Corrosion and Deposition:

The corrosion of metal heat-exchange surfaces in the interior of boilers being fired with fuel oil or coal is a potentially serious problem. It is caused by either an erosion effect of particles striking exposed surfaces or by the deposition of acid droplets on the cooled surfaces. These acid droplets are often formed by reaction of sulfur trioxide molecules with water in the flue gas. This reaction forms sulfuric acid, which acts as a corrosive agent upon metal surfaces.

It may be necessary to make tests on the rate of acid deposition on existing heat transfer surfaces to determine the types of materials needed in future installations. Recent studies by Jackson [14] and Alexander [15] have led to the development of an air-cooled probe which consists of a tube with air being purged through it into the flue gas. The temperature of the tube surface is controlled by varying the air flow rate, and is measured by use of a thermocouple placed at the end of the probe. The probe is placed in the duct for a given period at a controlled temperature. At the end of the exposure period, the probe is removed and the deposited material removed by washing. It is then analyzed for pH, sulfate ion concentration, and possible other tests. In a similar study, Haneef [16] exposed plates of various metals to flue gas and measured the rates of acid deposition and also corrosion. The plates could be exposed for extended periods to allow obtaining reliable results on a comparative basis.

C. Radioactivity:

Radioactive sources of air pollution must be carefully monitored as a warning regarding the possible release of dangerous materials. Harvey [17] developed a system for simultaneously monitoring four different isotopes. The gas stream was withdrawn from the duct through a tape sampling system followed by a continuous scrubber. One gamma- and one beta-radiation scintillation detector were used to monitor both the tape and the exit liquid on a continuous basis for the presence of these isotopes. The signals from each detector were connected to a computer and a readout system. The device is perhaps the most comprehensive continuous monitoring system constructed to date.

Hyatt [18] used a cascade impactor to measure the particle size distribution of particles from uranium manufacturing processes. A sheet of Whatman No. 41 filter paper was added as the final collection stage. Results showed the mass mean diameter of particles was from 2.0 to 4.5 microns, while the numerical mean diameter was 0.08 microns.

D. Bacteria:

Microbiological cells in the form of aerosols may be collected on membrane filters, as described by Goetz [19]. It is necessary that the bacterial cells be retained on the filter, and that their viability be preserved during collection. Goetz drew ambient air through a membrane filter followed by an impinger containing water. The liquid is then filtered through a second membrane filter, and both filters then cultured. The problem would be a difficult one for sources containing a large number of particles because of the difficulties in observing the cells, and possible toxic effects of the other particles.

Nelson and Ledbetter [20] measured the emissions from oxidation ponds including both gaseous and particulate materials. Air over the ponds was drawn into a membrane filter of 0.45 micron porosity for a given period. The filter was then removed and placed in one of several media and the types of organisms cultured were identified. Additional tests were run for methane, hydrogen sulfide, oxygen, and carbon dioxide being emitted from the pond.

An additional test method involves collection of bacterial cells from a flue gas stream by use of an Andersen sampler [21], where the device was originally developed for this purpose. The respective plates are coated with the media and used to collect the organisms, which may range in size from one to fifteen microns in size.

III. MOBILE SOURCES

Determining emissions from mobile sources presents a problem because it is normally necessary to simulate the test conditions at a stationary point. Exact duplication of field conditions then poses a serious problem in terms of obtaining representative measurements. The two sources studied to date have been automobiles and jet aircraft.

A. Automobiles:

Present discussion of emissions from automobile exhausts is limited to sampling systems, because the analytical procedures have already been described. Chipman and Massey [22] devised a proportional sampling system which removes a relative amount of gas from the exhaust, based on its ratio to the maximum flow rate. It is then possible to simplify calculations regarding emissions from the system. Hass and Brubacher [23] used a chassis dynamometer to simulate engine operating conditions at idle, cruise, accelerate, and decelerate. Emission rates could then be determined at each of the operating conditions.

Wallin [24] calibrated a smoke sampler used for measuring particulate emissions in diesel exhaust. The weight of material collected was compared to the optical density of the spot and was found to be significantly different from spots taken for ambient air. Calibration specifically for diesel smoke is then necessary.

The State of California employs a system for analysis of exhaust gas from automobiles as shown in Figure 102 [25]. It employs with-

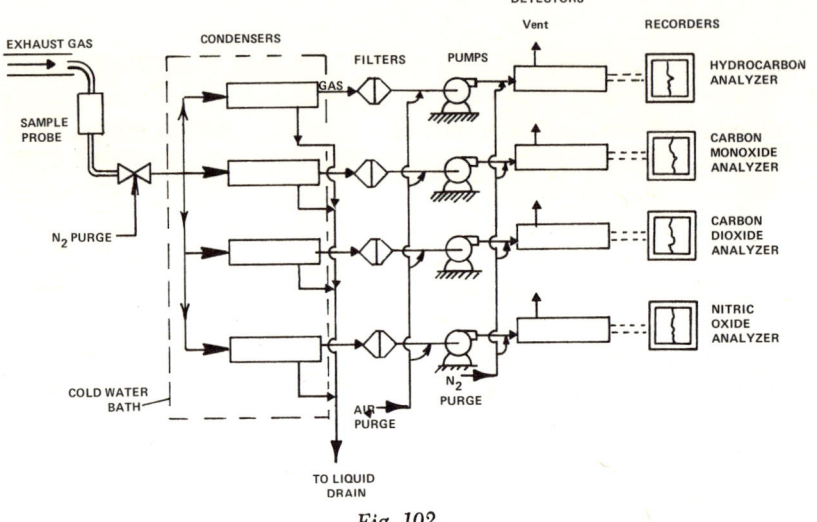

Fig. 102.

Courtesy of: California Air Resources Board, Sacramento, California, 1970 (25).

drawal of the exhaust gas through a series of parallel sampling trains, where flow is split and each of the gas streams is passed through a condenser for moisture removal and then filtered to remove particulates. Gas streams may then be diluted and then passed through a parallel arrangement of continuous analyzers, from which the hydrocarbon, carbon monoxide, carbon dioxide, and nitric oxide concentrations may be determined. In future installations, placement of the pumps downstream of the instruments would minimize the possibility of sample contamination by interaction with the pump. Analysis of the condensate water would allow for a complete material balance inventory. Placing a heated particulate filter at the sample probe prior to flow splitting would eliminate the need for having more than one filter and would allow separate determinations of both particulate and condensable constituent vapors to be made.

B. Jet Aircraft:

George and Burlin [26] studied the emissions of particulate matter, oxides of nitrogen, hydrocarbons, aldehydes, and carbon monoxide in the exhaust gases from commercial jet aircraft engines. The problem was made especially difficult because of gas velocities of up to 2,700 feet per second (162,000 feet per minute), high gas temperatures of up to 1000°F, the extremely high noise levels, and the need to keep the aircraft in a stationary position.

The irregular test conditions caused some unusual procedures to be used. A standard pitot tube and probes were bent into U-shaped arrangements, as shown in Figure 103, and mounted on one of the outer exhaust tubes. Results indicated that the standard pitot tube was sufficiently accurate for use at velocities greater than the speed of sound (1,600 feet per second or 97,000 feet per minute). The sampling period during each test was approximately five minutes because of temperature limitations on the engines when in a stationary position.

The particulate sampling system, illustrated in Figure 104, allowed withdrawal of approximately six cfm through a ¼ inch sample probe, where the large flow rate was necessary to maintain isokinetic conditions. The gas was then drawn through two parallel sampling trains, each consisting of three impingers followed by a paper thimble. The flow rate in each section was measured by two parallel lines, each employing a separate gas meter and vacuum pump in series. The entire system was fitted on a four-wheeled dolly and placed adjacent to and alongside the engine.

Gaseous components were monitored by means of withdrawal of the flue gas and splitting the flow to each of the instruments. There was no necessity for isokinetic sampling in these determinations, so gas was withdrawn at a constant rate.

Emissions were monitored during simulated engine operation for take-off, climb-out, approach, landing, and idle while the plane was at ground level by varying the operating conditions of the engine.

Additional findings on jet aircraft emissions were also presented in a recent study by the Los Angeles County Air Pollution Control District [27].

Fig. 103.

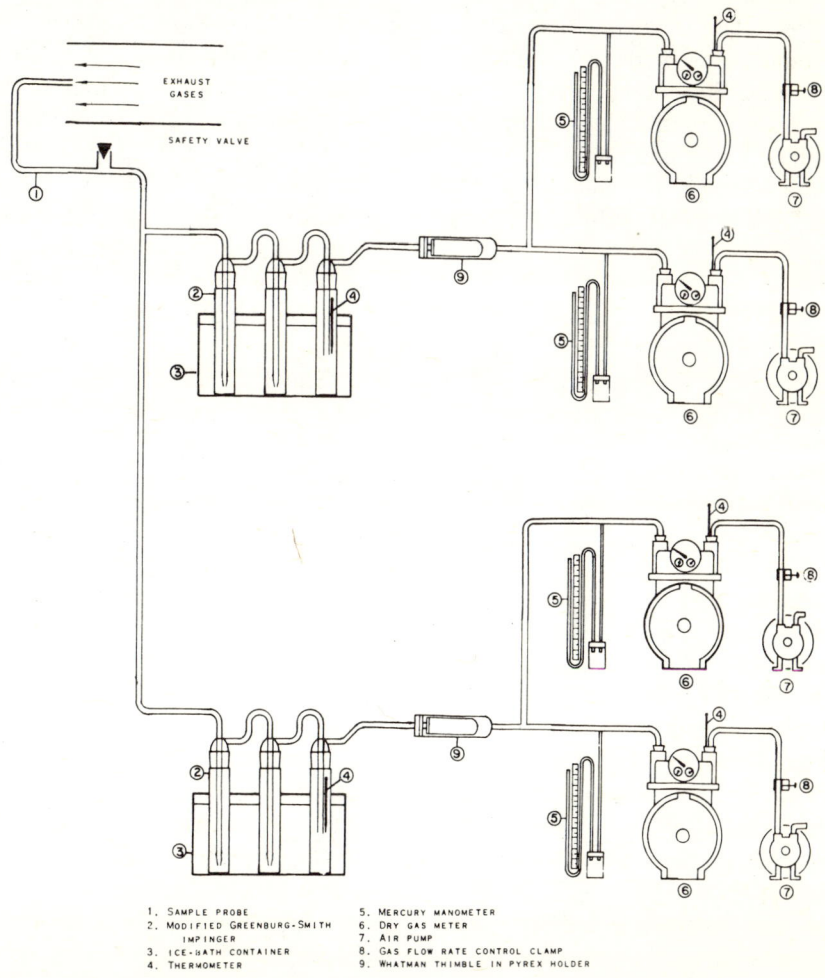

1. Sample probe
2. Modified Greenburg-Smith impinger
3. Ice-bath container
4. Thermometer
5. Mercury manometer
6. Dry gas meter
7. Air pump
8. Gas flow rate control clamp
9. Whatman thimble in Pyrex holder

Fig. 104.

Courtesy of: Los Angeles County Air Pollution Control District (26).

IV. ADDITIONAL SOURCES

A. Plume Measurements:

Measurement of materials in plumes at ground level is an application of ambient air monitoring techniques. The reasons for making these measurements are to determine the degree of dilution from the source or to trace the plume. It is also desirable to employ a remote monitoring of source gases as an aid in enforcement, so that direct measurement in the duct is unnecessary.

1. Particulate Measurements:

Meland [28] used a membrane filter, cascade impactor, and a rotating cylinder or "rotorod" sampler to measure the particulate concentration, size distribution, and types of materials present in the plume downwind from a source where it reaches ground level. The rotorod sampler was useful for observing the large particles greater than 20 microns in size because of its adhesive surface. Both the cascade impactor and membrane filter could be used for size analyses and give similar results. The membrane filter is preferable because its operation is simpler. Sources tested included an aluminum and brass smelter, a glass fiber factory, an oil-fired apartment heater vent, and a Kraft pulp mill. The major problem involved was to remain in the plume for a sufficient period to take a sample. A mobile vehicle with a generator aboard may be the best solution.

Noll [29] [30] recently developed a centrifugal sampler for measuring large particles by size, between 10 and 50 microns in size, in the ambient air downwind from sources. The device is illustrated in Figure 105, and consisted of a rotating shaft containing several cylindrical shafts of incrementally increasing diameter. The shafts were coated with adhesive so that particles would adhere to the surfaces. When air was caused to flow past the rapidly rotating cylinder, the smallest particles of highest mobility were caused to deposit out first on the upper cylinder of the smallest diameter. The larger particles of higher inertia, were subsequently caused to deposit on progressively lower cylinders of larger diameter. Extensive research in the laboratory and the field indicated the suitability of the device for measuring large particles in the atmosphere, including downwind from sources.

2. Visual Evaluation:

The Ringelmann chart [31] [32] [33] has been extensively used for visual observation of the density of black smoke plumes. The chart makes an arbitrary classification of six levels of smoke density from zero (completely transparent) to five (completely black). The density of smoke plumes is determined by visual comparison between the degree of visual obscuration by the plume, and the chart scale. Sample Ringelmann charts are illustrated in Figure 106. The technique has been extended to evaluate the equivalent opacity or degree of visual obscuration by non-black plumes, with varying degrees of success.

The Ringelmann chart has been extensively used as a means for evaluating the visual ob-

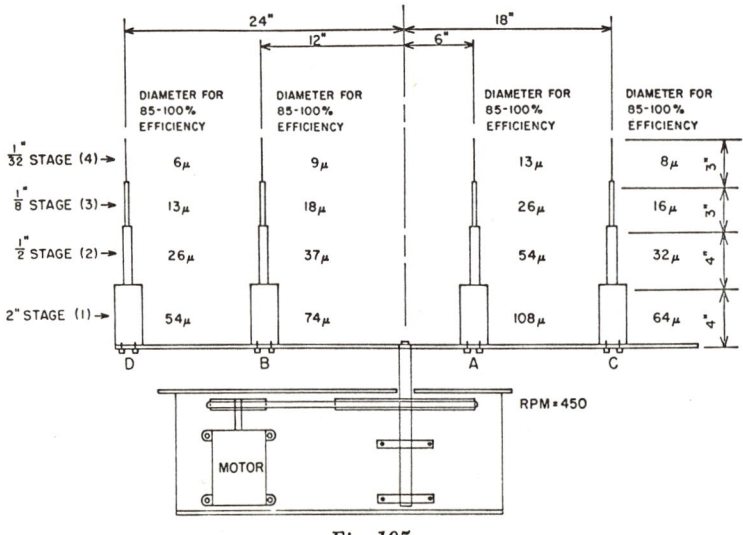

Fig. 105.

Courtesy of: Mr. Kenneth E. Noll, California Air Resources Board, Sacramento, California, 1969 (29), (30).

PLIBRICO SMOKE CHART
RINGELMANN TYPE

1. Circular

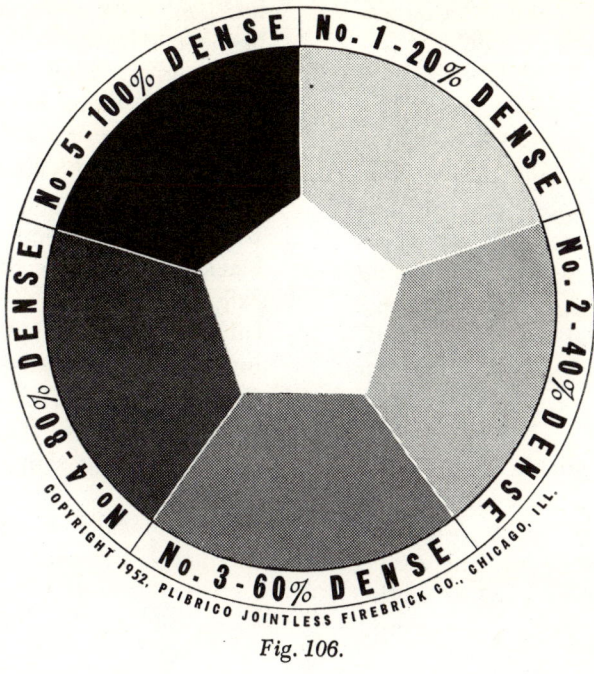

Fig. 106.

2. Linear

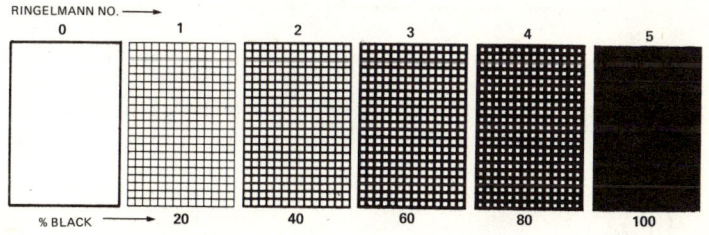

Fig. 106a.

Courtesy of: Mr. T. J. Barry, Plibrico Firebrick Co., Chicago, Ill., 1970.

scuration of both black and non-black plumes. It has been specified in numerous air pollution regulations as an enforcement tool for evaluating particulate source emission levels because of its low cost and simplicity. Use of the chart eliminates the need to enter the plant property, and the necessity for utilizing cumbersome and expensive source testing equipment. The Ringelmann chart has proved particularly useful for black smoke plumes from fossil-fuel burning sources as an indicator of combustion efficiency. However, the method should be used with considerable caution, particularly for sources with non-black plumes. For many sources, there is often no direct correlation between total particulate concentration and Ringelmann number.

Other problems may also interfere with the suitability of the Ringelmann chart for providing an index of particulate levels from sources, particularly for non-black plumes. These include atmospheric relative humidity, air temperature, angle of vision, color of the background sky, adjacent sources, particle size distribution, surface properties of the particles (such as color), stack diameter, and moisture content of the plume. Connor and Hodkinson [34] have presented an extensive discussion of factors affecting optical measurements of plume densities. The effect of particle size was particularly significant in attempting to correlate plume opacity as a function of particulate concentration. The results of a series of calculations for monodisperse aerosol systems of varying sizes is illustrated in Figure 107. It is noted that there is considerable variation (between white and black particles of varying sizes) in the relationship between particle concentration and optical transmittance. They concluded that the use of optical measuring devices could be very suitable for measurement of plume densities as an enforcement tool.

Evaluation of the equivalent opacity of a non-black plume following a scrubber is made especially difficult because of moisture condensation on cooling in the ambient air. The normal procedure is to observe the degree of obscuration at the point where the plume has just evaporated. An additional complicating factor is that plumes with higher particulate concentrations tend to retain their identity in condensed form for longer distances than plumes with lower particulate loadings.

Training of observers was described in recent publications by the Los Angeles County [35], and Bay Area Air Pollution Control Districts [36]. Smoke generating systems have been developed to produce source gases of varying Ringelmann numbers for both black and non-black plumes. Inspectors are required to periodically check their visual readings against the calibrated opacity meter on the smoke generating system. The system used by the Los Angeles County APCD is illustrated in Figure 108.

In summary, the Ringelmann chart has proved to be a simple, inexpensive, and useful method for providing an index of particulate losses for use by regulatory agencies. The method is very useful for providing an index of smoke densities for combustion efficiencies for black plumes. However, for non-black plumes,

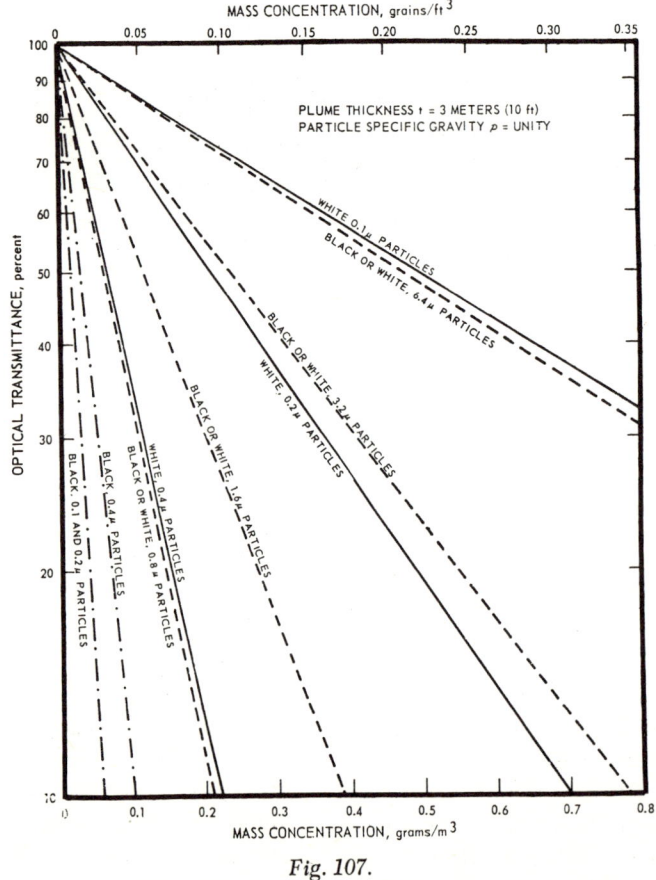

Fig. 107.

Courtesy of: W. D. Connor and J. R. Hodkinson, U.S. Public Health Service Publication No. 999-AP-30, 1967 (34).

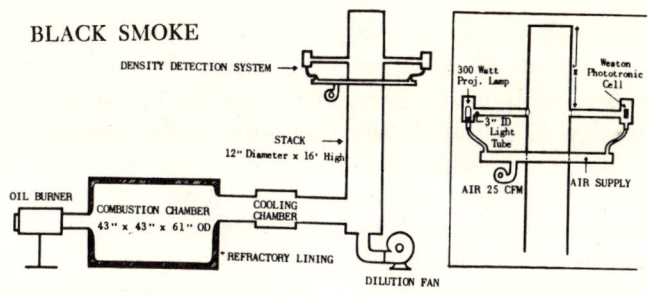
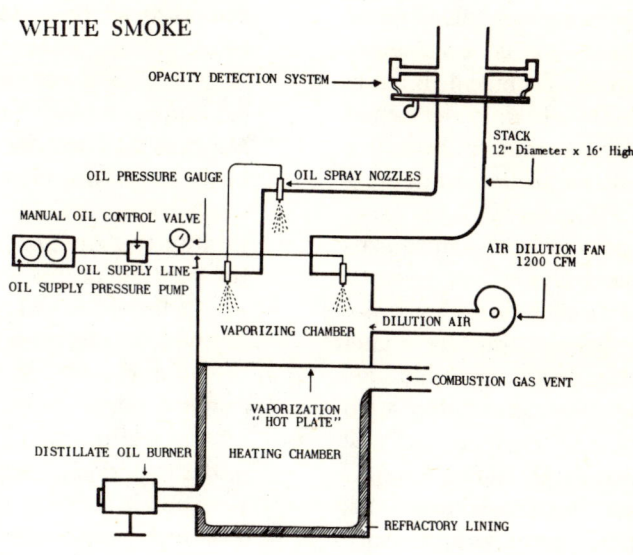

Fig. 108.
Courtesy of: Los Angeles County Air Pollution Control District and U.S. Public Health Service, 1962 (35).

and for any plume at Ringelmann numbers below 2.0, the technique should be exercised with considerable caution. Simultaneous Ringelmann number observation and particulate concentration measurements should be made over an extended period for sources where there is question regarding suitability of the chart. It is to be emphasized that under no circumstances does use of the Ringelmann chart supersede or eliminate the need for the more cumbersome, complicated, expensive, but more representative source testing methods previously described!

3. Remote Sensing:

It is also desired to measure remotely the concentrations of materials being emitted from sources, to avoid the many problems involved in conventional source testing and provide a useful enforcement tool. Two recent articles describe the use of optical radar (lidar) for measuring particulate concentrations in flue gases from a remote location [37], [38]. The technique may also be applied to sulfur dioxide and other gases. Much more work remains to be done in this area. In a similar study, Crider and Task [39] used a telephotometer to measure visual observation of a non-black plume caused by the presence of particles. The differences in signals between the light passing through the plume and the background sky were measured, and used as the basis for measuring optical density of the plume. Results were successful in a pilot test.

4. Airborne Sampling

The downwind dispersion characteristics of pollutants from the elevated sources can be measured directly in the ambient air as a means of evaluating the degree of dilution prior to reaching ground level. An early reported use of aircraft for atmospheric pollution measurements was a study of airborne bacterial microorganisms in the Arctic by Meier and Lindberg [40]. Gartrell [41] [42] used a helicopter to study the downwind dispersion of sulfur dioxide from the flue gases of a coal-fired power plant, and the rate of oxidation of sulfur dioxide to sulfur trioxide in ambient air. McCaldin and Johnson [43] used a single engine aircraft to study the downwind dispersion of particulate matter from a power plant plume, and a large forest fire. Adams and Koppe [44] used a twin-engine tandem aircraft for airborne measurements of gaseous and particulate matter for plume dispersion studies from stationary sources. The aircraft was equipped with continuous monitoring instruments for carbon dioxide, carbon monoxide, sulfur dioxide, and an integrating nephelometer to provide an index of particulate concentrations. A recent National Council Technical Bulletin [45] described the results of a recent study where the downwind dispersion of hydrogen sulfide and particulate matter from a Kraft pulp mill were measured on a continuous basis using a bromine coulometric detector and a light scattering photometer, respectively.

Use of instrumented aircraft provides a potentially useful tool for enforcement, and meteorological studies relating to dilution and dispersion of stack plumes. Instrumentation used must be light enough for easy transport and ease of installation and removal. Operating characteristics of the analytical instrumentation must be considered in terms of power requirements, sensitivity, response time, intereferences, specificity, background noise level, and electronic drift [44]. An electric power generation system of sufficient capacity for all instrumentation carried aloft must be powered off the plane's engines. It is also necessary to locate inlet sample parts in such a manner that possible contamination from the plane's engine exhausts is eliminated. Additional research is necessary in the area of using aircraft for source monitoring and plume dispersion studies to facilitate prediction of downwind ground level concentrations.

B. Mobile Laboratories:

Mobile laboratories provide the capability for bringing the necessary monitoring equipment to the field for extended studies of either gaseous or particulate materials. Walther and Amberg [46] built a mobile laboratory in a van for monitoring sulfur gas emissions from Kraft pulp mills. It was equipped with two gas chromatographs for determining the composition of flue gas streams, and an electrolytic titration unit for continuous monitoring of the total sulfur present in the gas. An electrically-heated line provided for transfer of the sample from the source to the instruments. A heated ceramic probe was used to prefilter the gas stream prior to entering the sample line to remove particulate matter. The use of the heated probe is especially important for wet saturated gas streams, where moisture droplets may be present, to avoid losses by condensation. Dilution of the gas stream just upstream of the instruments was useful for high concentration sources or those of high moisture contents. It was necessary to keep the system heated at all times to avoid condensation, however.

Megy [47] [48] recently designed a mobile laboratory which has been employed by the National Council for Air and Stream Improvement for measuring sulfur gas emissions in extensive field studies at several Kraft pulp mills. Continuous monitoring of the total reduced sulfur content of the gas stream is facilitated by use of bromine coulometric titration. Breakdown of the individual gaseous constituents in terms of sulfur dioxide, hydrogen sulfide, and organic reduced sulfur gas concentrations is provided for by the use of gas chromatography on an intermittent, periodic basis. A schematic diagram of the mobile laboratory is shown in Figure 109, where the major sections are the sample collection, chemical detection, and data readout, respectively. The system actually contains parallel sample lines for continuous total sulfur monitoring of two sources simultaneously, with valving of the gas chromatograph for alternate use with either line. However, only one sample line is shown in the diagram for purposes of simplicity.

1. Sample Collection

The sample collection system consists of an inlet probe, an extended sample line, and a sample conditioning box. It is designed to maintain the gas stream at above its dew point at all times when upstream of the detectors, to prevent gas losses by condensation, and resulting malfunctions in the system caused by liquid droplets. Heating is particularly important for Kraft pulp mill flue gases, where moisture contents of gas streams may range between 30 and 95 percent by volume, depending on the source. The two inch diameter and 12 inch long stainless steel probe is packed with glass wool for particulate removal. It is superheated to approximately 300°F to evaporate any water droplets which may enter the probe.

The gas stream then enters an electrically

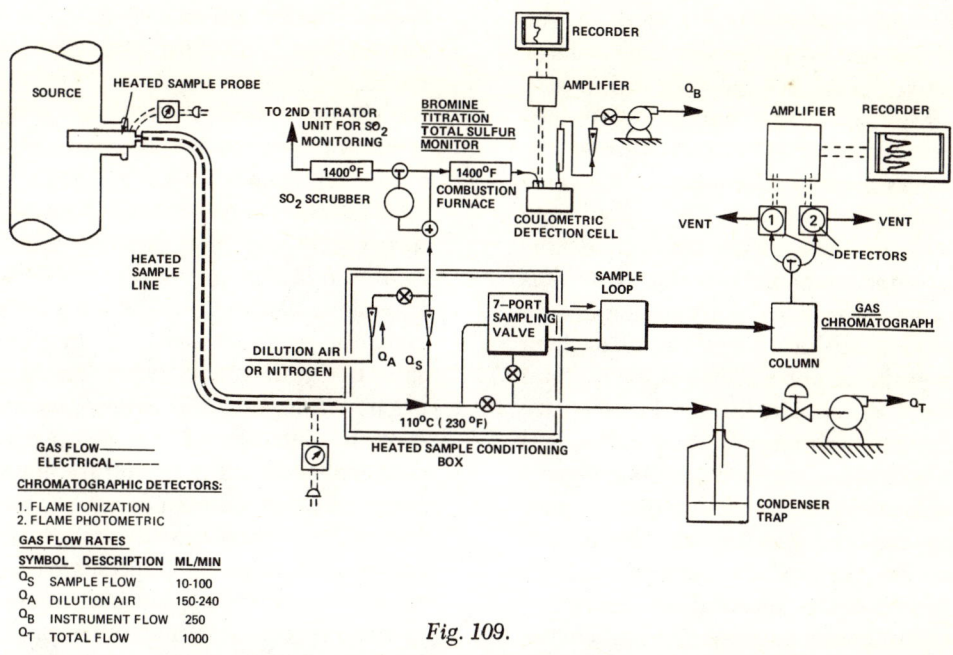

Fig. 109.

Courtesy of: Mr. Joseph A. Megy, National Council for Air and Stream Improvement, Corvallis, Oregon, 1970 (47) (48).

heated sample line maintained at approximately 225°F to be conveyed to the mobile laboratory. Either one or two 100 foot sections can be employed in the heated sample line.

The gas stream then enters the heated sample conditioning box maintained at 110°C. It is then split into three separate streams, going respectively to the continuous coulometric sulfur titrator, the gas chromatograph, and the vacuum source. The major portion of the total gas flow rate of approximately one liter per minute is drawn directly to the vacuum pump to minimize total retention time in the system. This high flow rate reduces the chances for gas losses by adsorption on the tubing walls. A moisture trap, consisting of a gallon bottle, is placed immediately upstream of the vacuum pump to reduce the chances of pump-flooding and malfunctions in highly moisture-laden gas streams.

2. Total Sulfur Monitoring:

Analyses of the sulfur gas concentrations present in the gas stream proceed in two different phases. A portion of the gas stream of between 10 and 100 milliliters per minute is removed and diluted with air or nitrogen to a total flow rate of 250 milliliters per minute. It is then passed through the total sulfur monitoring system, consisting of a combustion furnace and the bromine coulometric titration detection cell.

It is necessary to dilute gas streams, such as the highly moisture-laden digester noncondensable gases, by between 10 and 25 times to avoid excessive buildup of moisture in the detection cell. Purified air is used as the diluent medium for most sources because of its simplicity and because it provides sufficient oxygen for combustion in inlet gas streams which may be essentially devoid of oxygen.

The gas stream from the heated sample conditioning box is then passed through a specially designed furnace of NCASI design. The reduced sulfur gases are all converted to sulfur dioxide by thermal oxidation at 1400°F, where subsequent conversion to sulfur trioxide is less than five percent [49]. The organic constituents of carbon and hydrogen are converted to carbon dioxide and water at the same time that sulfur is being converted to sulfur dioxide. The potential interference of the detector-sensitive terpene and other unsaturated bond organic compounds is thus eliminated. The sulfur dioxide concentration is then measured continuously by means of a bromine coulometric titrator of commercial design. The system provides for measuring total reduced sulfur concentrations from less than 0.5 to more than 40,000 parts per million by volume with appropriate adjustment of dilution rates and instrument settings.

Sulfur dioxide and hydrogen sulfide plus other reduced gaseous components may also be monitored continuously by employing a sulfur dioxide-selective scrubber (using potassium acid phthalate) upstream of the combustion furnace and two titration units in parallel. One of the instruments then measures sulfur dioxide, hydrogen sulfide, plus the organic sulfur gases, while the other measures only hydrogen sulfide plus the organic sulfur compounds. Both gas streams are passed through the combustion furnace in separate parallel sampling lines after one of them passes through the scrubber. The respective total sulfur concentrations (as sulfur dioxide) are measured by two titration units in parallel, and the difference between the two resulting signals is used as a measure of the sulfur dioxide concentration in the flue gas stream. The normal configuration for the system employs a single instrument without the sulfur dioxide-selective prescrubber for continuous monitoring of the total reduced sulfur gases, including SO_2, H_2S, and the organic sulfur gases, all as sulfur dioxide.

Initial calibration of the system is performed with sulfur dioxide using the rotating syringe technique for producing known concentrations. Daily checks are made to assure continued accuracy.

3. Gas Chromatography:

Determination of the respective concentrations of the reduced sulfur gases present is made by means of gas chromatography, employing both flame ionization and flame photometric detectors. The flame ionization detector is particularly sensitive to organic sulfur compounds such as methyl mercaptan, dimethyl sulfide, dimethyl disulfide, and others. The flame photometric detector is useful in measuring hydrogen sulfide and sulfur dioxide concentrations directly. Gas is flushed through the seven-port sampling valve continuously, and samples injected periodically into the chromatograph through a sample loop of a given volume. The gas stream is separated into its constituent gases by a chromatographic column maintained at 100°C. The individual gas concentrations are then determined consecutively as the gases pass through the detectors.

Calibration of the chromatographic detectors is made for sulfur dioxide, hydrogen sulfide, methyl, mercaptan, dimethyl sulfide, and dimethyl disulfide using the rotating syringe technique. Daily checks are made with standard gas used to calibrate the titration cells to assure continued accuracy.

The system described above is useful for monitoring sulfurous gas emissions from a wide variety of sources over an extensive range of concentration levels. The method includes a measure of the total gaseous sulfur concentration in a gas stream, with the exception of sulfur trioxide, as well as a breakdown of its individual constituents. The potential interference of unsaturated bond and other bromine-consuming organic compounds can be eliminated by combustion upstream of the detector. This causes conversion of sulfur to sulfur dioxide, and carbon-hydrogen compounds to carbon dioxide and water. It is then possible to measure reduced sulfur gas emissions from such diverse sources as Kraft and sulfite pulp mill offgases, catalytic cracker regenerator and other oil refinery vent gases, coal-, oil-, and wood-fired power boilers, plus numerous other applications.

V. BIBLIOGRAPHIES

Several recent articles have been devoted to reviews of available literature on source sampling. The Coordinating Committee on Air Pollution of the Engineering Foundation has published four reviews of articles on particulate and gaseous source sampling methodology [50], [51] [52] [53]. The proceedings of a German symposium on technical measurements in dust systems was recently published by the Verein Deutscher Ingenieure [54]. Lindorf [55] has written a history of flue gas analyzers, while Wilson and Duff [56] present an extensive literature review of gaseous analytical methods. Ruch [57] has published an annotated bibliography of more than 500 different methods for analyzing gaseous components. Leithe [58] has recently authored a text on analytical methods for analysis of gaseous and particulate air pollutants. The above references are a valuable source of background information.

REFERENCES

1. Wilby, F. V., "Variation in Recognition Odor Threshold of a Panel," *Journal of the APCA*, Volume 19, Number 2, pp. 96-100, February 1969.
2. Cederlof, R., Edfers, M., Friberg, L., and Lindvall, T., "On the Determination of Odor Thresholds in Air Pollution Control—An Experimental Field Study of

Flue Gases from Sulfate Cellulose Plants," *Journal of the APCA*, Volume 16, Number 2, pp. 92-94, February 1966.
3. Mills, J. L., Walsh, R. T., Luedke, K. D., and Smith, L. K., "Quantitative Odor Measurement," *Journal of the APCA*, Volume 13, Number 10, pp. 467-475, October 1963.
4. Fox, E. A., and Gex, V. E., "Procedure for Measuring Odor Concentration in Air and Gases," *Journal of the APCA*, Volume 7, Number 1, pp. 60-61, May 1957.
5. "Standard Method for Measurement of Odors in Atmospheres (Dilution Method)," ASTM Standard D 139-57, American Society for Testing and Materials, Philadelphia, Pennsylvania, 1957.
6. Devorkin, H., Chass, R. L., Fudurich, A., and Kanter, C. V., ed., R. G., Holmes, "Source Testing Manual," Los Angeles County Air Pollution Control District, Los Angeles, California 1965.
7. Benforado, D. M., Rotella, W. J., and Horton, D. L., "Development of an Odor Panel for Evaluation of Odor Control Equipment," *Journal of the APCA*, Volume 19, Number 2, pp. 101-105, February 1969.
8. Nader, J. S., "An Odor Evaluation Apparatus for Field and Laboratory Use," *American Industrial Hygiene Association Journal*, Volume 19, Number 1, pp. 1-7, February 1958.
9. Hemeon, W. C. L., "Technique for Quantitative Measurement of Odor Emissions," *Journal of the APCA*, Volume 18, Number 3, pp. 166-170, March 1968.
10. Leonardos, G., Kendall, D., and Barnard, N., "Odor Threshold Determination of 53 Odorant Chemicals," *Journal of the APCA*, Volume 19, Number 2, pp. 91-95, February 1969.
11. Nickol, G. B., "Rank Odor Method for Evaluating Stack Gases," *Journal of the APCA*, Volume 7, Number 1, pp. 55, May 1955.
12. Turk, A., "Selection and Training of Judges for Sensory Evaluation of the Intensity and Character of Diesel Exhaust Odors," U.S. Public Health Service, National Air Pollution Control Administration, Cincinnati, Ohio, 1967.
13. Turk, A., "Measurement and Control of Community Malodors," Chapter 8 in Rossano, A. T., ed., *Air Pollution Control—Guidebook for Management*, Environmental Science Services Corp., Stamford, Connecticut, 1969.
14. Jackson, P. J., and Raask, E., "A Probe for Studying the Deposition of Solid Materials from Flue Gas at High Temperatures," *Journal of the Institute of Fuel*, Volume 34, Number 246, pp. 275-280, July 1961.
15. Alexander, P. A., Fielder, R. S., Jackson, P. J., and Raask, E., "An Air-Cooled Probe for Measuring Acid Deposition in Boiler Flue Gases," *Journal of the Institute of Fuel*, Volume 33, Number 228, pp. 31-37, January 1960.
16. Haneef, M., "Corrosion Tests on Materials Exposed in Flue Gases from Oil Firing," *Journal of the Institute of Fuel*, Volume 33, Number 233, pp. 285-294, June 1960.
17. Harvey, R. A., "A Stack Effluent Radioisotope Monitor," *Institute of Radio Engineers Transactions on Nuclear Science*, Volume NS-6, Number 4, pp. 20-28, December 1959.
18. Hyatt, E., Moss, W., and Schulte, H., "Particle Size Studies on Uranium Aerosols from Machining and Metallurgy Operations," *American Industrial Hygiene Association Journal*, Volume 20, Number 2, pp. 99-107, March-April 1965.
19. Goetz, A., "Application of Molecular Filter Membranes to the Analyses of Aerosols," *American Journal of Public Health*, Volume 43, Number 2, pp. 150-159, February 1953.
20. Nelson, R. Y., and Ledbetter, J. O., "Atmospheric Emissions from Oxidation Ponds," *Journal of the APCA*, Volume 14, Number 3, pp. 50-52, February 1964.
21. Andersen, A. A., "New Samples for the Collection, Sizing, and Enumeration of Viable Airborne Particles," *Journal of Bacteriology*, Volume 76, Number 10, pp. 471-490, November 1958.
22. Chipman, J. C., and Massey, M. T. "Proportional Sampling System for the Collection of an Integrated Auto Exhaust Sample," *Journal of the APCA*, Volume 10, Number 1, pp. 60-69, February 1960.
23. Hass, G. C., and Brubacher, M. L. "A Test Procedure for Motor Vehicle Exhaust Emissions," *Journal of the APCA*, Volume 12, Number 11, pp. 505-509, November 1962.
24. Wallin, S. C., "Calibration of the DSIR Standard Smoke Filter for Diesel Exhaust," *International Journal of Air and Water Pollution*, Volume 9, Number 6, pp. 351-356, June 1965.
25. "California Exhaust Emission Standards and Test Procedures for 1971 and Subsequent Model Gasoline Powered Vehicles under 6,001 pounds Gross Vehicle Weight," State of California Air Resources Board, Sacramento, California, November 20, 1968.
26. George, R. E., and Burlin, R. M. "Air Pollution Emissions from Commercial Jet Aircraft" Los Angeles County Air Pollution Control District, Los Angeles, California, April 1960.
27. George, R. E., Verssen, J. A., and Chass, R. L., "Jet Aircraft: A Growing Pollution Source," *Journal of the APCA*, Volume 19, Number 11, pp. 847-855, November 1969.
28. Meland, B. R., "A Comparative Study of Particulate Loadings in Plumes Using Multiple Sampling Devices," *Journal of the APCA*, Volume 18, Number 8, pp. 529-533, August 1968.
29. Noll, K. E., "A Rotary Inertial Impactor for Sampling Giant Particles in the Atmosphere," Ph.D. Dissertation Submitted to the Department of Civil Engineering, University of Washington, Seattle, Washington, July 1969.
30. Noll, K. E., "A Rotary Inertial Impactor for Sampling Giant Particles in the Atmosphere," *Atmospheric Environment*, Vol. 4, No. 1, 9-19, January 1970.
31. Developed by Professor Maximilian Ringelmann Institut National Agronomique and Directeur de los Station d'Essais die Machines, Paris, France, 1888.
32. Kudlich, R., and Burdick, L. R., "Ringelmann Smoke Chart," U.S. Dept. of the Interior, Bureau of Mines Information Circular No. 7718, Washington, D. C., August 1955.
33. Jacobs, M. B., *The Chemical Analysis of Air Pollutants*, Interscience Publishers, New York, New York, 1960.
34. Connor, W. D., and Hodkinson, J. R., "Optical Properties and Visual Effects of Smoke—Stack Plumes," U.S. Public Health Service Publication No. 999-AP-30, Cincinnati, Ohio, 1967.
35. Weisburd, M. I., ed., "Air Pollution Control Field Operations Manual, A Guide for Inspection and Enforcement," Compiled by the Los Angeles County Air Pollution Control District, U.S. Public Health Service Publication No. 937, Washington, D. C., 1962.

36. Coons, J. D., James, H. A., Johnson, H. C., and Walker, M. S., "Development, Calibration, and Use of a Plume Evaluation Training Unit," *Journal of the APCA*, Volume 15, Number 5, pp. 199-203, May 1965.
37. Johnson, W. B., "Lidar Applications in Air Pollution Research and Control," *Journal of the APCA*, Volume 19, Number 3, pp. 176-180, March 1969.
38. Hamilton, P. M., "The Use of Lidar in Air Pollution Studies," *International Journal of Air and Water Pollution*, Volume 10, Nos. 6/7, pp. 427-434, June—July 1968.
39. Crider, W. L., and Task, T. A., "Status Report: Study of Vision Obscuration by Non-black Plumes," *Journal of the APCA*, Volume 14, Number 5, pp. 161-167, May 1964.
40. Meier, F. C., and Lindberg, C. A., "Collecting Micro-Organisms from the Arctic Atmosphere," *The Scientific Monthly*, Vol. 40, No. 1, pp. 5-10, January 1935.
41. Gartrell, F. E., and Carpenter, S. B., "Aerial Sampling by Helicopter—A Method for Study of Diffusion Patterns, "*Journal of Meteorology*, Vol. 12, No. 3, 215-219, March 1955.
42. Gartrell, F. E., Thomas, J. W., and Carpenter, S. B., "An Interim Report on Full-Scale Study of Dispersion of Stack Gases," *Journal of the Air Pollution Control Association*, Vol. 11, No. 2, pp. 60-63, February 1961.
43. McCaldin, R. O., and Johnson, L. W., "The Use of Aircraft in Air Pollution Research," *Journal of the Air Pollution Control Association*, Vol. 19, No. 6, 405-409, June 1969.
44. Adams, D. F., and Koppe, R. K., "Instrumenting Light Aircraft for Air Pollution Research," *Journal of the Air Pollution Control Association*, Vol. 19, No. 6, pp. 410-415, June 1969.
45. Tuggle, M. L., McCaldin, R. O., Duncan, L., and Tucker, T. W., "Airborne Sampling of Gaseous and Particulate Emissions from Kraft Pulp Mills," Atmospheric Pollution Technical Bulletin No. 49, National Council of the Paper Industry for Air and Stream Improvement, New York, New York, September 1970.
46. Walther, J. E., and Amberg, H. R., "Experience with a Mobile Laboratory in Source Sampling Kraft Mill Emissions," *Tappi*, Volume 51, Number 14, pp. 126A-129A, November 1968.
47. Megy, J. A., "Design, Operation, and Use of a Mobile Laboratory for Continuous Monitoring of Kraft Mill Source Gases," Presented at the West Coast Regional Meeting of the National Council of the Paper Industry for Air and Stream Improvement, Seattle, Washington, October 15, 1969.
48. Blosser, R. O., Megy, J. A., Franklin, M. E., Caron, A. L., Duncan, L., and Tucker, T. W., "An Inventory of Miscellaneous Sources of Reduced Sulfur Emissions from the Kraft Pulping Process," Presented at the 7th Tappi Air and Water Conference, Minneapolis, Minnesota, June 7, 1970, Submitted for publication to *Tappi*, June 1970.
49. Lewis, W. K., Radasch, A. H., and Lewis, H. C., *Industrial Stoichiometry: Chemical Calculations of Manufacturing Processes*, McGraw-Hill Book Co., New York, New York, p. 171, 1954.
50. "Measurement of Stack Gas Velocity and Methods of Sampling Stack Gases and Their Particulates," Literature Search No. 6216, Engineering Foundation Coordinating Committee on Air Pollution, New York, New York, July 1962.
51. "Research into Sampling, Analysis or Monitoring of Gaseous Pollutant Emissions from Stacks," Literature Search 6269, Engineering Foundation Coordinating Committee on Air Pollution, New York, New York, August 1963.
52. "Research into Sampling, Analysis or Monitoring of Gaseous Pollutant Emissions from Stacks," Literature Search II, Engineering Foundation Coordinating Committee on Air Pollution, New York, New York, 1965.
53. "Stack Sampling and Monitoring, A Workshop Meeting," Engineering Foundation Coordinating Committee on Air Pollution, New York, New York, November 1963.
54. "Technical Measurements on Dust Systems," Papers Presented at the VDI Meeting in Augsburg, Germany, Verein Deutscher Ingenieure Reports (Dusseldorf), Volume 7, pp. 1-82, 1955.
55. Lindorf, H., "History of Flue Gas Analyzers," *Archiv fur Technisches Messen*, Number 274, pp. 233-234, November 1958, and Number 282, pp. 137-140, July 1959.
56. Wilson, H. N., and Duff, G., "Industrial Gas Analysis—A Literature Review," *The Analyst*, Volume 92, Number 1101, pp. 723-758, December 1967.
57. Ruch, W., *Chemical Detection of Gaseous Pollutants*, Ann Arbor Science Publishers, Inc., Ann Arbor, Michigan, 1967.
58. Leithe, W., *The Analysis of Air Pollutants*, Ann Arbor Science Publishers, Inc., Ann Arbor, Michigan, 1970.

CHAPTER 12
APPENDICES

APPENDIX A.

PERTINENT SOURCE TEST DATA

1. Vapor Pressures of Water in Inches of Mercury.

Temp. Deg. F.	0	1	2	3	4	5	6	7	8	9
−20	.0126	.0119	.0112	.0106	.0100	.0095	.0089	.0084	.0080	.0075
−10	.0222	.0209	.0199	.0187	.0176	.0168	.0158	.0150	.0142	.0134
−	.0376	.0359	.0339	.0324	.0306	.0289	.0275	.0259	.0247	.0233
0	.0376	.0398	.0417	.0441	.0463	.0489	.0517	.0541	.0571	.0598
10	.0631	.0660	.0696	.0728	.0768	.0810	.0846	.0892	.0932	.0982
20	.1025	.1080	.1127	.1186	.1248	.1302	.1370	.1429	.1502	.1567
30	.1647	.1716	.1803	.1878	.1955	.2035	.2118	.2203	.2292	.2383
40	.2478	.2576	.2677	.2782	.2891	.3004	.3120	.3240	.3364	.3493
50	.3626	.3764	.3906	.4052	.4203	.4359	.4520	.4686	.4858	.5035
60	.5218	.5407	.5601	.5802	.6009	.6222	.6442	.6669	.6903	.7144
70	.7392	.7648	.7912	.8183	.8462	.8750	.9046	.9352	.9666	.9989
80	1.032	1.066	1.102	1.138	1.175	1.213	1.253	1.293	1.335	1.378
90	1.422	1.467	1.513	1.561	1.610	1.660	1.712	1.765	1.819	1.875
100	1.932	1.992	2.052	2.114	2.178	2.243	2.310	2.379	2.449	2.521
110	2.596	2.672	2.749	2.829	2.911	2.995	3.081	3.169	3.259	3.351
120	3.446	3.543	3.642	3.744	3.848	3.954	4.063	4.174	4.289	4.406
130	4.525	4.647	4.772	4.900	5.031	5.165	5.302	5.442	5.585	5.732
140	5.881	6.034	6.190	6.350	6.513	6.680	6.850	7.024	7.202	7.384
150	7.569	7.759	7.952	8.150	8.351	8.557	8.767	8.981	9.200	9.424
160	9.652	9.885	10.12	10.36	10.61	10.86	11.12	11.38	11.65	11.92
170	12.20	12.48	12.77	13.07	13.37	13.67	13.98	14.30	14.62	14.96
180	15.29	15.63	15.98	16.34	16.70	17.07	17.44	17.82	18.21	18.61
190	19.01	19.42	19.84	20.27	20.70	21.14	21.59	22.05	22.52	22.99
200	23.47	23.96	24.46	24.97	25.48	26.00	26.53	27.07	27.62	28.18
210	28.75	29.33	29.92	30.52	31.13	31.75	32.38	33.02	33.67	34.33
220	35.00	35.68	36.37	37.07	37.78	38.50	39.24	39.99	40.75	41.52
230	42.31	43.11	43.92	44.74	45.57	46.41	47.27	48.14	49.03	49.93
240	50.84	51.76	52.70	53.65	54.62	55.60	56.60	57.61	58.63	59.67
250	60.72	61.79	62.88	63.98	65.10	66.23	67.38	68.54	69.72	70.92
260	72.13	74.36	74.61	75.88	77.16	78.46	79.78	81.11	82.46	83.83
270	85.22	86.63	88.06	89.51	90.97	92.45	93.96	95.49	97.03	98.61
280	100.2	101.8	103.4	105.0	106.7	108.4	110.1	111.8	113.6	115.4
290	117.2	119.0	120.8	122.7	124.6	126.5	128.4	130.4	132.4	134.4
300	136.4	138.5	140.6	142.7	144.8	147.0	149.2	151.4	153.6	155.9
310	158.2	160.5	162.8	165.2	167.6	170.0	172.5	175.0	177.5	180.0
320	182.6	185.2	187.8	190.4	193.1	195.8	198.5	201.3	204.1	206.9
330	209.8	212.7	215.6	218.6	221.6	224.6	227.7	230.8	233.9	237.1
340	240.3	243.5	246.8	250.1	253.4	256.7	260.1	263.6	267.1	270.6
350	274.1	277.7	281.3	284.9	288.6	292.3	296.1	299.9	303.8	307.7
360	311.6	315.5	219.5	323.5	327.6	331.7	335.9	340.1	344.4	348.7
370	353.0	357.4	361.8	366.2	370.7	375.2	379.8	384.4	389.1	393.8
380	398.6	403.4	408.2	413.1	418.1	423.1	428.1	433.1	438.2	443.4
390	448.6	453.9	459.2	464.6	470.0	475.5	481.0	486.6	492.2	497.9
400	503.6	509.3	515.1	521.0	526.9	532.9	538.9	545.0	551.1	557.3

Courtesy of: Western Precipitation Corp., Los Angeles, California.

2. Flue Gas Velocity Tables (Feet per Minute) for a Standard Pitot Tube with Dry Air.

Pressure Differential in. H$_2$O	Flue Gas Temperature—°F					
	70	100	150	200	250	300
0.0	0.0	0.0	0.0	0.0	0.0	0.0
0.05	897	921	963	1010	1040	1077
0.1	1263	1300	1360	1412	1463	1518
0.15						
0.2	1790	1840	1920	2000	2067	2144
0.3	2200	2255	2360	2455	2540	2635
0.4	2530	2600	2720	2825	2925	3040
0.5	2830	2900	3040	3160	3270	3400
0.6	3100	3180	3325	3460	3580	3720
0.7	3360	3350	3600	3740	3880	4020
0.8	3590	3680	3840	4000	4140	4290
0.9	3800	3900	4075	4240	4390	4560
1.0	4000	4100	4300	4470	4630	4800
1.1	4210	4310	4510	4690	4860	5040
1.2	4380	4490	4710	4890	5060	5250
2.0	5620	5760	6020	6270	6490	6730
3.0	7000	7180	7510	7820	8090	8390
4.0	8000	8210	8600	8940	9250	9600
5.0	8950	9180	9620	10,000	10,340	10,720
6.0	9800	10,070	10,530	10,950	11,350	11,780
7.0	10,600	10,870	11,380	11,820	12,240	12,700
8.0	11,320	11,620	12,170	12,650	13,100	13,600
9.0	12,000	12,320	12,900	13,420	13,900	14,420
10.0	12,620	12,980	13,580	14,120	14,620	15,180
11.0	13,330	13,620	14,270	14,830	15,380	15,920
12.0	13,850	14,220	14,900	15,500	16,000	16,620

Flue Gas Temperature—°F						
400	500	600	700	800	900	1000
0.0	0.0	0.0	0.0	0.0	0.0	0.0
1140	1210	1270	1327	1385	1435	1490
1617	1703	1792	1870	1950	2020	2100
2285	2410	2540	2640	2760	2860	2970
2805	2960	3110	3340	3390	3510	3640
3235	3410	3580	3740	3900	4040	4200
3620	3810	4010	4180	4360	4530	4700
3960	4170	4390	4580	4770	4960	5140
4280	4510	4750	4950	5170	5360	5560
4570	4820	5070	5290	5520	5720	5950
4850	5120	5380	5610	5860	6070	6300
5120	5390	5670	5910	6170	6400	6640
5370	5660	5960	6210	6480	6720	6970
5600	5900	6210	6480	6760	7010	7280
7170	7560	7950	8290	8660	8970	9320
8940	9420	9910	10,330	10,800	11,190	11,620
10,240	10,800	11,350	11,830	12,350	12,800	13,300
11,450	12,050	12,650	13,220	13,800	14,300	14,850
12,530	13,220	13,900	14,500	15,150	15,700	16,300
13,530	14,280	15,000	15,650	16,350	16,930	17,600
14,480	15,280	16,050	16,720	17,480	18,120	18,800
15,360	16,200	17,000	17,750	18,530	19,200	19,950
16,150	17,030	17,910	18,700	19,500	20,600	21,000
16,980	17,900	18,800	19,600	20,450	21,200	22,050
17,700	18,680	19,630	20,440	21,400	22,200	23,000

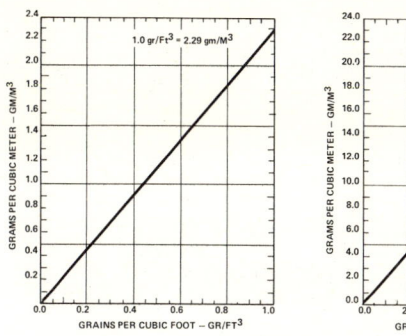 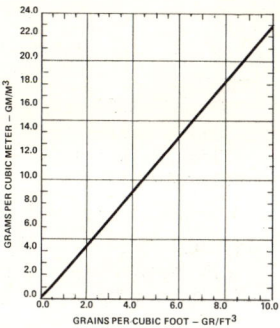 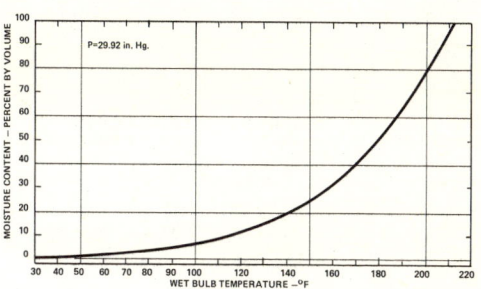

Item 3. *Item 4.*

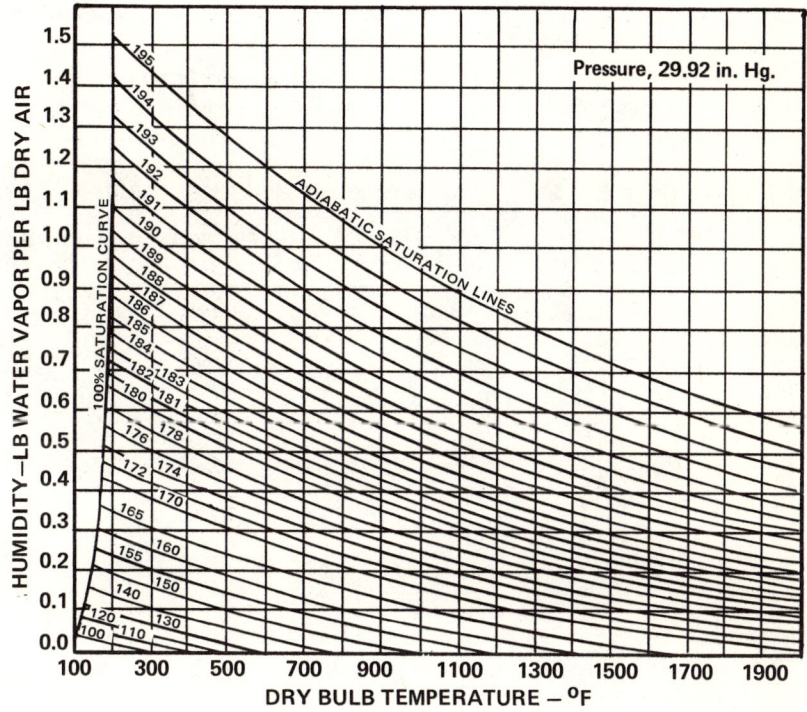

Item 5.

Courtesy of: Koch Engineering Company, Wichita, Kansas.

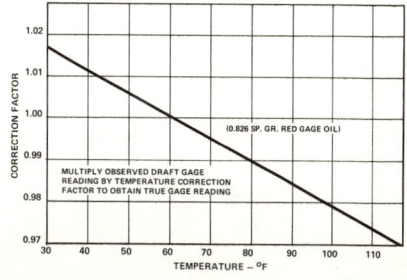

Item 6.

Courtesy of: Bay Area Air Pollution Control District, "Source Test Methods," 1961.

7. Conversion Factors between Metric and English Systems of Units.

Mass
1.0 pound = 7,000 grains
1.0 gram = 15.43 grains
1.0 kilogram = 2.2046 pounds
1.0 metric ton = 2,204.6 pounds
1.0 short ton = 2,000 pounds

Concentration
1.0 grains/ft^3 = 2.29 grams/meter^3
1.0 gram/meter^3 = 0.437 grains/ft^3

Velocity
1.0 meter/hour = 0.0547 ft/minute
1.0 meter/second = 3.28 feet/second

Length
1.0 cm = 0.394 inch
1.0 inch = 2.54 cm
1.0 foot = 30.5 cm
1.0 meter^2 = 10.76 ft^2
1.0 meter^3 = 35.31 ft^3

Emission Rate
1.0 kilogram/hour = 2.2046 pounds/hour
1.0 kg/met. ton = 2.0 lb/short ton

Flow Rate
1.0 meter^3/hour = 0.585 ft^3/minute
1.0 meter^3/minute = 35.3 ft^3/minute

8. Conversion of Particulate Concentrations in Parts per Million by Weight to Grains per Standard Dry Cubic Foot.

ppm by wt	gr/SDCF
1.0	0.0005
10.0	0.0053
20.0	0.0107
50.0	0.0266
100.0	0.0533
200.0	0.1067
500.0	0.266
800.0	0.426
1,000.0	0.533
5,000.0	2.665
10,000.0	5.33
50,000.0	26.65
100,000.0	53.33
500,000.0	266.5
1,000,000.0	533.3
2,000,000.0	1,067.0
5,000,000.0	2,665.0

Basis: 1. Dry gas at 60°F, 29.92 in. Hg.
2. Gas density is 0.076 lb/ft^3.
3. Gr/SDCF = (5.33 × 10^{-4}) (ppm by wt.).

9. Cross-sectional Areas for Round Sampling Probes.

Diameter-Inches			Cross-sectional Area	
Nominal	Decimal	Inch^2	Feet^2	Cm^2
1/16	0.0625	0.00307	0.213 × 10^{-4}	0.0198
1/8	0.1250	0.0123	0.854 × 10^{-4}	0.0795
3/16	0.1875	0.0276	1.92 × 10^{-4}	0.1788
1/4	0.2500	0.0491	3.41 × 10^{-4}	0.317
5/16	0.3125	0.0767	5.33 × 10^{-4}	0.496
3/8	0.3750	0.1102	7.65 × 10^{-4}	0.712
7/16	0.4375	0.1501	1.043 × 10^{-3}	0.970
1/2	0.5000	0.196	1.362 × 10^{-3}	1.270
5/8	0.6250	0.307	2.135 × 10^{-3}	1.985
3/4	0.7500	0.442	3.07 × 10^{-3}	2.86
7/8	0.8750	0.602	4.18 × 10^{-3}	3.89
1.0	1.0000	0.785	5.45 × 10^{-3}	5.07
1-1/4	1.2500	1.228	8.54 × 10^{-2}	7.95
1-1/2	1.5000	1.768	0.1228	11.42
1-3/4	1.7500	2.40	0.1665	15.50
2.0	2.0000	3.14	0.218	20.30
2.5	2.5000	4.91	0.341	31.75
3.0	3.0000	7.07	0.491	45.70

10. Cross-sectional Areas for Round Ducts.

Diameter			Cross-sectional Area	
Inches	Feet	Meters	Feet^2	Meter^2
6	0.5	0.152	0.392	0.036
12	1.0	0.305	0.785	0.073
18	1.5	0.457	1.768	0.164
24	2.0	0.610	3.141	0.292
30	2.5	0.763	4.910	0.456
36	3.0	0.915	7.070	0.657
48	4.0	1.200	12.560	1.167
60	5.0	1.525	19.650	1.827
72	6.0	1.830	28.300	2.625
84	7.0	2.135	38.450	3.570
96	8.0	2.440	50.250	4.665
108	9.0	2.745	63.600	5.900
120	10.0	3.050	78.560	7.300
132	11.0	3.355	95.000	8.825
144	12.0	3.660	113.200	10.520
180	15.0	4.575	176.800	16.400
240	20.0	6.100	314.160	29.200
300	25.0	7.625	491.000	45.600
360	30.0	9.150	707.070	65.700
420	35.0	10.675	962.500	89.400
480	40.0	12.000	1,256.000	116.700
540	45.0	13.725	1,592.500	148.000
600	50.0	15.250	1,965.000	182.700

APPENDIX B

DERIVATIONS

1. CORRECTION FACTOR FOR 12% CARBON DIOXIDE

Terminology:

V_0 = Volume of gas at standard conditions.
$V_{12\%}$ = Volume of gas at standard conditions corrected to 12% CO_2 by volume.
C_0 = Concentration at standard conditions.
$C_{12\%}$ = Concentration at standard conditions corrected to 12% CO_2 by volume.

Volumetric:

$$\frac{\% CO_2}{100}(V_0) = \frac{12.0}{100.0}(V_{12\%})$$

$$(\% CO_2)(V_0) = (12.0)(V_{12\%})$$

Concentration:

$$\frac{1}{(\% CO_2)}\frac{Wt}{(V_0)} = \frac{1}{(12.0)}\frac{Wt}{(V_{12\%})}$$

$$\frac{1}{(\% CO_2)}(C_0) = \frac{1}{(12.0)}(C_{12\%})$$

$$C_{12\%} = \frac{12.0}{(\% CO_2)}(C_0)$$

$$F_{CO_2} = \frac{C_{12\%}}{C_0} = \frac{12.0}{(\% CO_2)}$$

2. CORRECTION FACTOR FOR 6.0% OXYGEN

$$\frac{\% O_2}{100}(1 \text{ mole}) + X(0.21) = (1+X)(0.06)$$

X = lb-mole of air added
$1 + X$ = total flue gas after air

$$X = \frac{0.06 - \% O_2/100}{0.15}$$

$$F_{O_2} = \frac{1}{1+X} = \frac{1.00}{1.00 + \dfrac{0.06 - (\% O_2/100)}{0.15}}$$

$$F_{O_2} = \frac{1.00(0.15)}{0.15 + 0.06 - (\% O_2/100)}$$

$$F_{O_2} = \frac{0.15}{0.21 - (\% O_2/100)}$$

3. CORRECTION FACTORS FOR 50% EXCESS AIR

$$F_{EA} = \frac{(N_2/O_2)_{50\% EA}}{(N_2/O_2)_{Orsat}} = \frac{(11.30)}{(N_2/O_2)}$$

$$C_{50\%} = (F_{EA})(C_0)$$

F_{EA} = Correction factor to 50% excess air.
N_2/O_2 = Nitrogen-to-oxygen ratios on a dry volumetric basis from Orsat analysis.
C_0 = Concentration at standard conditions.
$C_{50\%}$ = Concentration at standard conditions corrected to 50% excess air.

Correction Factors for 50% Excess Air (Volumetric Fractions).

%EA	Combustion Air		Excess Air		Total Air		N_2/O_2	F_{EA}
	N_2	O_2	N_2	O_2	N_2	O_2		
0.00	0.791	0.000	0.000	0.000	0.791	0.000	∞	0.000
5.0	0.791	0.000	0.039	0.010	0.830	0.010	83.0	0.136
10.0	0.791	0.000	0.079	0.021	0.870	0.021	41.4	0.273
15.0	0.791	0.000	0.119	0.032	0.910	0.032	28.4	0.398
20.0	0.791	0.000	0.158	0.043	0.949	0.043	22.1	0.512
25.0	0.791	0.000	0.198	0.052	0.989	0.052	19.00	0.595
30.0	0.791	0.000	0.238	0.063	1.029	0.063	16.50	0.684
35.0	0.791	0.000	0.277	0.073	1.068	0.073	14.63	0.773
40.0	0.791	0.000	0.316	0.084	1.107	0.084	13.18	0.857
45.0	0.791	0.000	0.356	0.094	1.147	0.094	12.22	0.924
50.0	0.791	0.000	0.395	0.105	1.186	0.105	11.30	1.000
55.0	0.791	0.000	0.435	0.115	1.226	0.115	10.65	1.062
60.0	0.791	0.000	0.474	0.125	1.265	0.125	10.12	1.116
70.0	0.791	0.000	0.554	0.146	1.345	0.146	9.21	1.230
80.0	0.791	0.000	0.633	0.167	1.424	0.167	8.52	1.328
90.0	0.791	0.000	0.712	0.188	1.503	0.188	8.00	1.412
100.0	0.791	0.000	0.791	0.209	1.582	0.209	7.57	1.493
125.0	0.791	0.000	0.989	0.261	1.780	0.261	6.83	1.655
150.0	0.791	0.000	1.187	0.314	1.978	0.314	6.29	1.800
175.0	0.791	0.000	1.383	0.366	2.174	0.366	5.93	1.910
200.0	0.791	0.000	1.582	0.418	2.373	0.418	5.67	1.990
250.0	0.791	0.000	1.978	0.522	2.769	0.522	5.30	2.135
300.0	0.791	0.000	2.375	0.627	3.166	0.627	5.05	2.240
350.0	0.791	0.000	2.770	0.731	3.561	0.731	4.87	2.320
400.0	0.791	0.000	3.160	0.836	3.951	0.836	4.73	2.390
450.0	0.791	0.000	3.560	0.940	4.351	0.940	4.63	2.440
500.0	0.791	0.000	3.950	1.045	4.741	1.045	4.53	2.50
600.0	0.791	0.000	4.740	1.255	5.531	1.255	4.41	2.56
700.0	0.791	0.000	5.540	1.465	6.331	1.465	4.32	2.62
800.0	0.791	0.000	6.330	1.672	7.121	1.672	4.26	2.66
900.0	0.791	0.000	7.120	1.882	7.911	1.882	4.20	2.69
1000.0	0.791	0.000	7.910	2.092	8.701	2.090	4.17	2.71
∞							3.76	3.01

APPENDIX C

SOURCES OF EQUIPMENT

1. LISTING BY TYPE OF EQUIPMENT

EQUIPMENT   PAGE

    Alundum Thimbles
    Bolometric Particulate Monitors
    Bubble Tube Flow Meters
    Draft Gauges
    Dry Gas Meters
    Electrostatic Precipitators
    Flexible Fabric Bag Material
    Flue Gas Composition Analyzers
    Gas Detector Tubes
    Glass Fiber Filter Paper
    Glass Syringes
    Heat Traced Sample Lines
    Impingers
    Indicating Pyrometers
    Industrial Gases
    Liquid Pumps
    Manometers
    Membrane Filters
    Metallic Thermometers
    Needle Valves
    Orifice Flow Meters
    Particle Size Analyzers and Counters
    Particle Size Measurement
    Particulate Sampling Trains
    Pitot Tubes
    Plastic Tubing
    Recorders
    Rotameters
    Rotating Vane Anemometers
    Sample Flow Meters
    Sequential Samplers
    Specific Ion Electrodes
    Stack Opacity Monitors
    Swinging Vane Anemometers
    Thermal Anemometers
    Thermal Precipitators
    Tubing Fittings
    Vacuum Pumps
    Wet Test Meters

Note: It has been the desire of the authors to make this list of suppliers as complete as possible. However, they can assume no responsibility for information of which they were unaware. All listings are alphabetical.

Alundum Thimbles
1. American Hospital Supply Corp.
   Scientific Products Division
   1210 Leon Place
   Evanston, Illinois   60201
2. Research-Cottrell, Inc.
   P.O. Box 750
   Bound Brook, New Jersey   08805

3. Van Waters & Rogers, Inc.
   P.O. Box 3200
   Rincon Annex
   San Francisco, California   94119
4. Western Precipitation Co.
   1000 W. Ninth Street
   Los Angeles, California   90015

Bolometric Particulate Monitors
1. Bailey Meter Company
   Instrument Division
   Wickliffe, Ohio   44092
2. Leeds and Northrup Co.
   Instrument Division
   North Wales, Pennsylvania   19454

Bubble Tube Flow Meters
1. ITT—Barton Instrument Co.
   580 Monterey Pass Road
   Monterey Park, California   91754
2. SKC, Inc.
   P.O. Box 8538
   Pittsburgh, Pennsylvania   15220
3. Varian Associates, Inc.
   611 Hansen Way
   Palo Alto, California   94303

Draft Gauges
1. Alnor Instrument Company
   420 N. LaSalle Street
   Chicago, Illinois   60610
2. F. W. Dwyer Mfg. Co.
   P.O. Box 373
   Michigan City, Indiana   46360
3. Meriam Instrument Company
   10920 Madison Avenue
   Cleveland, Ohio   44102
4. Research-Cottrell, Inc.
   P.O. Box 750
   Bound Brook, New Jersey   08805
5. Western Precipitation Co.
   1000 W. Ninth Street
   Los Angeles, California   90015

Dry Gas Meters
1. American Meter Company, Inc.
   991 Broadway
   Albany, New York   12201
2. American Standard, Inc.
   Industrial Products Dept.
   Detroit, Michigan   48232
3. Research-Cottrell, Inc.
   P.O. Box 750
   Bound Brook, New Jersey   08805
4. Rockwell Manufacturing Co.
   400 N. Lexington Avenue
   Pittsburgh, Pennsylvania   15208
5. Sprague Meter Co.
   35 South Avenue
   Bridgeport, Connecticut   06601

6. Western Precipitation Corp.
   1000 W. Ninth Street
   Los Angeles, California   90015

Electrostatic Precipitators
1. Del Electronics, Inc.
   250 E. Sandford Blvd.
   Mount Vernon, New York   10550
2. Litton Industries, Inc.
   Applied Science Division
   2003 E. Hennepin Avenue
   Minneapolis, Minnesota   55413
3. Mine Safety Appliance Co.
   201 N. Braddock Ave.
   Pittsburgh, Pennsylvania   15208
4. Thermo-Systems, Inc.
   2500 Cleveland Ave. No.
   St. Paul, Minnesota   55113
5. Western Precipitation Corp.
   1000 W. Ninth Street
   Los Angeles, California   90015

Flexible Fabric Bag Material
1. 3M Company, Inc.
   5201 Hudson Road
   St. Paul, Minnesota   55119
2. The G. T. Schjeldahl Co.
   Sales Division
   Northfield, Minnesota   55057
3. Robert H. Wager Co.
   423 Valley Street
   South Orange, New Jersey   07079

Flue Gas Composition Analyzers
1. Bacharach Instrument Company
   RID Industrial Park
   Pittsburgh, Pennsylvania   15238
2. Burrell Technical Supply Co.
   1936 Fifth Avenue
   Pittsburgh, Pennsylvania   15219
3. F. W. Dwyer Mfg. Co.
   P.O. Box 373
   Michigan City, Indiana   46360
4. Hays Corporation
   742 E. Eighth Street
   Michigan City, Indiana   46360

Gas Detector Tubes
1. Metronics Associates, Inc.
   3201 Porter Drive
   Stanford Industrial Park
   Palo Alto, California   94304
2. Mine Safety Appliance Co.
   201 N. Braddock Avenue
   Pittsburgh, Pennsylvania   15208
3. Unico Environmental Instruments, Inc.
   150 Cove Street
   Fall River, Massachusetts   02720

Glass Fiber Filter Paper
1. Gelman Instrument Company
   600 S. Wagner Road
   Ann Arbor, Michigan   48106
2. Mine Safety Appliance Co.
   201 N. Braddock Avenue
   Pittsburgh, Pennsylvania   15208

Glass Syringes
1. Cole-Parmer Instrument Co.
   7425 N. Oak Park Avenue
   Chicago, Illinois   60648
2. East Rutherford Syringes, Inc.
   480 Patterson Avenue
   East Rutherford, New Jersey   07073
3. The Hamilton Company
   P.O. Box 300
   Whittier, California   90601
4. Perco Supplies, Inc.
   P.O. Box 201
   San Gabriel, California   91778
5. Precision Sampling Corp.
   P.O. Box 15119
   Baton Rouge, Louisiana   70815
6. SKC, Inc.
   P.O. Box 8538
   Pittsburgh, Pennsylvania   15220

Heat Traced Sample Lines
1. Samuel Moore & Co.
   Dekoron Division
   Mantua, Ohio   44255

Impingers
1. Bel-Art Industries
   Sales Division
   Pequannock, New Jersey   07440
2. Gelman Instrument Company
   600 S. Wagner Road
   Ann Arbor, Michigan   48106
3. Mine Safety Appliance Co.
   201 N. Braddock Avenue
   Pittsburgh, Pennsylvania   15208
4. Van Waters and Rogers, Inc.
   P.O. Box 3200
   Rincon Annex
   San Francisco, California   94119
5. Weather Measure Corp.
   P.O. Box 41257
   Sacramento, California   95841

Indicating Pyrometers
1. Cole-Parmer Instrument Co.
   7425 North Oak Park Avenue
   Chicago, Illinois   60648
2. W. H. Curtin Co.
   P.O. Box 1546
   Houston, Texas   77001
3. Fisher Scientific Co.
   711 Forbes Avenue
   Pittsburgh, Pennsylvania   15219
4. The Pyrometer Instrument Co.
   Sales Division
   Bergenfield, New Jersey   07621

Industrial Gases
1. Airco Air Reduction Co.
   575 Mountain Ave.
   Murray Hill, New Jersey   07974
2. Air Products & Chemicals, Inc.
   P.O. Box 538
   Allentown, Pennsylvania   18101
3. Matheson Company
   Gas Products Division
   P.O. Box 85
   East Rutherford, New Jersey   07073
4. Scott Research Laboratories, Inc.
   P.O. Box No. 66
   Perkosie, Pennsylvania   18944
5. Union Carbide Corporation
   Linde Division

270 Park Avenue
New York, New York   10017

Liquid Pumps
1. American Instrument Co.
   8030 Georgia Avenue
   Silver Spring, Maryland   20901
2. Ideal Precision Glass Co.
   P.O. Box 287
   Industrial Road
   Carlstadt, New Jersey   07072
3. Jabsco Pump Company
   1485 Dale Way
   Costa Mesa, California   92626
4. Zero-Max Company
   2845 Harriet Avenue S.
   Minneapolis, Minnesota   55408

Manometers
1. Alnor Instrument Co.
   420 N. LaSalle Street
   Chicago, Illinois   60610
2. Cole-Parmer Instrument Company
   7425 N. Oak Park Avenue
   Chicago, Illinois   60648
3. F. W. Dwyer Mfg. Co.
   P.O. Box 373
   Michigan City, Indiana   46360
4. Meriam Instrument Co.
   10920 Madison Ave.
   Cleveland, Ohio   44102

Membrane Filters
1. Gelman Instrument Co.
   600 S. Wagner Road
   Ann Arbor, Michigan   48106
2. Millipore Corporation
   Ashby Road
   Bedford, Massachusetts   01730

Metallic Thermometers
1. Cole-Parmer Instrument Co.
   7425 North Oak Park Avenue
   Chicago, Illinois   60648
2. Fisher Scientific Co.
   711 Forbes Avenue
   Pittsburgh, Pennsylvania   15219
3. Marshalltown Mfg. Co.
   P.O. Box 400
   Marshalltown, Iowa   50128
4. Palmer Instruments, Inc.
   2501 Norwood Ave.
   Cincinnati, Ohio   45212

Needle Valves
1. American Instrument Co.
   8030 Georgia Avenue
   Silver Spring, Maryland   20901
2. Cajon Company
   32550 Old South Miles Road
   Solon, Ohio   44139
3. Hoke Manufacturing Co.
   P.O. Box 501
   Tenafly, New Jersey   07670
4. Nupro Company
   15635 Saranac Road
   Cleveland, Ohio   44110
5. Whitey Research Tool Co.
   5679 Landregan Street
   Emeryville, California   94608

Orifice Flow Meters
1. Gelman Instrument Co.
   600 S. Wagner Road
   Ann Arbor, Michigan   48106
2. Meriam Instrument Co.
   10920 Madison Ave.
   Cleveland, Ohio   44102

Particle Size Analyzers and Counters
1. Air Technology Inc.
   1717 Whitehead Road
   Baltimore, Maryland   21207
2. Bausch & Lomb, Inc.
   635 St. Paul Street
   Rochester, New York   14602
3. Climet Instruments, Inc.
   1240 Birchwood Drive
   Sunnyvale, California   94086
4. Coulter Electronics Corp.
   Industrial Division
   590 W. 20th Street
   Hialeah, Florida   33010
5. Particle Technology, Inc.
   P.O. Box 265
   Elmhurst, Illinois   60126
6. Phoenix Precision Instrument Co.
   3805 N. Fifth Street
   Philadelphia, Pennsylvania   19140
7. Royco Instruments, Inc.
   141 Jefferson Drive
   Menlo Park, California   94025
8. Thermo-Systems, Inc.
   2500 Cleveland Ave. No.
   St. Paul, Minnesota   55113

Particle Size Measurement
1. A. A. Andersen & Co.
   1423-S. 2nd Street
   Salt Lake City, Utah   84115
2. C. F Cassella & Co., Ltd.
   Regent House, Fitzroy Square
   London W. 1, England
3. Harry W. Dietert Co. (Bahco)
   9330 Roselawn Avenue
   Detroit, Michigan   48204
4. Mine Safety Appliance Co.
   201 N. Braddock Ave.
   Pittsburgh, Pennsylvania   15208
5. Unico Environmental Instruments, Inc.
   150 Cove Street
   Fall River, Massachusetts   02720
6. Vickers Instruments, Inc.
   15 Waite Court
   Malden, Massachusetts   02148

Particulate Sampling Trains
1. Airflow Developments, Ltd.
   244 Newkirk Road
   Richmond Hill, Ontario, Canada
2. American Precision Industries, Inc.
   Dustex Division
   2777 Walden Avenue
   Buffalo, New York   14225
3. Gelman Instrument Co.
   600 S. Wagner Road
   Ann Arbor, Michigan   48106
4. Research Appliance Corp.
   Route 8 and Craighead Road
   Allison Park, Pennsylvania   15101
5. Universal Oil Products Co.

Aerotec Industries Division
Greenwich, Connecticut   06830

Pitot Tubes

1. Airflow Developments, Ltd.
   244 Newkirk Road
   Richmond Hill, Ontario, Canada
2. Alnor Instrument Co.
   420 N. LaSalle Street
   Chicago, Illinois   60610
3. Research-Cottrell, Inc.
   P.O. Box 750
   Bound Brook, New Jersey   08805
4. Weather Measure Corp.
   P.O. Box 41275
   Sacramento, California   95841
5. Western Precipitation Co.
   1000 W. Ninth Street
   Los Angeles, California   90015

Plastic Tubing

1. Barton Instrument Co.
   580 Monterey Pass Road
   Monterey Park, California   91754
2. Cole-Parmer Instrument Co.
   7425 N. Oak Park Avenue
   Chicago, Illinois   60648
3. D & G Plastics Co.
   P.O. Box 209
   Kent, Ohio   44240
4. Pennsylvania Fluorocarbon Co.—Shrinkable Tubing
   Penntube Plastics Division
   Holley Street & Madison Avenue
   Clifton Heights, Pennsylvania   19018
5. The Polymer Corp.
   Polypreneco Division
   Reading, Pennsylvania   19603
6. Portland Valve & Fitting Co.
   112 NE Holladay Street
   Portland, Oregon   97232
7. Rex Valve & Control, Inc.—Polyethylene
   928 NW Tualatin Valley Highway
   Beaverton, Oregon   97005
8. Universal Plastics Co.
   3663 First Avenue S.
   Seattle, Washington   98104
9. U.S. Stoneware Co.—Tygon
   Plastics & Synthetics Division
   Akron, Ohio   44309

Recorders

1. Beckman Instruments, Inc.
   Scientific Instruments Division
   2500 Harbor Boulevard
   Fullerton, California   92634
2. The Foxboro Company
   38 Neponset Street
   Foxboro, Massachusetts   02035
3. Leeds and Northrup Co.
   Instrument Division
   North Wales, Pennsylvania   19454
4. Minneapolis-Honeywell, Inc.
   Industrial Division
   1100 Virginia Drive
   Fort Washington, Pennsylvania   19034
5. Varian Associates
   611 Hansen Way
   Palo Alto, California   94303
6. Weather Measure Corp.
   P.O. Box 41257
   Sacramento, California   95841

Rotameters

1. Brooks Instrument Co.
   407 W. Vine Street
   Hatfield, Pennsylvania   19440
2. F. W. Dwyer Mfg. Co.
   P.O. Box 373
   Michigan City, Indiana   46360
3. Fischer & Porter, Inc.
   County Line Road
   Warminster, Pennsylvania   18974
4. Ideal Precision Glass Co.
   Manostat Division
   P.O. Box 287
   Carlstadt, New Jersey   07072
5. Meriam Instrument Co.
   10920 Madison Avenue
   Cleveland, Ohio   44102
6. Schutte & Koerting, Inc.
   2239 State Road
   Cornwells Heights
   Bucks County, Pennsylvania   19020

Rotating Vane Anemometers

1. Airflow Developments (Canada) Ltd.
   244 Newkirk Street
   Richmond Hill, Ontario, Canada
2. Taylor Instrument Co.
   95 Ames Street
   Rochester, New York   14601
3. Weather Measure Corp.
   P.O. Box 41257
   Sacramento, California   95841

Sample Flow Meters

1. Electro-Neutronics, Inc.
   94710 Dwight Way
   Berkeley, California   94710
2. Flow Corporation
   127 Coolidge Hill Road
   Watertown, Massachusetts   02172
3. Matheson Company
   P.O. Box 85
   East Rutherford, New Jersey   07073

Sequential Samplers

1. Gelman Instrument Co.
   600 S. Wagner Road
   Ann Arbor, Michigan   48106
2. International Chemical & Nuclear Corp.
   630-20th Street
   Oakland, California   94612
3. Research Appliance Corp.
   Route 8 and Craighead Road
   Allison Park, Pennsylvania   15101

Specific Ion Electrodes

1. Calgon Corporation
   P.O. Box 1346
   Pittsburgh, Pennsylvania   15230
2. Orion Research, Inc.
   11 Blackstone Street
   Cambridge, Massachusetts   02139

Stack Opacity Monitors

1. Airflow Developments, Ltd.
   244 Newkirk Road
   Richmond Hill, Ontario, Canada
2. Bailey Meter Company
   Instrument Division
   Wickliffe, Ohio   44092

3. Leeds and Northrup Co.
   Instrument Division
   North Wales, Pennsylvania   19454
4. Reliance Instrument Mfg. Co.
   143 Lawrence Street
   Hackensack, New Jersey   07602

Swinging Vane Anemometers

1. Alnor Instrument Company
   420 N. LaSalle Street
   Chicago, Illinois   60610
2. F. W. Dwyer Mfg. Co.
   P.O. Box 373
   Michigan City, Indiana   46360

Thermal Anemometers

1. Airflow Developments, Ltd.
   244 Newkirk Road
   Richmond Hill, Ontario, Canada
2. Alnor Instrument Co.
   420 N. LaSalle Street
   Chicago, Illinois   60610
3. Gelman Instrument Co.
   600 S. Wagner Road
   Ann Arbor, Michigan   48106
4. Hastings-Raydist, Inc.
   Sales Division
   Hampton, Virginia   23360
5. Weather Measure Corp.
   P.O. Box 41257
   Sacramento, California   95841
6. Willson Products Company
   P.O. Box 622
   Reading, Pennsylvania   19603

Thermal Precipitators

1. American Instrument Co., Inc.
   8030 Georgia Avenue
   Silver Spring, Maryland   20901
2. C. F. Cassella & Co., Ltd.
   Regent House, Fitzroy Square
   London W. 1, England
3. Joseph B. Ficklen, III
   1848 E. Mountain Street
   Pasadena, California   91107

Tubing Fittings

1. Cajon Company
   32550 Old South Miles Road
   Solon, Ohio   44139
2. Crawford Fittings Co.
   29500 Solon Road
   Solon, Ohio   44139
3. D & G Plastics Co.
   P.O. Box 209
   Kent, Ohio   44240
4. Hoke Manufacturing Co.
   P.O. Box 501
   Tenafly, New Jersey   07670

Vacuum Pumps

1. Cole-Parmer Instrument Co.
   7425 North Oak Park Avenue
   Chicago, Illinois   60648
2. Gast Manufacturing Corp.
   P.O. Box 117
   Benton Harbor, Michigan   49002
3. Gelman Instrument Co.
   600 S. Wagner Road
   Ann Arbor, Michigan   48106

4. International Telephone & Telegraph Co.
   Bell & Gossett Division
   8200 N. Austin Avenue
   Morton Grove, Illinois   60053
5. Leiman Bros., Inc.
   140 E. Union Avenue
   East Rutherford, New Jersey   07071
6. Research Appliance Corp.
   Route 8 and Craighead Road
   Allison Park, Pennsylvania   15101

Wet Test Meters

1. American Meter Company, Inc.
   991 Broadway
   Albany, New York   12201
2. Precision Scientific Company
   3737 W. Cortland Street
   Chicago, Illinois   60647

2. ALPHABETICAL LISTING OF SUPPLIERS

1. A. A. Andersen & Co.
   1423 S. 2nd Street
   Salt Lake City, Utah   84115
2. Airco Air Reduction Co.
   575 Mountain Avenue
   Murray Hill, New Jersey   07974
3. Air Flow Developments (Canada) Ltd.
   244 Newkirk Road
   Richmond Hill, Ontario, Canada
4. Air Products and Chemicals, Inc.
   P.O. Box 538
   Allentown, Pennsylvania   18101
5. Air Technology, Inc.
   1717 Whitehead Road
   Baltimore, Maryland   21207
6. Alnor Instrument Company
   420 N. LaSalle Street
   Chicago, Illinois   60610
7. American Hospital Supply Corp.
   Scientific Products Division
   1210 Leon Place
   Evanston, Illinois   60201
8. American Instrument Company
   8030 Georgia Avenue
   Silver Spring, Maryland   20901
9. American Meter Company, Inc.
   991 Broadway
   Albany, New York   12201
10. American Precision Industries, Inc.
    Dustex Division
    2777 Walden Avenue
    Buffalo, New York   14225
11. American Standard, Inc.
    Industrial Products Dept.
    Detroit, Michigan   48232
12. Bacharach Instrument Company
    RID Industrial Park
    Pittsburgh, Pennsylvania   15238
13. Bailey Meter Co.
    Instrument Division
    Wickliffe, Ohio   44092
14. Barton Instrument Co.
    580 Monterey Pass Road

Monterey Park, California   91754
15. Bausch and Lomb, Inc.
    635 St. Paul Street
    Rochester, New York   14602
16. Beckman Instruments, Inc.
    Scientific Instruments Division
    2500 Harbor Boulevard
    Fullerton, California   92634
17. Bel-Art Industries, Inc.
    Sales Division
    Pequannock, New Jersey   07440
18. Brooks Instrument Company
    407 W. Vine Street
    Hatfield, Pennsylvania   19440
19. Burrell Technical Supply Company
    1936 Fifth Avenue
    Pittsburgh, Pennsylvania   15219
20. Cajon Company
    32550 Old South Miles Road
    Solon, Ohio   44139
21. Calgon Corporation
    P.O. Box 1346
    Pittsburgh, Pennsylvania   15230
22. C. F. Cassella & Co., Ltd.
    Regent House, Fitzroy Square
    London W. 1, England
23. Climet Instruments, Inc.
    1240 Birchwood Drive
    Sunnyvale, California   94806
24. Cole-Parmer Instrument Co.
    7425 N. Oak Park Avenue
    Chicago, Illinois   60648
25. Coulter Electronics Corp.
    Industrial Division
    590 W. 20th Street
    Hialeah, Florida   33010
26. Crawford Fittings Co.
    29500 Solon Road
    Solon, Ohio   44139
27. W. H. Curtin Co.
    P.O. Box 1546
    Houston, Texas   77001
28. Del Electronics, Inc.
    250 E. Sandford Avenue
    Mount Vernon, New York   10550
29. D & G Plastics Co.
    P.O. Box 209
    Kent, Ohio   44240
30. Harry W. Dietert Co. (Bahco)
    9330 Roselawn Avenue
    Detroit, Michigan   48204
31. F. W. Dwyer Mfg. Co.
    P.O. Box 373
    Michigan City, Indiana   46360
32. East Rutherford Syringes, Inc.
    480 Patterson Avenue
    East Rutherford, New Jersey   07073
33. Electro-Neutronics, Inc.
    9471 Dwight Way
    Berkeley, California   94710
34. Joseph B. Ficklen, III
    1848 E. Mountain Street
    Pasadena, California   91107
35. Fischer & Porter, Inc.
    County Line Road
    Warminster, Pennsylvania   18974
36. Fisher Scientific Company
    711 Forbes Avenue
    Pittsburgh, Pennsylvania   15219
37. Flow Corporation
    127 Coolidge Hill Road
    Watertown, Massachusetts   02172
38. The Foxboro Company
    38 Neponset Street
    Foxboro, Massachusetts   02035
39. Gast Manufacturing Company
    P.O. Box 117
    Benton Harbor, Michigan   49002
40. Gelman Instrument Company
    600 S. Wagner Road
    Ann Arbor, Michigan   48106
41. The Hamilton Company
    P.O. Box 300
    Whittier, California   90601
42. Hastings-Raydist, Inc.
    Sales Division
    Hampton, Virginia   23360
43. Hays Corporation
    742 E. Eighth Street
    Michigan City, Indiana   46360
44. Hoke Manufacturing Co.
    P.O. Box 501
    Tenafly, New Jersey   07670
45. Ideal Precision Glass Co.
    P.O. Box 287
    Industrial Road
    Carlstadt, New Jersey   07072
46. International Chemical & Nuclear Corp.
    630-20th Street
    Oakland, California   94612
47. ITT—Barton Instrument Corp.
    580 Monterey Pass Road
    Monterey Park, California   91754
48. ITT—Bell & Gossett Division
    8200 N. Austin Avenue
    Morton Grove, Illinois   60053
49. Jabsco Pump Company
    1485 Dale Way
    Costa Mesa, California   92626
50. Leeds and Northrup Company
    Instrument Division
    North Wales, Pennsylvania   19454
51. Leiman Brothers, Inc.
    140 E. Union Avenue
    East Rutherford, New Jersey   07071
52. Litton Industries, Inc.
    Applied Science Division
    2003 E. Hennepin Avenue
    Minneapolis, Minnesota   55413
53. Marshalltown Mfg. Co.
    P.O. Box 400
    Marshalltown, Iowa   50128
54. The Matheson Company
    Gas Products Division
    P.O. Box 85
    East Rutherford, New Jersey   07073

55. Meriam Instrument Company
    10920 Madison Avenue
    Cleveland, Ohio   44102
56. Metronics Associates, Inc.
    3201 Porter Drive
    Stanford Industrial Park
    Palo Alto, California   94304
57. Millipore Corp.
    Ashby Road
    Bedford, Massachusetts   01730
58. Mine Safety Appliance Co.
    201 N. Braddock Ave.
    Pittsburgh, Pennsylvania   15208
59. Minneapolis-Honeywell, Inc.
    Industrial Division
    1100 Virginia Drive
    Fort Washington, Pennsylvania   19034
60. Samuel Moore and Co.
    Dekoron Division
    Mantua, Ohio   44255
61. Nupro Company
    15635 Saranac Road
    Cleveland, Ohio   44110
62. Orion Research Inc.
    11 Blackstone Street
    Cambridge, Massachusetts   02139
63. Palmer Instruments, Inc.
    2501 Norwood Avenue
    Cincinnati, Ohio   45212
64. Particle Technology, Inc.
    P.O. Box 265
    Elmhurst, Illinois   60126
65. Pennsylvania Fluorocarbon Co.
    Penntube Plastics Division
    Holley Street & Madison Ave.
    Clifton Heights, Pennsylvania   19018
66. Perco Supplies, Inc.
    P.O. Box 201
    San Gabriel, California   91778
67. Phoenix Precision Instrument Co.
    3805 N. Fifth Street
    Philadelphia, Pennsylvania   19140
68. The Polymer Corp.
    Polyprenco Division
    Reading, Pennsylvania   19603
69. Portland Valve and Fitting Co.
    112 N.E. Holladay Street
    Portland, Oregon   97232
70. Precision Sampling Corp.
    P.O. Box 15119
    Baton Rouge, Louisiana   70815
71. Precision Scientific Co.
    3737 W. Cortland St.
    Chicago, Illinois   60647
72. The Pyrometer Instrument Co.
    Sales Division
    Bergenfield, New Jersey   07621
73. Reliance Instrument Mfg. Co.
    143 Lawrence Street
    Hackensack, New Jersey   07602
74. Research Appliance Corp.
    Route 8 and Craighead Road
    Allison Park, Pennsylvania   15101
75. Research-Cottrell, Inc.
    P.O. Box 750
    Bound Brook, New Jersey   08805
76. Rex Valve and Control, Inc.
    928 N.W. Tualatin Valley Highway
    Beaverton, Oregon   97005
77. Rockwell Mfg. Co.
    400 N. Lexington Avenue
    Pittsburgh, Pennsylvania   15208
78. Royco Instruments, Inc.
    141 Jefferson Drive
    Menlo Park, California   94025
79. G. T. Schjeldahl Company
    Sales Division
    Northfield, Minnesota   55057
80. Schutte & Koerting, Inc.
    2239 State Road
    Cornwell Heights
    Bucks County, Pennsylvania   19020
81. Scott Research Laboratories
    P.O. Box 66
    Perkosie, Pennsylvania   18944
82. SKC, Inc.
    P.O. Box 8538
    Pittsburgh, Pennsylvania   15220
83. Sprague Meter Co.
    35 South Avenue
    Bridgeport, Connecticut   06601
84. Taylor Instrument Co.
    95 Ames Street
    Rochester, New York   14601
85. Thermo-Systems, Inc.
    2500 Cleveland Avenue N.
    St. Paul, Minnesota   55113
86. Three M. Company, Inc.
    2501 Hudson Road
    St. Paul, Minnesota   55119
87. Unico Environmental Instruments, Inc.
    150 Cove Street
    Fall River, Massachusetts   02720
88. Union Carbide Corp.
    Linde Division
    270 Park Avenue
    New York, New York   10017
89. Universal Oil Products Co.
    Aerotec Industries Division
    Greenwich, Connecticut   06830
90. Universal Plastics Co.
    3663 First Avenue S.
    Seattle, Washington   98104
91. U.S. Stoneware Co.
    Plastics & Synthetics Division
    Akron, Ohio   44309
92. Van Waters and Rogers, Inc.
    P.O. Box 3200
    Rincon Annex
    San Francisco, California   94119
93. Varian Associates, Inc.
    611 Hansen Way
    Palo Alto, California   94303
94. Vickers Instruments, Inc.
    15 Waite Court
    Malden, Massachusetts   02148

95. Robert H. Wager Co.
    423 Valley Street
    South Orange, New Jersey   07079
96. Weather Measure Corp.
    P.O. Box 41257
    Sacramento, California   95811
97. Western Precipitation Corp.
    1000 W. Ninth Street
    Los Angeles, California   90015

98. Whitey Research Tool Co.
    5679 Landregan Street
    Emeryville, California   94608
99. Willson Products Co.
    P.O. Box 622
    Reading, Pennsylvania   19603
100. Zero-Max Company
    2845 Harriet Avenue
    Minneapolis, Minnesota   55408

APPENDIX D

ADDRESSES

1. GOVERNMENT AGENCIES

    1. Bay Area Air Pollution Control District
       939 Ellis Avenue
       San Francisco, California   94109
    2. County of Los Angeles
       Air Pollution Control District
       434 S. San Pedro Street
       Los Angeles, California   90013
    3. Dade County Pollution Control Authority
       864 NW 23rd Street
       Miami, Florida   33127
    4. Federal Clearinghouse for Scientific and
       Technical Information
       National Bureau of Standards
       U. S. Department of Commerce
       Springfield, Virginia   22151
    5. U. S. Public Health Service
       National Air Pollution Control Administration
       4211 West Chapel Hill Street
       Durham, North Carolina   27706
    6. U.S. Public Health Service
       National Air Pollution Control Administration
       801 N. Randolph Street
       Arlington, Virginia   22201

2. TECHNICAL AND PROFESSIONAL ORGANIZATIONS

    1. Air Pollution Control Association
       4400 Fifth Avenue
       Pittsburgh, Pennsylvania   15213
    2. American Industrial Hygiene Association
       25711 Southfield Road
       Southfield, Michigan   48075
    3. American Petroleum Institute, Inc.
       1271 Avenue of the Americas
       New York, New York   10020
    4. American Society of Mechanical Engineers
       United Engineering Center
       345 East 47th Street
       New York, New York   10017
    5. American Society for Testing and Materials
       1916 Race Street
       Philadelphia, Pennsylvania   19103
    6. British Coal Utilization Research Association
       Randalls Road, Leatherhead
       Surrey, England
    7. British Standards Institution
       British Standards House
       2 Park Street
       London, W., 1 England
    8. Incinerator Institute of America
       60 East 42nd Street
       New York, New York   10017
    9. Industrial Gas Cleaning Institute, Inc.
       P.O. Box No. 448
       Rye, New York   10580
    10. National Council of the Paper Industry for
        Air and Stream Improvement
        260 Madison Avenue
        New York, New York   10016
    11. Oregon State University
        Forest Products Research Laboratory
        Corvallis, Oregon   97330

3. PRIVATE FIRMS AND PUBLISHERS

    1. Academic Press, Inc.
       111 Fifth Avenue
       New York, New York   10003
    2. Buffalo Forge Company
       490 Broadway
       Buffalo, New York   14204
    3. Environmental Science Services Corp.
       24 Danbury Road
       Wilton, Connecticut   06897
    4. Wiley-Interscience, Inc.
       605 Third Avenue
       New York, New York   10016
    5. McGraw-Hill Book Company
       330 West 42nd Street
       New York, New York   10036
    6. Research-Cottrell, Inc.
       P.O. Box 750
       Bound Brook, New Jersey   08805
    7. Western Precipitation Corp.
       1000 W. Ninth Street
       Los Angeles, California   90013
    8. John Wiley and Sons, Inc.
       605 Third Avenue
       New York, New York   10016

LIST OF FIGURES

Figure	Title	Page
1.	Approach to Source Testing	2
2.	Correction of Stack Concentrations to 12 Percent Carbon Dioxide	10
3.	Correction of Stack Concentrations to 6 Percent Oxygen	11
4.	Correction of Stack Concentrations to 50 Percent Excess Air	12
5.	Characteristics of Particles and Particle Dispersoids	19
6.	Traverse Positions in Round Ducts	23
7.	Traverse Positions in Rectangular Ducts	23
8.	Typical Sample Port Arrangements	24
9.	Sampling Platform on a Stack	24
10.	Sample Port Assembly for Positive Pressure Ducts	24
11.	Sample Port Assembly for Inclined Ducts Utilizing a Water Trap	25
12.	Determination of Static Pressure in a Duct	25
13.	Moisture Content Determination by Direct Wet and Dry Bulb Temperature Measurement in Hot Ducts	27
14.	Moisture Content Determination in Hot Ducts by Remote Measurement of Wet and Dry Bulb Temperatures	27
15.	Moisture Content by the Condensation Method	29
16.	Orsat Analyzer for Measuring CO_2, O_2 and CO Contents of Flue Gas Streams	30
17.	Standard Pitot Tube	34
18.	Stausscheibe (S-Type) Pitot Tube	34
19.	Calibration Curve for a Stausscheibe (S-Type) Pitot Tube	34
20.	Operation of a Double Pitot-Venturi Flow Meter	34
21.	Operation of a Differential Pitot Tube	35
22.	Alternative Arrangement for Traverse Positions in Rectangular Ducts	37
23.	Low Velocity Measurement by Chemical Addition	40
24.	Velocity Measurement by Reducing the Cross-Sectional Area of a Duct	40
25.	Velocity Traverse with an S-Type Pitot Tube	41
26.	Tracer Dilution Technique for Gas Flow Measurement	42
27.	Elements of a Source Sampling Train	47
28.	Calibration of an Orifice Flow Meter	50
29.	Use of a Bubble Tube Flow Meter for Flow Calibration	51
30.	Calibration of a Dry Gas Meter with a Standard Orifice	51
31.	Flow Meter Calibration with a Spirometer	51
32.	Generation of a Known Gas Volume by Liquid Displacement	52
33.	Nonrepresentative Sampling Caused by Nonisokinetic Conditions	53
34.	Parameter for Particle Size in Isokinetic Sampling	54
35.	Effect of Anisokinetic Sampling Velocity on Measured Particle Concentration	54
36.	Effect of Subisokinetic Sampling Velocity on Measured Particle Concentration	55
37.	Sampling Errors Created by Misalignment of Probe in the Flow Stream	56
38.	Effect of Sampling Nozzle Configuration on Measured Particle Concentrations	57
39.	Estimation of Sampling Flow Rates for Maintaining Isokinetic Conditions	59
40.	Estimation of Isokinetic Sampling Rates	59
41.	Maintaining Isokinetic Sampling Conditions by Parallel Location of S-Type Pitot Tube and Sampling Probe	60
42.	Operating Nomograph for Sampling Rate Estimation	60
43.	Nomograph for Moisture Content, Temperature, and Pressure Corrections	60
44.	Nomograph for Estimating Isokinetic Sampling Rates	62
45.	Nomograph for Estimating Isokinetic Sampling Rates	63
46.	Automatic Isokinetic Sampling Control System Using Differential Pressure Relays	63
47.	Automatic Isokinetic Sampling Control System Using Pressure Actuated Transducers	65
48.	Pressure Null Balance Nozzles for Estimating Isokinetic Sampling Rates a. Parallel Tube b. Concentric Tube	65
49.	Sampling Errors Caused by Departure from Pressure Null Conditions	65
50.	Thermal Null Balance System for Estimating Isokinetic Sampling Conditions	66
51.	Variation of Particulate Emissions with Firing Rate for a Coal-Fired Power Boiler	69
52.	Effect of Duct Configuration on Particle Emission and Gas Velocity Profiles	69
53.	Water Cooled Probe for Hot Ducts	80
54.	Double Needle Valve Bleed System	82
55.	Impingers Used in Particulate Sampling Trains	87
56.	Filtration Devices Used in Particulate Sampling	90
57.	Effect of Flow Rate on Pressure Drop Across Membrane Filters	92
58.	Continuous "Cegrit" Sampler for Maintaining Isokinetic Conditions	94

Figure	Title	Page
59.	Effect of Flow Rate on Pressure Drop Across Single Impinger Units	99
60.	Low Rate Particulate Sampling Trains Employed by the Los Angeles County APCD	106
61.	Particulate Sampling Train Employed by the National Air Pollution Control Administration	106
62.	Particulate Sampling Train Employed by the Dade County Pollution Control Authority	106
63.	High Rate Sampling Train Employed by the Los Angeles County APCD	106
64.	High Rate Particulate Sampling Train Employed by the British Standards Institution	106
65.	BCURA Internal Cyclone-Filter Probe for Particulate Sampling	107
66.	Sampling Train for Polynuclear Aromatic Hydrocarbons	107
67.	Sampling Train for Gaseous and Particulate Fluoride Emissions	108
68.	Sampling Train for Acid Mist Droplets	109
69.	Particulate Sampling Train for Refuse Incinerators Employing a Pressure Null Balance System	111
70.	Particulate Sampling Train for a Wigwam Waste Burner Employing a Thermal Null Balance System	112
71.	Particulate Sampling Train for Kraft Pulp Mill Flue Gas Streams with High Moisture Contents	113
72.	Sampling Train for Particulate and Gaseous Emissions from Magnesium Base Sulfite Recovery Furnaces	115
73.	Particulate Sampling Trains for Flue Gases from Metallurgical Operations	117
74.	Particulate Sampling Trains for Determining Collection Efficiency of a Wet Scrubber	118
75.	Particulate Sampling System Employed in the Soviet Union	119
76.	Graphical Presentation of Particle Size Distribution	120
77.	Logarithmic Probability Representation of Particle Size Distribution	121
78.	Effect of Particle Size and Density on Terminal Settling Velocity	122
79.	Particle Size Measurement in Flue Gases with a Modified Cascade Impactor	124
80.	Particle Size Measurement with an Internally Located Cascade Impactor	124
81.	Particle Size Measurement with an Externally Located Andersen Multiple Plate Sampler	126
82.	Centrifugal Devices Used for Particle Size Measurements	127
83.	Particle Size Measurement with a Membrane Filter Assembly	128
84.	Effect of Flow Rate on Pressure Differential and Static Pressure for an Orifice Flow Meter	138
85.	Effect of Duct Velocity on Sampling Rate for Proportional Sampling	139
86.	Gaseous Sampling Train Employing Wet Impingers	140
87.	Impingers Used in Gaseous Sampling	141
88.	Gaseous Sampling Train Employing Consequtive Freeze Out Traps	145
89.	Effect of Sulfur Trioxide Concentration on Sulfuric Acid Dew-point Temperature	147
90.	Gaseous Sampling Trains for Simultaneous Collection of Sulfur Dioxide and Sulfur Trioxide	148
91.	Arrangement of a Typical Gas Chromatograph	152
92.	Photometric Smoke Density Meter	162
93.	Continuous Monitoring of Particulate Emissions with a Conductivity Cell	166
94.	Continuous Monitoring of Particulate Emissions with a Specific Ion Electrode	166
95.	Continuous Monitoring of Reduced Sulfur Gases from a Kraft Recovery Furnace	172
96.	Continuous Monitoring System for Total Reduced Sulfur and Sulfur Dioxide from a Kraft Recovery Furnace	172
97.	Nondispersive Infrared Analyzer for Carbon Dioxide Measurement	174
98.	Dynamic Dilution System Employing a Rotating Syringe	177
99.	Dynamic Dilution System Employing a Pressurized Stainless Steel Cylinder	179
100.	Continuous Velocity Measurement in Flue Gases	180
101.	Evaluating the Odor Threshhold Level of Flue Gas Samples	190
102.	Schematic Automobile Exhaust Gas Sampling and Analytical System Used by the State of California	192
103.	Sample Probe and Pitot Tube Inserted into a Jet Engine Exhaust	193
104.	Particulate Sampling Train for Aircraft Jet Engine Exhaust Gases	193
105.	Rotary Inertial Impactor for Sampling Large Particles in the Atmosphere	194
106.	Ringelmann Charts Used for Evaluating Optical Densities of Smoke Plumes	195
107.	Relationship of Transmittance and Mass Concentration for Plumes Containing Particles of Various Diameters and Irregular Shapes	196
108.	Smoke Generating Units for Plume Opacity Training of Inspectors	197
109.	Mobile Laboratory for Source Monitoring of Sulfur Gas Emissions from Kraft Pulp Mills	199

LIST OF TABLES

Table	Title	Page
1.	Summary of Air Pollution Sources	5
2.	Emission Factors for Refuse Incineration	5
3.	Summary of Standard Conditions	12
4.	Changes in Density and Barometric Pressure of Air with Altitude	13
5.	Nitrogen/Oxygen Ratios for Varying Excess Air Levels	13
6.	Volume per Mole for Gases	21
7.	Determination of Flue Gas Density	44
8.	Number of Traverse Points in Ducts	44
9.	Equal Area Zones for Velocity Traverses in Round Ducts	44
10.	Equal Area Zones for Velocity Traverses in Rectangular Ducts	44
11.	Velocity Traverse Points in Rectangular Ducts with Perpendicular Ports	45
12.	Dimensionless Constants for Gas Flow Measuring Devices	45
13.	Effect of Departure from Isokinetic Conditions on Measured Concentrations	73
14.	Effect of Nozzle Size on Efficiency of Particle Collection	73
15.	Confidence Levels for Minimum Probe Size Expression	73
16A.	Sampling Nozzle Selection for Different Gas Velocities: For Thimble-Impinger Sampling Trains, Where $Q = 0.5$-2.00 cfm	73
16B.	Sampling Nozzle Selection for Different Gas Velocities: Cyclone-Filter Sampling Train, Where $Q = 4.0$-15.0 cfm	74
17.	Estimation of Isokinetic Sampling Rates	74
18.	Corrections in Sampling Rate with Meter Pressure Changes	74
19.	Magnitude of Systematic Errors in Particulate Sampling Measurements	74
20.	Estimated Accuracy of Particulate Emission Rate Measurements	74
21.	Magnitude of Errors in Measuring Particulate Emission Rates	75
22.	Distribution of Particle Collection in a Sampling Train	101
23.	Maximum Operating Temperatures for Filter Media	102
24.	Particle Retention Characteristics of Clean Alundum Thimbles	102
25.	Effect of Flow Rate on Pressure Drop and DOP Smoke Penetration for Various Air Sampling Media	102
26.	Operating Characteristics of Electrostatic Samples	103
27.	Operating Characteristics of Thermal Precipitators	103
28.	Properties of Collection Devices Used in Particulate Sampling	103
29A.	Particulate Sampling Trains in Use. A. Low Flow Rate Systems where Q is Less than 3.0 cfm	132
29B.	Particulate Sampling Trains in Use. B. High Flow Rate Systems where Q is Greater than 3.0 cfm	134
30.	Distribution of Particle Collection in a Sampling Train on Gases from a Furnace for Burning Insulated Copper Wire	134
31.	Minimum Particle Sizes Collected by Final Stage Devices	134
32.	Techniques for Particle Size Determination	135
33.	Microscopy in Particle Size Analysis	135
34.	The Effect of Particulate Concentration on Sampling Period When Using a Cascade Impactor	135
35.	Properties of the Anderson Sampler	135
36.	Cooling Agents Used for Freezeout Traps	158
37.	Analytical Methods for Sulfur Dioxide and Sulfur Trioxide	159
38.	Analytical Methods for Hydrogen Sulfide and Mercaptans	159
39.	Analytical Methods for Oxides of Nitrogen	159
40.	Calibration of a Photometric Meter for Measuring Particulate Concentrations on a Kraft Recovery Furnace	183
41.	Operating Characteristics of Continuous Gaseous Analyzers	184
42.	Selective Prescrubbing Solutions for Sulfur Gas Analyzers with Coulometric Titration	184
43.	Response of a Coulometric Titrator to Different Gases	184
44.	Oxidation of Nitric Oxide to Nitrogen Dioxide	185
45.	Relative Magnetic Susceptibilities of Flue Gas Constituents	185

Author Index

Abrams, R.—57, 110, 169
Achinger, W. C.—67
Adams, D. F.—148, 153, 167, 170, 198
Adley, F. E.—170
Aitken, J.—97, 101
Alexander, P. A.—190
Allen, D.—43
Allner, W.—110
Altshuller, A. P.—146, 156
Amberg, H. R.—5, 153, 175, 198
Amtower, R. E.—179
Andersen, A. A.—126, 191
Anderson, E.—90, 100, 156
Anderson, M. S.—182
Arbogast, A. H.—130
Austin, R. R.—171

Badzioch, S.—21, 42, 55, 56, 94, 100
Bailey, D.—162
Baker, R. A.—146
Balestrieri, S.—156
Banciu, I.—100
Barkov, N. N.—168
Barnes, E. C.—96
Baum, F.—100, 130
Belcher, R.—129
Bellack, E.—129
Benforado, D. M.—189
Bent, R.—147, 173
Bergshoeff, G.—156
Bewick, H. A.—157
Bialkowsky, H. W.—149
Binek, B.—96, 101
Blackett, J. H.—21, 42, 94, 99
Bloomfield, B. D.—5, 48, 55, 58, 62, 99
Bloomfield, J.—99, 121
Blosser, R. O.—153, 171
Bonelli, E. J.—152
Bonnet, M.—55
Boothroyd, R. G.—66
Bosch, J. C.—123, 124
Bostrom, C. E.—142
Boubel, R. W.—112, 113
Bracewell, J. M.—142
Brady, W.—116
Brawn, J.—57, 110, 169
Bredl, J.—97, 101
Breitling, K.—167
Brief, R. S.—145
Briglio, A.—171
Brockman, J.—171
Brown, J. E.—169
Brubacher, M. L.—191
Budd, M. S.—157
Bulba, E.—164, 167
Burlin, R. M.—192
Bye, W. E.—130

Cadle, R. D.—21, 99, 120, 131
Caldwell, W. E.—110

Calvert, S.—56, 140, 156
Campau, R. M.—177
Caplan, K. J.—12, 73, 100
Carrier, W. H.—18
Cederlof, R.—187
Chass, R. L.—5, 12, 43, 72, 99, 129, 156
Chojnowski, B.—48
Cholak, J.—150
Chory, J. P.—147, 157
Christman, R. F.—183
Clayton, G. D.—5, 42, 72, 99, 128, 156
Collins, K. E.—181
Collins, T. T.—130
Colombo, P.—149
Connor, W. D.—196
Coons, J. D.—202
Cooper, H. B. H.—5, 153, 171, 177
Cooper, S. R.—163
Corn, M.—43
Cravitt, S.—101
Crider, W. L.—202
Crosse, P. A.—181
Crummett, W. B.—151
Cuffe, S.—73
Curley, L. C.—43

Dalla Valle, J. M.—21, 120, 131
Darley, E. F.—112, 113
Davis, D. S.—73
Decker, C. E.—166
DeHaas, G. G.—149
Dennis, R.—57, 64
Detwiler, C. G.—100
Devorkin, H.—5, 12, 43, 72, 99, 106, 129, 156
DiGiovanni, H.—101
Doerr, R. C.—146
Donoso, J. J.—164
Dooley, A.—129
Dorsey, J. A.—108
Doubek, J.—158
Drinker, P. A.—100, 121, 145
Duckworth, S.—178
Duff, G.—200
Duncan, L.—202
Durst, D.—72
Duwel, L.—168

Einstein, A.—97
Elfers, L. A.—166
Engdahl, R.—101, 167
Ensor, D. S.—123
Epstein, P.—97
Ettre, L. S.—153
Evans, D G.—52, 99

Fahrenbach, W.—53
Feldstein, M.—144, 156, 176
Felicetta, V. F.—148
Fielder, R. S.—157
First, M. V.—100

Fitzgerald, J. J.—100
Flesch, J.—126
Flint, D.—157
Fox, E. A.—187
Franklin, M. E.—202
Fudurich, A.—5, 12, 42, 72, 99, 111, 129, 156

Gaeke, G. C.—157
Gallaer, C. A.—62
Gansler, N. R.—162, 163
Gartrell, F. E.—197, 202
George, R. E.—192
Gerstle, R.—73
Gex, V. E.—187
Giuntini, J.—101
Godard, L.—101
Goetz, A.—100, 127, 191
Goksoyr, H.—157
Goodeve, C. F.—129
Gordon, M.—101, 123, 124
Greenburg, L. G.—99, 121
Grieve, T.—97
Grindell, D. H.—101, 167
Grondona, A.—147
Groutsch, E. R.—164
Gruber, C. W.—99, 164
Grune, W. N.—153
Guse, W.—169
Gussman, R. A.—123, 124

Haaland, H. H.—5, 43, 52, 100
Hagen, J. E.—131
Haines, G. F.—54, 57, 64, 100, 164
Halliday, D.—167
Hama, G. M.—35
Haneef, M.—191
Hangebrauck, R. P.—105, 107
Hardie, P. H.—72, 99, 110
Harding, C. I.—43, 73, 149
Harkens, W. D.—117
Harkins, J.—174
Harvey, R. A.—191
Haskell, G. F.—163
Hass, G. C.—191
Hatch, T.—100, 121
Hatchard, R. E.—108
Hawksley, P. G. W.—18, 42, 50, 56, 68, 69, 72, 94, 100
Hays, A. D.—21
Hayek, H.—153
Heimke, W.—163
Hein, G. M.—111, 144
Hemeon, W. C. L.—54, 57, 64, 100, 164, 189
Hendrickson, E. R.—43, 141
High, M. D.—130
Hissink, M.—157
Himmelblau, D. M.—8, 43
Hochheiser, S.—157
Hodgson, A. E. M.—142
Hodkinson, J. R.—196
Holmes, R. G.—5, 12, 43, 72, 99, 129, 156
Holton, W. C.—94, 100
Horsley, R.—43
Hougen, O. A.—18

Hudson, R. G.—52
Hughes, A. D.—43
Hull, R.—20
Hultz, J. A.—52, 99
Hurley, T. F.—162, 181
Hyatt, E.—191
Hyde, P. E.—66, 112
Hyland, R.—72

Ide, H. M.—164
Izmailov, G. A.—167

Jackson, P. J.—157, 170, 189
Jacobs, M. B.—5, 43, 52, 72, 100, 148, 152
James, H. A.—202
Johnson, H. C.—202
Johnson, L. W.—198, 202
Johnson, W. B.—202
Jorgensen, R.—12
Jutze, G. A.—179

Kaiser, E. R.—118
Kallai, T.—127
Kanter, C. V.—5, 12, 43, 72, 99, 111, 129, 156
Kast, W.—131
Katz, J.—98
Katz, M.—156
Katz, S. H.—128
Kemnitz, D. A.—108
Kennedy, E. D.—43
Kethley, T.—101
King, W. J.—154
Klemperer, H.—164
Knudsen, V. O.—182
Kogan, L.—100, 129
Kolp, P.—43
Koppe, R. K.—198
Kuczynski, E. R.—178

Ladner, W.—173
Lamb, A.—101
LaMer, V. K.—97
Landry, J. E.—73
Lapple, C. E.—21, 131, 168
Laxton, J. W.—170
Ledbetter, J. O.—191
Lees, B.—98
Leithe, W.—200
Leonard, J. S.—96, 114, 130, 145, 165, 166, 169
Leonardos, G.—189
Levaggi, D.—156
Levin, E. D.—100, 129
Lewis, R. L.—179
Lilienfeld, P.—101
Lindberg, C. A.—197, 202
Lindorf, H.—200
Lippman, M.—96, 123, 131
Lisle, E. S.—147, 157
Lodding, W.—152
Lucas, D. H.—163, 181
Lunche, R. G.—111
Lundgren, D.—56

McCaldin, R. O.—198
McDonald, W.—163
McKee, H. C.—141, 142
McLean, J. D.—151
McNair, H. M.—152
McNary, R.—150
McPhee, R. D.—153
McShane, W. P.—164, 167
Magill, P. L.—42, 156
Mandl, M.—158
Marcucci, G. P.—157
Markovs, J.—172
Martin, R. M.—72
Matoi, H. J.—96, 116
May, K. R.—123, 131
Meeker, J. E.—107
Megy, J. A.—149, 171, 172, 178, 198, 199
Meier, F. C.—197, 202
Meland, B. R.—112, 194
Messer, P.—156
Miller, A. M.—57, 110, 169
Mills, J. L.—150
Mitchell, R.—101, 167
Moody, D.—40
Moore, A. S.—63
Moore, H.—148
Morley, M.—98
Mukai, M.—129
Mullin, W.—173
Murphy, D. E.—43
Murray, F. E.—157

Nacovsky, W.—157, 170
Nader, J. S.—189
Naumann, A.—169
Neerman, J. C.—177
Nelson, H. W.—191
Nestell, R. J.—156
Nelson, R. Y.—111
Nickol, G. B.—189
Nicksic, S. W.—174
Nieuwenhuizen, J. K.—42, 43
Noll, K. E.—194
Northcraf, M.—112
Noss, P.—110

Okita, T.—127
Orning, A. A.—52, 73, 99
Orr, C.—21, 101, 120, 131
Ott, R. R.—108
Ower, E.—41
Ozolins, G.—1

Parker, G. J.—53, 56, 57, 64, 99
Parkes, W. J. S.—58, 64
Patty, F. A.—158
Paulus, H. J.—168
Pengelly, A. E. S.—35, 43
Penney, G. W.—96
Perry, J. H.—15, 25, 41, 52, 62
Peter, L. J.—20
Peterson, C. M.—168
Petty, G. M.—158
Pilat, M. J.—123, 124

Pixova, J.—101
Pohlman, R.—182
Posthumus, H.—43
Preining, O.—100
Price, T. D.—164
Pritchard, W.—101
Purdon, P. W.—181

Raask, E.—157
Radwanska, A.—157
Rammler, E.—94, 121
Ranz, W. E.—97, 123, 131
Razbegaeva, A. P.—30, 52
Reed, L. E.—56
Rehm, F. R.—52, 57, 72, 99, 110, 129
Reichardt, I.—101
Rendle, L. K.—100
Ringelmann, M.—201
Rivera, M. F.—7, 12, 58, 72, 99
Roberts, L. R.—141, 142
Robinson, M.—95
Roesler, J. F.—131
Rose, A. H.—43
Rosenblatt, P.—97
Rosin, P.—94, 121
Ross, K.—157
Rossano, A. T.—5, 100, 177
Roth, H.—40, 42
Rounds, G. L.—96, 116
Rozsa, J. T.—164
Ruch, W.—14, 156, 200
Rushing, D. E.—152, 153

Saltzman, B. E.—150, 157
Samples, W.—57
Saxton, R. L.—97
Sawyer, K. F.—127
Schadt, C.—96, 99
Schell, N. E.—5, 100
Schoeboe, P. J.—129
Schneider, R. L.—116
Schuck, E. A.—113
Schueneman, J. J.—130
Schultz, E. J.—94, 100
Schumann, C. E.—99, 164
Schutz, A.—167
Schwartz, C.—73
Sehmel, G. A.—55
Seidman, E. B.—157
Shaffer, P. A.—171
Shepherd, M.—154
Shevchenko, N. S.—100, 129
Shigehara, R. T.—67
Sholtes, R. S.—43
Shuck, E. A.—174
Silverman, L.—100
Skillern, C. P.—170
Sleik, H.—156
Sleva, S. F.—156
Smith, G. W.—128
Smith, J. F.—52, 99
Smith, R.—1
Smith, S. H.—116
Smith, W. J.—91

Smith, W. S.—61, 72, 81, 82, 94, 105, 106, 129
Snowsill, W. L.—163, 181
Sonkin, L. S.—131
Spaite, P. W.—131
Sparks, R. E.—43
Spurny, K.—101
Stairmand, C. J.—93, 96, 99
Steele, D. J.—181
Stein, J.—110
Stenburg, R. L.—43
Stern, A. C.—94, 95, 100, 152
Stevenson, H.—100
Streeter, V. L.—21, 43
Sullivan, F.—39
Surprenant, N. F.—91
Swain, R. E.—117
Sweeney, M. P.—174

Task, T. A.—202
Taylor, C. E.—166
Teague, M. C.—158
Tebbens, B. D.—107
Thoen, G. N.—171, 172
Thoenes, H. W.—169
Thomas, J. F.—129
Thomas, M. D.—179
Tipping, F.—176
Todd, W. F.—131
Tolciss, J.—130
Touzalin, L. A.—116
Toynbee, P. A.—58, 64
Tretter, V. J.—166
Tucker, T. W.—202
Turk, A.—144, 189, 190
Tuttle, W. N.—153
Twiss, S. B.—153

Uzhov, V. N.—119, 131

Verrochi, W. A.—98
Vitols, V.—55
Walker, A. B.—111
Walker, C. G.—131
Walker, M. S.—202
Wallin, S. C.—191
Walters, K.—182
Walther, J. E.—5, 39, 153, 175, 198
Walton, W. H.—127
Wang, G. K. M.—157
Warner, T. B.—166
Warren, H.—121
Warshavsky, T. P.—92, 100, 129
Wartburg, A. F.—156
Wasser, R. W.—110
Watson, H. H.—54, 56, 72, 101
Watson, K. M.—18
Weber, H. J.—116, 128
Wendt, G.—101
West, P. W.—157
West, T. S.—129
White, H. J.—95
Whiteley, A. B.—56
Wilby, F. V.—187
Wilcox, J. D.—99, 123, 131
Willard, H. H.—152, 158
Wilson, A. C.—94
Wilson, H. N.—200
Wilson, J. V.—182
Wilson, R.—101
Wise, K. R.—112
Wolfe, E. A.—43, 72, 99, 129
Wong, J. B.—123, 131
Workman, W.—140, 156

Yamashita, S.—127
Yocom, J. E.—111, 744, 163

Zimmerman, E.—94, 100

Subject Index

Adsorption Method of Moisture Content Determination—29, 30
Aerosol—7
Agricultural Field Burning—110-113
Airborne Sampling—197, 198
Air Contaminants—7
Air Pollution—1, 7
Air Pollution Control Association—93, 95
Alundum Thimbles—90, 91
American Petroleum Institute—52, 72, 93
American Society of Mechanical Engineers—7, 34, 35, 43, 57, 58, 62, 71, 72, 83, 84, 89, 93, 102, 110, 118, 122, 147
American Society for Testing and Materials—102, 189, 201
Ammonia—150, 151
Andersen Sampler—126

Approaches to Source Testing—1-2
Arrangement of Sampling Ports—23
Automobiles—191, 192

Bacteria—191
Bahco Particle Classifier—127
Balloons for Low Velocity Measurement—39
Bay Area Air Pollution Control District—7, 40, 58, 92, 93, 196
Beta Ray Attenuation—167
Bibliographies—200
Bluford's Law—20
British Coal Utilization Research Association—62, 74, 75, 107
British Standards Institution—58, 62, 92, 93, 106, 110, 162
Bubble Tube Meter—50, 51

Calibration Techniques—177, 178, 179
California Air Resources Board—191, 192
Carbon Monoxide—151, 152
Cascade Impactor—123, 124, 125, 126
Cegrit Sampler—94
Central Electricity Generating Board—98
Centrifugal Particle Collectors—93, 94
Chemical Addition Techniques for Flow Measurement—40, 42
Chlorine Compounds—151
Coal-Fired Power Boilers—110
Collection Devices—7, 81, 84, 99, 121-128, 139-146
Colored Smokes for Low Velocity Measurement—39
Combustion Sources—110-113
Components of Sampling Trains—48-49
Condensation Method for Moisture Content—28, 29
Conductivity Probes—165, 166
Conifuge—127
Conservation of Energy—17
Conservation of Mass—17
Continuous Flow Measurement—179, 180, 181
Continuous Monitoring—161-185
Continuous Gaseous Monitoring—168-181
Continuous Particulate Monitoring—161-168
Converging Nozzles—41
Corrosion—190, 191
Cross-Sectional Area Adjustment—40
Cyclone Flow Meter—50

Dade County Pollution Control Agency—7, 58, 88, 93, 111
Deposition—190, 191
Droplets—109
Dry Gas Meter—50
Dust—7

Elements of Sampling Trains—47
Electrostatic Precipitation Collectors—94-97
Emissions—7
Emission Factors—1, 5
Engineering Foundation—200
Errors in Particulate Sampling—69-72
Evacuated Flask Method for Moisture Content—30

Ferrous Metallurgy—116
Filtration Collection Devices—89-93
Flexible Wall Containers—178, 179
Flow Calibration Devices—51, 52
Flame Ionization—154, 175, 176
Flow Corrections—52
Flue—7
Fluoride Compounds—108
Flyash—7
Freeze-Out Collection—144, 145
Fumes—7
Fluid Motion—17, 18

Gas Chromatography—152, 153, 155
Gas Composition and Density—30-33
Gaseous Sampling—137-159
Gaseous Collection Methods—139-146
Gases—7

Gas Flow Conditions—24-33
Gas Flow Measurement—4, 23-45
Gas Laws—15, 16, 17
Generalities—20
Goetz Spectrometer—127
Grab Sampling—145, 146
Greenburg-Smith Impinger—121, 123, 141, 142

Hydrogen Sulfide—148, 149

Ideal Gas Laws—15, 16, 17
Incinerator Institute of America—93, 111
Industrial Gas Cleaning Institute—118
Instrumental Analyses—152, 153, 154
Ion-Specific Electrodes—166
Isokinetic Sampling—53-67

Jet Aircraft—192, 193

Kraft Recovery Furnaces—114, 115

Lime Kilns—113-114
Liquid—7
Liquid Absorption—139-143
Liquid Displacement for Flow Meter Calibration—52
Location of Sampling Ports—23
Los Angeles County Air Pollution Control District—7, 57, 87, 88, 89, 93, 95, 106, 110, 111, 145, 152, 189, 190, 192, 193, 196, 197
Low Velocity Measurement—38, 39, 40

Mass Spectrometry—154, 176, 177
Material Balance—1
Membrane Filters—91
Mercaptans—148, 149
Metallurgical Operations—116, 117, 118
Methodology of Particulate Sampling—77-103
MgO Recovery Furnaces—115, 116
Millipore Filter Corp.—92
Mists—7, 109
Mobile Laboratories—198, 199, 200
Mobile Sources—191, 192, 193
Moisture Content Determination—25-30
Motor-Driven Syringes—178
Murphy's Law—20

National Air Pollution Control Administration—88, 93, 106
National Council for Air and Stream Improvement—56, 58, 72, 88, 93, 128, 143, 198, 199
NIIOGAZ—119
Nitrogen Compounds—149, 150, 151

Odor Threshold Levels—187-190
Oil-Fired Power Boilers—110
Oregon State University—112
Organic Materials—152
Orifice Flow Meters—41, 49
Orsat Analyzer—30, 31, 32
Oxides of Nitrogen—149, 150
Oxides of Sulfur—146, 147, 148

Paramagnetism—176, 185
Particle Dynamics—18, 19, 20
Particle Size Analysis—120-128
Particulate Collection Devices—84-99
Particulate Matter—7
Particulate Sampling—53-135
Particulate Sampling Trains—83, 84, 105-135
Performance Testing—12, 118, 119, 120
Permeation Tubes—179
Peter Principle—20
Photometric Detectors—161, 162, 163
Pitot Tubes—33, 34, 35
Plibrico Company—195
Polynuclear Hydrocarbons—105, 106, 107
Pressure Null Balance—64, 65
Prime Mover—7, 49
Principles of Particulate Sampling—53-75
Principles of Source Testing—15-21
Procedures for Source Testing—2, 3
Psychrometry—17, 18
Pulp and Paper Industry—113-116
Pulsating Sources—41, 42

Radioactive Tracers—40
Radioactivity Measurement—191
Reasons for Source Testing—1
Reference Conditions—9-12
Refuse Incinerators—110, 111
Remote Sensing—197
Reports for Source Testing—4, 5
Requirements for Particulate Collection—84, 85
Requirements for Sampling Trains—47, 48
Requirements of Source Testing—2
Research-Cottrell, Inc.—24, 58, 82, 99
Ringelmann Chart—194-197
Rotameter—50
Rotating Syringes—177, 178
Rotating Vane Anemometer—39

Sample Flow Measurement—49, 50, 5
Sampling Ports—23, 24
Sampling Trains—7, 47-52
Smoke—7
Solid Adsorption—143, 144
Solids—7
Soot—8
Sound Attenuation—167
Source—8
Sources of Air Pollution—5
Source Testing Report—4, 5

Specified Conditions—10, 11, 12
Spherical Pitometer—35
Spirometer—51
Stack Gas—8
Stainless Steel Cylinders—178
Standard Conditions—8-12
Standard Pitot Tube—33, 34
Stanford Research Institute—18
State of California—191, 192
Static Pressure Determination—25
Statistical Considerations in Particulate Sampling—67-72
Stausscheibe (S-Type) Pitot Tube—33, 34, 35
Sulfur Dioxide—147, 148
Sulfur Trioxide—146, 147
Swinging Vane Anemometer—39
Symbols Used in Calculations—8, 9
Syringes—177, 178

Tape Samplers—163, 164, 165
Tars—109
Temperature Measurement—24, 25
Terminology—7-13
Thermal Null Balance—64-67
Thermal Precipitation Collectors—97, 98
Thermistor Anemometer—39
Thermo-Anemometer—38
Total Flow Rate in Ducts—37, 38, 40, 41
Total Sulfur—149

U.S. Bureau of Mines—73, 87
U.S. Public Health Service—58, 72, 87, 109, 110, 111, 129,

Vapor—8
Variable Flow Measurement—42
Velocity Determination—33-40
Velocity Tables—40
Velocity Traverses—35, 36, 37
Venturi Meters—41
Venturi Pitot Tube—35
Verein Deutschen Ingenieure—119, 200

Water Vapor—8
Western Precipitation Corp.—34, 52, 67, 101
Wet and Dry Bulb Measurements—25, 26, 27, 28
Wet Impingement—85, 86, 87, 88
Wet Test Meter—51
Wigwam Burners—111, 112
Wood-Fired Power Boilers—110